Analogue and D[illegible]echniques

Analogue and Digital Communication Techniques

Grahame Smillie
Lecturer, Hull College

OXFORD AMSTERDAM BOSTON LONDON NEW YORK PARIS
SAN DIEGO SAN FRANCISCO SINGAPORE SYDNEY TOKYO

Newnes
An imprint of Elsevier Science
Linacre House, Jordan Hill, Oxford OX2 8DP
200 Wheeler Road, Burlington, MA 1803

First published 1999
Transferred to digital printing 2002

British Library Cataloguing in Publication Data
A catalogue record for this book is available from the British Library

ISBN 0 340 73125 7

For information on all Newnes publications
visit our website at www.newnespress.com

Contents

- Filters & Frequency Responses
- Signals in Time & Frequency Domain
-

About the author

Grahame Smillie started out as a telecommunications technician in South Africa, working on transmission equipment for TELKOM SA.

In 1978 he joined the TELKOM SA training college and lectured on the National Certificate programme. After further study he gained an HND in Telecommunications and a teaching qualification. Subsequently he lectured on the HND programme as well as running training courses for the transmission staff in the company.

In 1993 he came to the UK where he is currently the BTEC EDEXCEL Higher National Diploma Engineering (Electronics & Communication) course tutor at Hull College, where the idea for the book was born. He also lectures on the BTEC EDEXCEL National Certificate and national Diploma as well as the City & Guilds programmes in electronics and telecommunications.

To my wife Irene and my three children Donna, Darren and Warren.
I will always cherish their support and encouragement

Preface

The idea for this book was born out of the notes that I made while lecturing on the HND Engineering (Electronics & Communication) Programme. From my experiences as a lecturer I very soon realised that the text books that the students consulted were either too advanced and too mathematical, more suitable to 3-year and 4-year degree courses, or too simple in that they were orientated to the BTEC/EDEXCEL National Diploma and National Certificate programmes.

In this book I have tried to bridge the gap between the two educational levels. Hopefully, students taking advanced GNVQ, HND and degree courses will find the information given useful and helpful in their studies. I have also tried to bridge some of the gaps between purely descriptive text with mathematical descriptions.

The knowledge gained and experiences whilst working as a technician and training officer in the telecommunications industry, together with the experience of lecturing at a further education college, equipped me to write this book

I sincerely hope that all students who use this book will benefit from the information contained within its covers.

Acknowledgements

I would like to thank my colleague and friend Jane Hutchinson for proofreading the manuscript and making valuable suggestions. I also want to thank Arnold Publishers for giving me the opportunity to have this book published and for the support that they have given me.

1 Definition of terms

Knowledge is the foundation of civilisation.

1.1 INTRODUCTION

Before starting, it is necessary to discuss some fundamentals that are important in understanding analogue and digital transmission. The first area that is discussed is what is meant by frequency. Also the make-up of complex waveforms such as square and sawtooth waveforms is described. In order to understand a complex waveform a knowledge of harmonic frequencies is necessary, and this is explained.

Units of measurement are also very important, so an explanation of the gain or loss of a network is given. The discussion then includes an explanation of the decibel. Other common units such as dBm, dBr and dBmO are also discussed.

1.2 FREQUENCIES

Frequency is measured in hertz (Hz), equivalent to cycles per second: thus 1 Hz is 1 c/s. The SI measurement system is used to designate frequency ranges. The common frequency ranges with their SI symbols are given in Table 1.1.

Table 1.1 Common frequency ranges

Frequency	SI symbol
1 Hz	1 Hz
1 000 Hz	1 kHz
1 000 000 Hz	1 MHz
1 000 000 000 Hz	1 GHz

1.3 TYPES OF SIGNAL

There are only two different types of signal that are processed, transmitted and received by telecommunication equipment. These are *analogue* and *digital.* Each type has individual characteristics.

1.4 ANALOGUE SIGNAL

An analogue signal is defined as a continuous waveform having a positive peak and a negative peak and having an infinite range of levels. This infinite range is due to the fact that if two discrete points on the waveform are chosen then a point halfway between these two points will yield a different amplitude. If the distance between this new point and one of the previous points is halved then a new point having a different amplitude will be obtained. This process will continue an infinite number of times.

An analogue signal has no discontinuous points, i.e. it follows an unbroken curve for its full duration. Typical analogue waveforms are shown in Fig. 1.1.

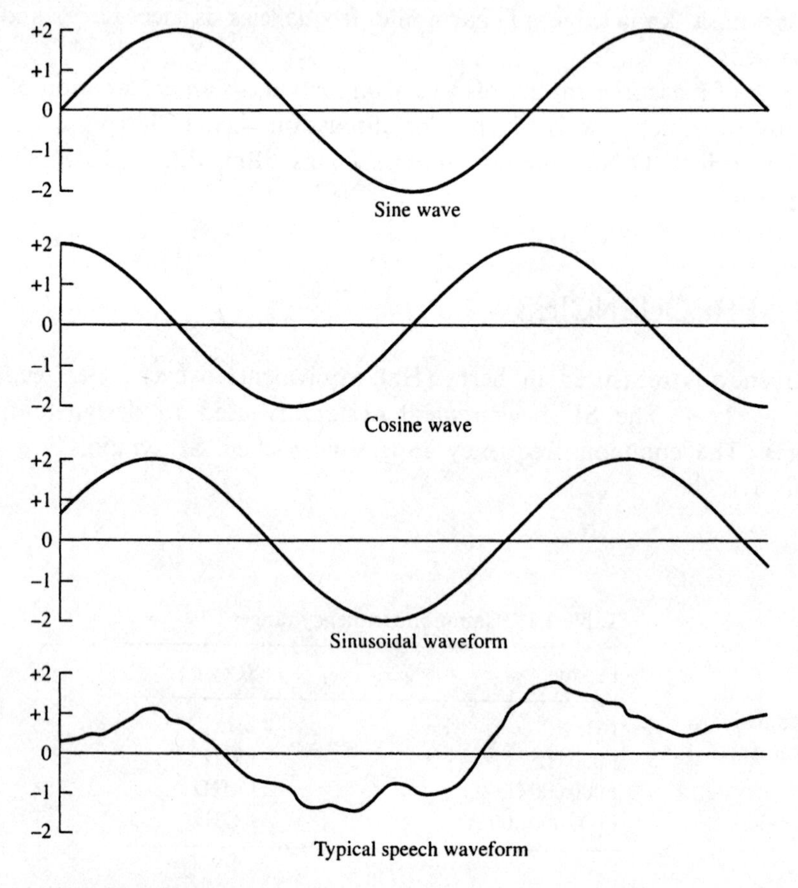

Fig. 1.1 Analogue waveforms

The frequency f of the analogue wave is determined by the following formula:

$$f = \frac{1}{t} \text{ Hz}$$

where t = time in seconds

The time t is the duration of one complete cycle, i.e. one wavelength.

The formula used to determine the wavelength λ in free space (vacuum) is as follows:

$$\lambda = \frac{c}{f} \text{ m}$$

where c = velocity of light in a vacuum
$= 3 \times 10^8$ m/s

The waveforms shown in Fig. 1.1 are shown in the time domain, which means that the amplitude is plotted against time.

A sine wave and a cosine wave consist of a single frequency. A sine wave can be described as having a start phase of 0° and an initial amplitude of zero. The wave then starts to move towards the positive maximum amplitude. A cosine wave can be described as having a start phase of 90° and an initial positive maximum amplitude. The wave then starts to move towards the zero amplitude point. A sinusoidal wave has a start phase anywhere from 0° to 360° and an initial amplitude anywhere between the positive maxima and the negative maxima. However, the sinusoidal wave follows the shape of a sine wave.

A complex wave is a wave that consists of a number of different frequencies.

1.5 DIGITAL SIGNAL

A digital signal is a complex waveform and can be defined as a discontinuous waveform having a finite range of levels.

A theoretical digital signal is shown in Fig. 1.2. At times t_1, t_2, t_3, t_4 and t_5, the signal assumes two states. These states are a logic 0 and a logic 1. It can be seen that at these times the signal is discontinuous (i.e. there is a break).

A practical digital waveform is not a discontinuous waveform, but has a leading edge, which is also referred to as the positive edge, and a trailing edge, which is referred to as a negative edge or lagging edge. A typical practical digital wave is shown in Fig. 1.2.

In practice the leading edge has a finite rise time and the trailing edge has a finite fall time. The rise time of the leading edge is the time taken for the output amplitude to rise from 10% of the final steady-state value to 90% of the final steady-state value. The fall time or decay time of the trailing edge is defined as the time taken to fall from 90% of the initial output amplitude to 10% of the initial output amplitude. The rise and decay times are shown in Fig. 1.2.

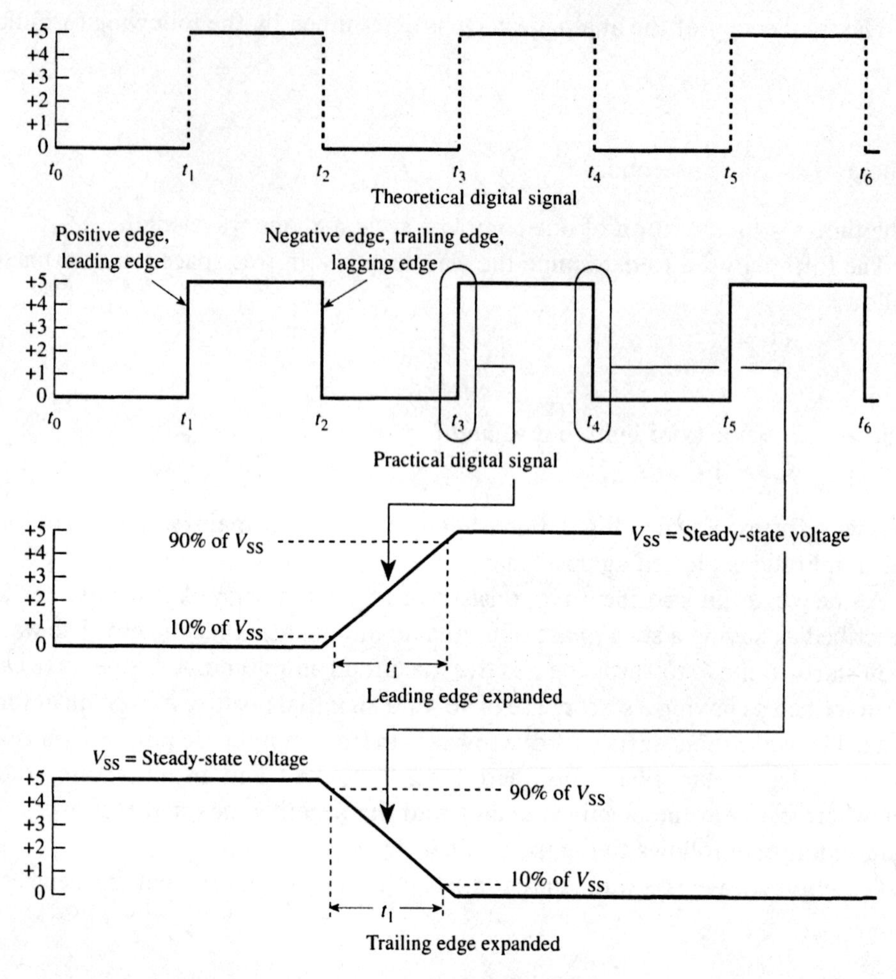

Fig. 1.2 Digital waveforms

A digital signal has a pulse repetition time (PRT) which is the duration of one full cycle (one wavelength). The fundamental frequency (i.e. the lowest frequency in the wave) is determined by means of the following formula:

$$f = \frac{1}{\text{PRT}} \text{ Hz}$$

1.6 WAVEFORMS

Different waveforms are produced using different harmonically related frequencies. The harmonic frequencies are those frequencies which are directly related to the fundamental frequency.

A fundamental frequency is the lowest frequency that exists in the complex wave and its frequency is determined by the inverse of the duration of one cycle. Some of these harmonic frequencies are given in Table 1.2.

Table 1.2 Harmonic frequencies

Fundamental frequency	Harmonic frequency	Relationship
f_1	2nd	$2f_1$
	3rd	$3f_1$
	4th	$4f_1$
	5th	$5f_1$
	50th	$50f_1$
	200th	$200f_1$

1.6.1 Square wave

A square wave is made up of a fundamental frequency and all the odd harmonic frequencies. The amplitude relationship between the fundamental frequency and the harmonic frequencies is very important, as well as the initial phase relationship between the different frequencies. The fundamental frequency and all the odd harmonic frequencies must be sine waves and the amplitude relationship

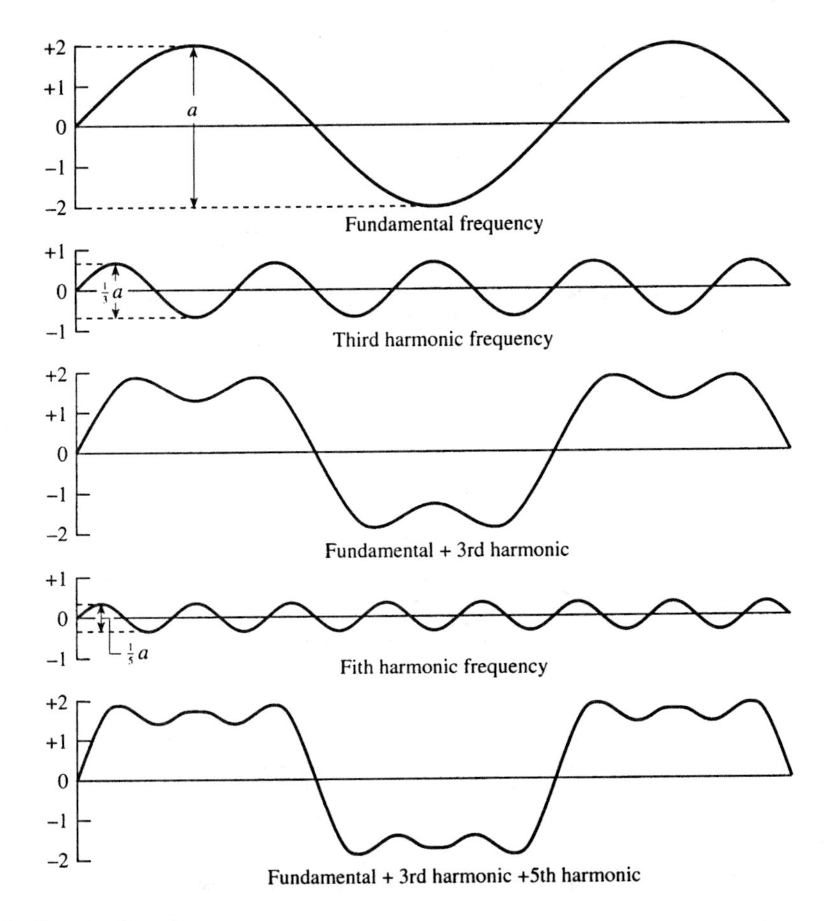

Fig. 1.3 Construction of a square wave

must be as follows:

$$n\text{th harmonic amplitude} = \frac{\text{max. amplitude of } f_1}{n}$$

where f_1 = fundamental frequency
n = harmonic number (1, 3, 5, 7 etc.)

The construction of a square wave is shown in Fig. 1.3. This figure shows the resultant of the fundamental frequency and the third harmonic frequency as well as the resultant of the fundamental frequency, the third harmonic frequency and the fifth harmonic frequency. By adding more and more odd harmonic frequencies, eventually a true square wave will result.

1.6.2 Sawtooth wave

This wave is also a complex wave consisting of a fundamental frequency and all the harmonic frequencies. Again the fundamental frequency and the harmonic frequencies are all sine waves. The amplitude relationship between the fundamental and the harmonic frequencies is given by the equation in Section 1.6.1. The only difference is that n = 1, 2, 3, 4, 5 etc.

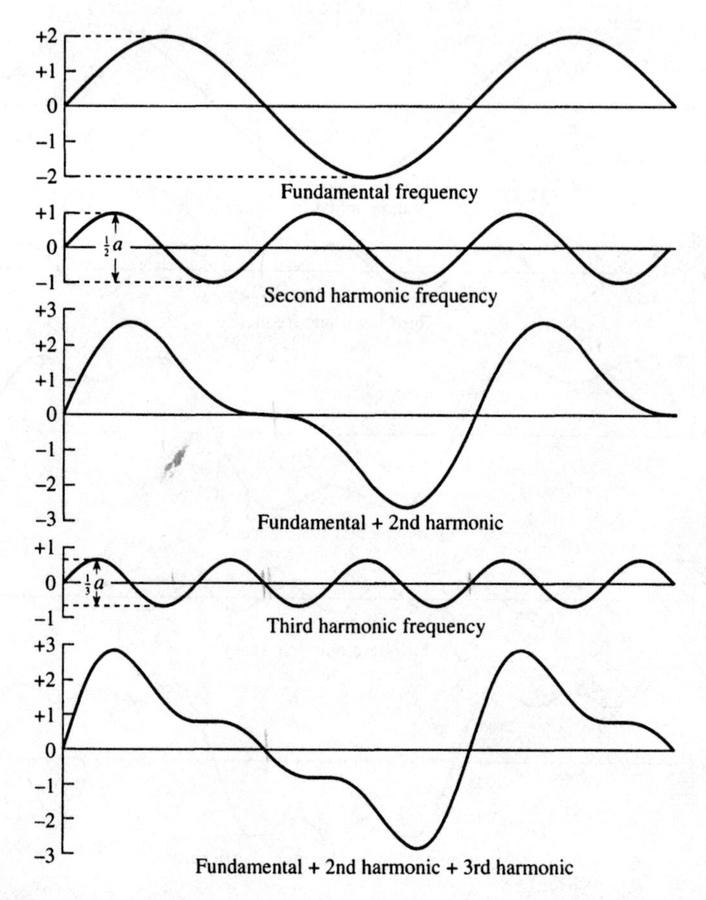

Fig. 1.4 Construction of a sawtooth wave

The construction of a sawtooth wave is shown in Fig. 1.4. This figure shows the resultant of a fundamental frequency and the second harmonic frequency as well as the resultant of a fundamental frequency, a second harmonic frequency and a third harmonic frequency.

If more and more harmonic frequencies are added then a true sawtooth wave results.

1.6.3 Noise spikes

In any communication system one of the biggest problems is noise. Noise can be caused by sudden current surges due to drastically changing loads in the supply leads. A typical noise spike is created by a fundamental frequency and all the harmonic frequencies. Again all the frequencies are sine waves but all the harmonic frequencies have the same amplitude as the fundamental frequency. This results in very narrow spikes of large amplitude.

Figure 1.5 shows the fundamental frequency, and the second, third and fourth harmonic frequencies. In this figure the fundamental and harmonic frequencies all

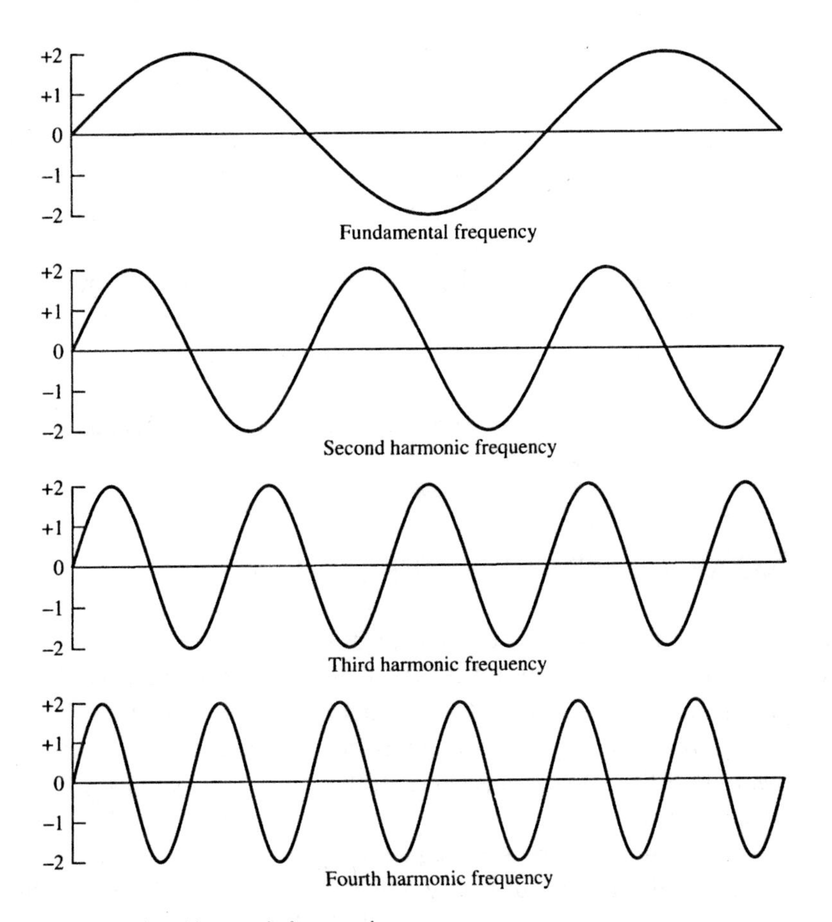

Fig. 1.5 Fundamental and harmonic frequencies

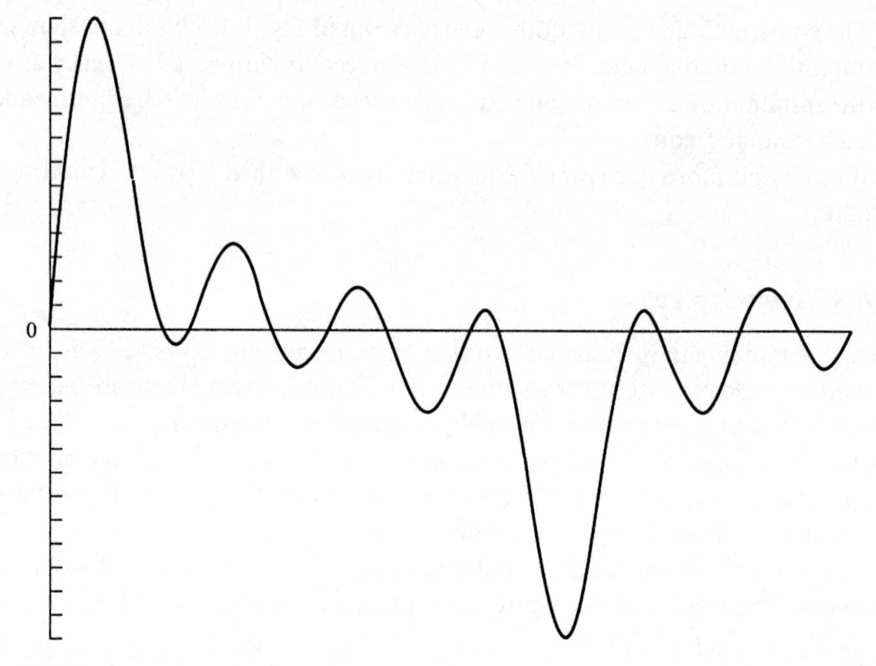

Fig. 1.6 Resultant noise spikes

have the same amplitude. Figure 1.6 shows the resultant wave when these frequencies are added together. By adding more and more harmonic frequencies the spike increases in amplitude and it is reduced in width; also the ripple between the positive and negative spikes is reduced.

These noise spikes can occur in the electricity supply in the early morning when workers in heavy industry switch on their equipment, causing a sudden surge of current in the supply, and in the late afternoon or early evening when the equipment is switched off, creating a sudden drain in the current.

Other causes of noise spikes are induced voltages and currents due to lightning strikes.

1.7 MEASUREMENT OF SIGNAL LEVEL

The gain or loss of a network can be expressed as a ratio of output power over input power. If the output power is greater than the input power then the network has a power ratio >1, which means that the network has a gain. If the output power is less than the input power then the network has a power ratio <1, which means that the network has a loss. If the power ratio equals 1 then the network has a unity gain.

Instead of measuring the power a signal dissipates in watts, it is easier to measure the signal level. The unit of measurement is the bel, but this is a rather large unit for practical purposes, so instead the decibel is used.

1.7.1 The decibel

A decibel (dB) is one-tenth of a bel and can be defined as the loss or gain of a network in a logarithmic form.

Power in (P_{in})—[N]—Power out (P_{out})

The loss or gain N_{Ratio} of the network shown above is given as follows:

$$N_{Ratio} = \frac{P_{out}}{P_{in}}$$

The above equation yields a ratio as both the output power and the input power are in watts.

To enable a unit of measurement to be attached to this ratio, the following formula is used:

$$N = 10\log_{10}\frac{P_{out}}{P_{in}}\ \text{dB}$$

The advantage of using the decibel as a unit of measurement is that the individual gains of the individual networks can be added together instead of being multiplied together.

Example 1.1

Determine a formula for the overall gain of the three cascaded networks shown below as a ratio and in dB.

P_{in} —[N_1]— P_1 —[N_2]— P_2 —[N_3]— P_{out}

$$N_{Ratio} = \frac{P_1}{P_{in}} \times \frac{P_2}{P_1} \times \frac{P_{out}}{P_2}$$

This yields the following as a ratio:

$$N = \frac{P_{out}}{P_{in}}$$

To convert this into dB the following results:

$$N = 10\log_{10}\left(\frac{P_1}{P_{in}} \times \frac{P_2}{P_1} \times \frac{P_{out}}{P_2}\right)$$

$$N = 10\log_{10}\frac{P_1}{P_{in}} + 10\log_{10}\frac{P_2}{P_1} + 10\log_{10}\frac{P_{out}}{P_2}\ \text{dB}$$

Example 1.2

Calculate the gain or loss of each of the following networks in dB:

1. Input power is 4 μW and output power is 16 μW.
2. Input power is 4 mW and output power is 16 mW.
3. Input power is 4 W and output power is 16 W.

Solution

1. $$N = 10\log_{10}\frac{P_{out}}{P_{in}} = 10\log_{10}\frac{16\,\mu W}{4\,\mu W}$$

$$N = 10\log_{10} 4 = 10 \times 0.602 = 6\,dB$$

2. $$N = 10\log_{10}\frac{P_{out}}{P_{in}} = 10\log_{10}\frac{16\,mW}{4\,mW}$$

$$N = 10\log_{10} 4 = 10 \times 0.602 = 6\,dB$$

3. $$N = 10\log_{10}\frac{P_{out}}{P_{in}} = 10\log_{10}\frac{16\,W}{4\,W}$$

$$N = 10\log_{10} 4 = 10 \times 0.602 = 6\,dB$$

In this example it can be seen that the equation yields the same answer for each of the three networks. Hence the dB gives no indication of the input power or output power. In all three examples the network has a gain of 6 dB.

If the input voltage to a network and output voltage from the network are given then the equation becomes

$$N = 20\log_{10}\frac{V_o}{V_i} + 10\log_{10}\frac{Z_i}{Z_o}\ dB$$

where V_o = output voltage
V_i = input voltage
Z_o = output impedance
Z_i = input impedance

If the input and output impedance of the network is the same then the equation becomes

$$N = 20\log_{10}\frac{V_o}{V_i}\ dB$$

where V_o = output voltage
V_i = input voltage

If the input current to a network and output current from the network are given then the equation becomes

$$N = 20\log_{10}\frac{I_o}{I_i} + 10\log_{10}\frac{Z_o}{Z_i}\ dB$$

where I_o = output voltage
I_i = input voltage
Z_o = output impedance
Z_i = input impedance

If the input and output impedance of the network is the same then the equation becomes

$$N = 20\log_{10}\frac{I_o}{I_i}\ dB$$

where I_o = output voltage
I_i = input voltage

1.7.2 The dBm

This unit of measurement is used to measure the actual power at a point in a system relative to 1 mW. This indicates whether the power at the point is greater than 1 mW or less than 1 mW. The dBm is defined as the gain or loss of a network, where the reference signal power is 1 mW, and the power ratio is expressed in a logarithmic form. This is shown by

$$N = 10\log_{10}\frac{P}{1\,\text{mW}}\,\text{dBm}$$

Example 1.3

Determine the signal level for each of the following signal powers in dBm:

1. 4 μW
2. 16 μW
3. 4 mW
4. 16 mW

1.
$$N = 10\log_{10}\frac{P}{1\,\text{mW}} = 10\log_{10}\frac{4\,\mu\text{W}}{1\,\text{mW}}$$
$$N = 10\log_{10}0.004 = -23.98\,\text{dBm}$$

2.
$$N = 10\log_{10}\frac{P}{1\,\text{mW}} = 10\log_{10}\frac{16\,\mu\text{W}}{1\,\text{mW}}$$
$$N = 10\log_{10}0.016 = -17.96\,\text{dBm}$$

3.
$$N = 10\log_{10}\frac{P}{1\,\text{mW}} = 10\log_{10}\frac{4\,\text{mW}}{1\,\text{mW}}$$
$$N = 10\log_{10}4 = 6\,\text{dBm}$$

4.
$$N = 10\log_{10}\frac{P}{1\,\text{mW}} = 10\log_{10}\frac{16\,\text{mW}}{1\,\text{mW}}$$
$$N = 10\log_{10}16 = 12\,\text{dBm}$$

As can be seen in the above example the level in dBm indicates the actual power in the signal relative to 1 mW. Table 1.3 illustrates this relationship more clearly.

From this it can be seen that the dBm is used to measure the amount of power at a point in a system.

Refer to Fig. 1.7; the input signal power at the origin of the system is 100 μW, which is a level of −10 dBm. The origin of the system is also referred to as the zero test level point (0 TLP). The total loss of the local telephone exchange and the

Table 1.3 Relationship between power and dBm

Power (mW)	Level (dBm)
16	+12
8	+9
4	+6
2	+3
1	0
0.5	−3
0.25	−6
0.125	−9
0.0625	−12

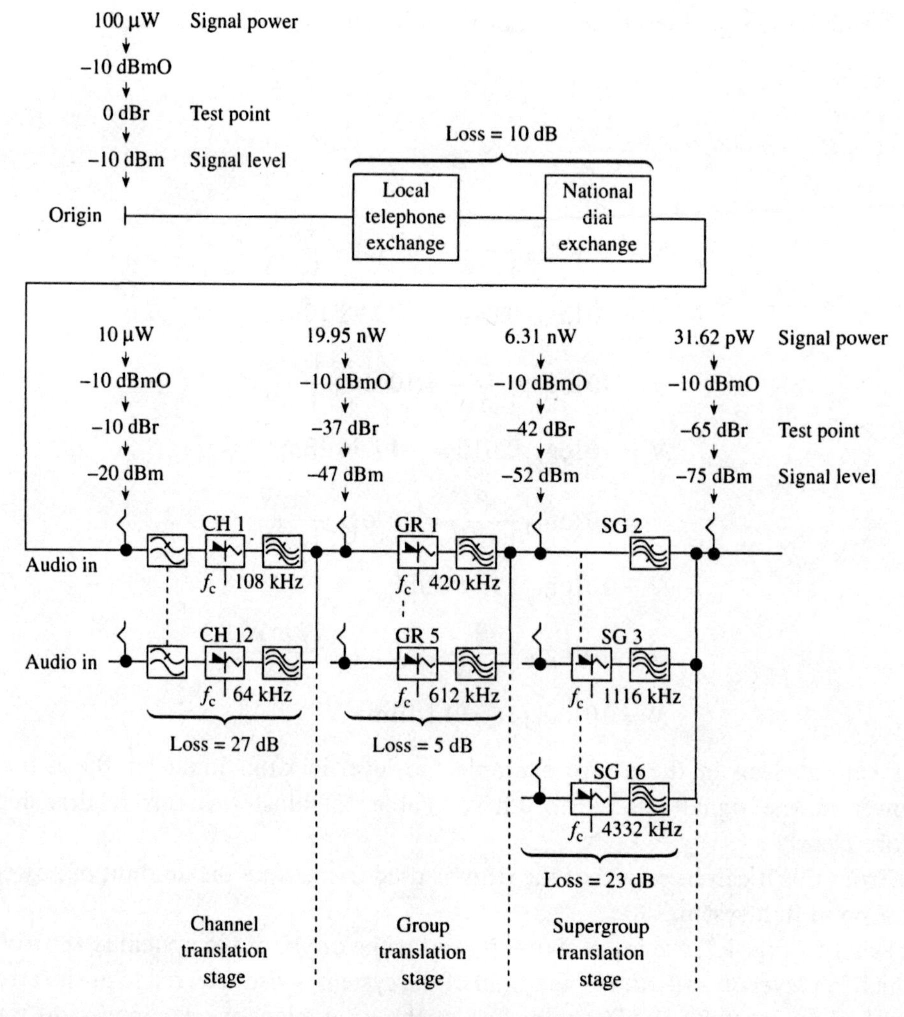

Fig. 1.7 Power losses through a telephone system

national dial exchange is 10 dB. The resultant signal power at the audio-in point is 10 μW, which is equivalent to a signal level of −20 dBm.

The channel presents a loss of 27 dB and as a result the signal power at the output of the channel translation stage is 19.95 nW, which equates to a signal level of −47 dBm. The group translation stage presents a loss of 5 dB and as a result the signal power at the output of the group translation stage is 6.31 nW, which equates to a signal level of −52 dBm. The supergroup translation stage presents a loss of 23 dB, which means that the output signal power from this stage is 31.62 pW, which equates to a signal level of −75 dBm.

1.7.3 The dBr

This unit of measurement is used to indicate the level at a point in a system relative to a level of 0 dBm being sent at the origin of the system or 0 TLP. Refer to Fig. 1.7; the total loss that is caused by the local telephone exchange and the national dial exchange is 10 dB. As can be seen in the figure, the level at the audio-in point relative to the 0 TLP is −10 dBr.

The loss presented to the signal by the channel is 27 dB, hence the level at the output of the channel translation stage relative to the 0 TLP is −37 dBr. The group translation stage presents a loss of 5 dB to the signal, hence the level at the output of the group translation stage relative to the 0 TLP is −42 dBr.

The loss presented to the signal by the supergroup translation stage is 23 dB, hence the level at the output of this stage relative to the 0 TLP is −65 dBr.

1.7.4 The dBmO

This signal level is determined by

$$N = (N\,\text{dBm} - N\,\text{dBr})\,\text{dBmO}$$

As can be seen in the above equation, it is the difference between the level in dBm and the level in dBr. It should be the same at every test point in a system for the same signal. A practical example of its usage is the half-power points of a filter. These points are often referred to as the 3 dB down points; a more correct way to designate these points is as the −3 dBmO points.

By referring to Fig. 1.7 it can be seen that the dBmO level at each test point is exactly the same. This unit of measurement is often used in fault finding to isolate a faulty piece of equipment in a system.

1.8 REVIEW QUESTIONS

1.1 Describe the relationships that must exist between a fundamental frequency and the harmonic frequencies for a square wave.
1.2 Describe, with sketches, the construction of a sawtooth wave.

1.3 Define each of the following:
 (a) A sine wave.
 (b) An analogue signal.
 (c) A digital signal.

1.4 Describe why noise spikes are so troublesome to telecommunication signals.

1.5 Calculate the gain or loss in dB for each of the following:
 (a) Input RMS voltage is 3.4 mV, output RMS current is 2.5 mA, input impedance is 550 Ω and the output impedance is 220 Ω.
 (b) The input power is 25 mW and the output power is 2.5 mW.
 (c) The input RMS current is 2.4 mA and the output peak current is 5 mA. The input and output impedance of the network is the same.

1.6 Calculate the signal level in dBm for each of the following:
 (a) 5.5 mW.
 (b) 6.4 V RMS across 600 Ω.
 (c) 0.8 mA RMS into 350 Ω.

2 Analogue modulation principles

Technology can help improve the quality of life.

2.1 INTRODUCTION

In order to understand how transmission systems translate the input frequency band to a higher frequency band it is necessary to understand the fundamentals of modulation and modulators. It is also necessary to understand the demodulation process and demodulators.

Firstly, the products of modulation for amplitude modulators are shown very clearly by means of line graphs in this chapter. The line graph is also used to describe the concepts behind frequency division multiplexing. Practical modulators produce products of modulation that are different to the theoretical model. The necessity for things such as single sideband transmission with suppressed carrier on high-quality media is explained and the resultant advantages that this gives.

Secondly, the principle of frequency modulation is necessary as this modulation technique is used extensively in microwave and satellite systems. A brief introduction to frequency modulation is given in this chapter.

Thirdly, the principle of phase modulation is necessary. Adaptations of phase modulation are used extensively in digital systems. This chapter also gives a brief introduction to the concept of phase modulation.

2.2 FREQUENCY BAND CLASSIFICATIONS

The electromagnetic spectrum consists of frequencies from just above 0 Hz to ∞ Hz. The frequency spectrum is broken up into different segments which indicate the usage for particular frequency range. Table 2.1 gives the classification of the different frequency bands as used in telecommunications.

Table 2.1 Classification of frequency bands

Frequency band	Classification	Abbreviation
20 Hz–21 kHz	Audio frequency band	–
300 Hz–3.4 kHz	Commercial speech band	AF
10 kHz–30 kHz	Very low frequencies	VLF
30 kHz–300 kHz	Low frequencies	LF
300 kHz–3 MHz	Medium frequencies	MF
3 MHz–30 MHz	High frequencies	HF
30 MHz–300 MHz	Very high frequencies	VHF
300 MHz–3 GHz	Ultra high frequencies	UHF
3 GHz–30 GHz	Super high frequencies	SHF
30 GHz–300 GHz	Extra high frequencies	EHF

The frequency band of 20 Hz to 21 kHz is the total band that the human ear can hear. However, very few people are capable of hearing this full range. In modern society there is a lot of manmade noise. The amplitude of this noise is partially responsible for desensitising our ears to the full range of frequencies that we should be able to hear. Ageing causes the ear to hear less of the frequency range. This means that as people age the bandwidth of their ears decreases, such that people stop hearing the higher frequencies in the frequency band.

The frequency band 300 Hz to 3400 Hz is referred to as the commercial speech band because this is the band that ITU recommended for use for the conveyance of speech over commercial telephony circuits; it is also referred to as the audio frequency (AF) band. When the CCITT[1] conducted tests it was found that speech could be conveyed and still be totally intelligible in a limited band of 3.1 kHz (300 Hz to 3400 Hz). However, for the conveyance of music either a 10 kHz or preferably a 15 kHz band is necessary. The CCITT discovered that the fundamental frequency of the average male voice was around 800 Hz and that of the average female was around 1 kHz.

The VLF band is used for communication under the sea and in deep mines. The sizes of the aerials to transmit these frequencies are too large to make it viable for communication through the air.

The LF, MF, HF and VHF bands are used by commercial radio stations to convey programmes from the studio to listeners.

The VHF and UHF bands are used by commercial TV stations to convey television pictures from the studio to viewers.

The SHF band is used for direct line-of-sight communication by means of microwaves. This is used by companies such as BT. This band is also used for satellite communication.

The EHF band is still largely experimental, except for the frequency range of light which is used in optical fibres.

[1] The Consultative Committee of International Telegraphy and Telephony (see Section 12.7).

2.3 MODULATION TECHNIQUES

The fundamental reason for modulation is to enable efficient use of the available frequency spectrum, on a particular medium, that is used for the communication. As a result many different modulation techniques have been developed. Each method has advantages and disadvantages, and some methods are easier to implement than others. Some modulation techniques favour one type of medium rather than another type.

All the different modulation techniques can be classified under the following main headings:

1. Amplitude modulation schemes.
2. Angle modulation schemes composed of:
   (a) Frequency modulation.
   (b) Phase modulation.
3. Composite modulation schemes composed of:
   (a) Pulse position modulation.
   (b) Pulse width modulation.
   (c) Pulse amplitude modulation.
   (d) Pulse code modulation.
   (e) Delta modulation.
   (f) Phase shift key modulation.
   (g) Frequency shift key modulation.
   (h) Quadrature amplitude modulation.

2.4 AMPLITUDE MODULATION

In amplitude modulation, the amplitude of the carrier frequency is made to vary in sympathy with the amplitude of the modulating signal. A true amplitude modulator produces a waveform that is shown in Fig. 2.1. The carrier frequency is designated as f_c and the modulating frequency is designated as f_a. Notice the variation of the amplitude of the carrier frequency. The shape of the resultant envelope is the same as that of the modulating frequency. Figure 2.1 shows a single modulating frequency. In practice the modulating signal will be a band of frequencies. This band of frequencies could be the commercial speech band which is used over a telephone network.

The mathematical relationship is given by

$$v_i = [f_a(t) + V_c] \sin n\omega t$$

where V_c is the peak amplitude of the carrier frequency, the instantaneous amplitude of the modulated signal is v_i and the instantaneous amplitude of the carrier frequency v_c is given by

$$v_c = V_c \sin n\omega t$$

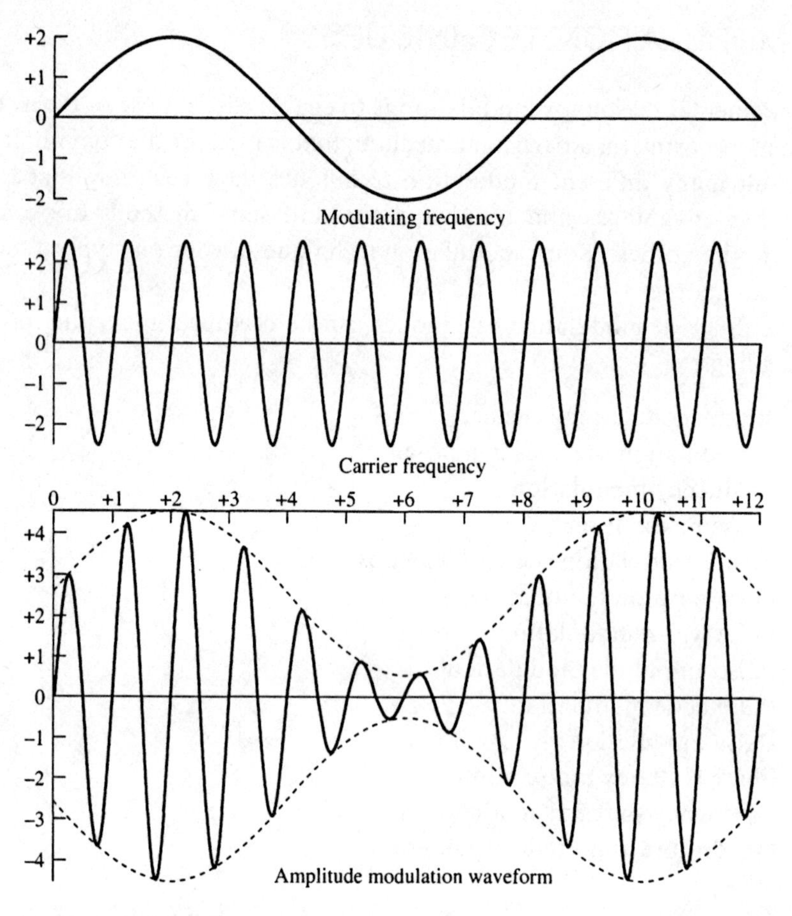

Fig. 2.1 Amplitude modulation

The instantaneous amplitude of the modulating frequency is given by

$$v_a = f_a(t)$$

As $[f_a(t) + V_c]$ is time dependent, this results in the amplitude of the carrier frequency being a function of the amplitude of the modulating frequency.

The complex waveform shown in Fig. 2.1 contains the following products of modulation:

- The carrier frequency.
- The lower sideband.
- The upper sideband.

The symbol for the modulator is shown in Fig. 2.2. Notice that the arrow points in the direction of the propagation of the information.

2.4.1 Sidebands

The upper sideband is determined by means of the following:

$$\text{USB} = f_c + f_a$$

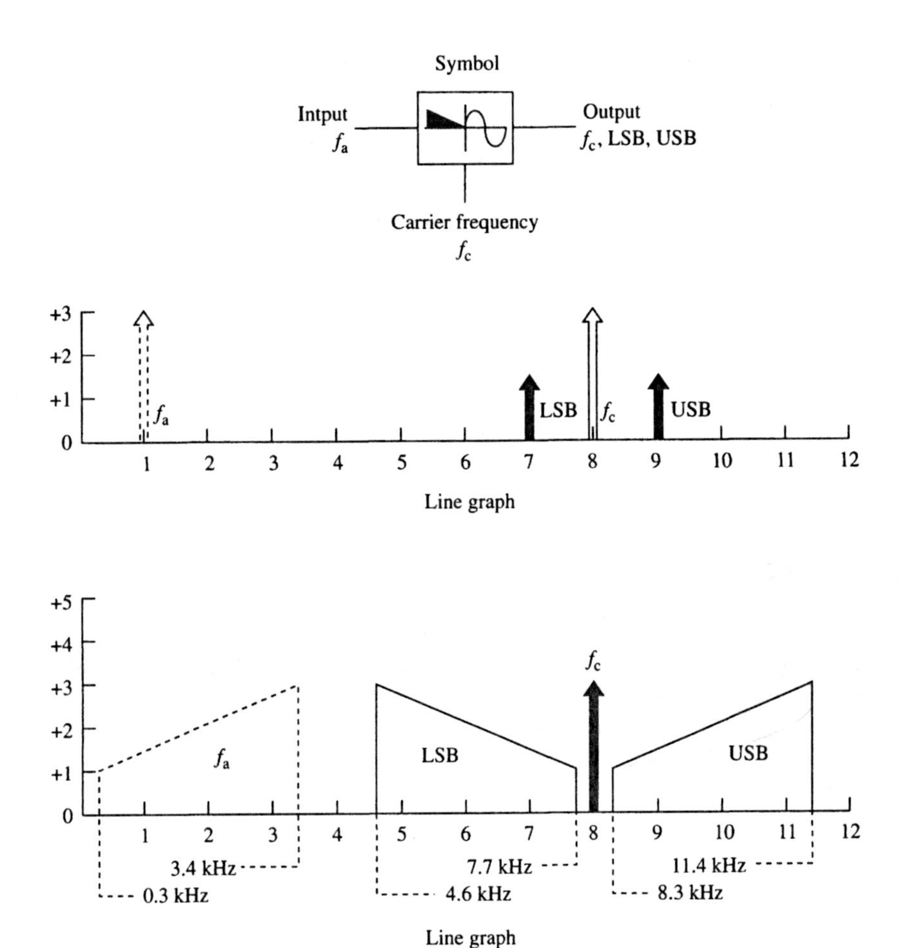

Fig. 2.2 Line graphs for double sideband transmission

The lower sideband is determined by the following:

$$\text{LSB} = |f_c - f_a|$$

The modulus is taken so as to produce a positive result. At a demodulator the modulating frequency is sometimes higher than the carrier frequency. To ensure that the answer is positive the modulus is taken as there are no negative frequencies. The frequency starts at just above 0 Hz and increases in a positive direction.

Example 2.1

Determine the frequency of the lower sideband and upper sideband for an amplitude modulator where the carrier frequency is 8 kHz and the modulating frequency is 1 kHz.

Solution

The upper sideband:

$$\begin{aligned}\text{USB} &= f_c + f_a \\ &= 8\,\text{kHz} + 1\,\text{kHz} \\ &= 9\,\text{kHz}\end{aligned}$$

The lower sideband:

$$\begin{aligned}\text{LSB} &= |f_c - f_a| \\ &= |8\,\text{kHz} - 1\,\text{kHz}| \\ &= 7\,\text{kHz}\end{aligned}$$

Example 2.2

Determine the frequency band of the upper sideband and the lower sideband for an amplitude modulator having a carrier frequency of 8 kHz and a modulating signal of 300 Hz to 3400 Hz.

Solution

The upper sideband:

$$\begin{aligned}\text{USB} &= f_c + f_a \\ &= 8\,\text{kHz} + (0.3\,\text{kHz} \leftrightarrow 3.4\,\text{kHz}) \\ &= 8\,\text{kHz} + 0.3\,\text{kHz} \leftrightarrow 8\,\text{kHz} + 3.4\,\text{kHz} \\ &= 8.3\,\text{kHz} \leftrightarrow 11.4\,\text{kHz}\end{aligned}$$

The lower sideband:

$$\begin{aligned}\text{LSB} &= |f_c - f_a| \\ &= |(8\,\text{kHz} - 0.3\,\text{kHz} \leftrightarrow 8\,\text{kHz} - 3.4\,\text{kHz})| \\ &= 7.7\,\text{kHz} \leftrightarrow 4.6\,\text{kHz} \\ &= 4.6\,\text{kHz} \leftrightarrow 7.7\,\text{kHz}\end{aligned}$$

2.4.2 Line graph

A line graph is a tool that is used to graphically show the products of modulation. Figure 2.2 shows the line graphs for Examples 2.1 and 2.2. Notice that the arrowhead points in the direction of the lowest frequency that produced that particular frequency. In Example 2.2 the frequencies of 7.7 kHz and 8.3 kHz are produced by the 300 Hz component. The arrowhead indicates this fact. The frequencies of

4.6 kHz and 11.4 kHz are produced by the 3.4 kHz component. The backs of the arrows indicate this.

2.4.3 Sideband transmission

The example shown in Fig. 2.2 is for double sideband (DSB) transmission. Double sideband transmission is wasteful of the frequency spectrum as the same information is carried in the USB and the LSB. In practice, on very good media only one of the sidebands is sent, and this is known as single sideband (SSB) transmission. On large-capacity point-to-point systems the carrier frequency is suppressed at the transmitter and reintroduced at the receiver. This means that only the one sideband is transmitted over the media.

This method of transmission is illustrated in Fig. 2.3. Notice that this filter at the output of the modulator either selects the USB or the LSB.

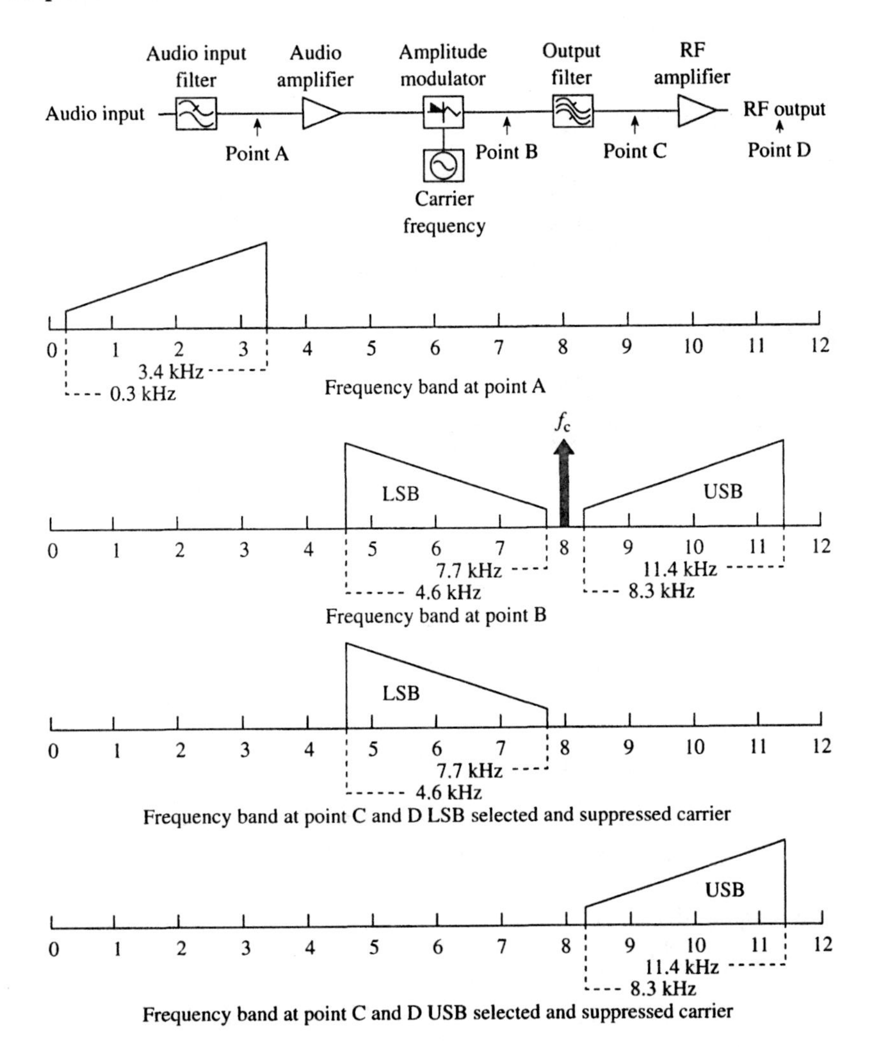

Fig. 2.3 Line graphs for single sideband transmission

Example 2.3

Transmitter: a carrier frequency of 108 kHz and a modulating frequency of 0.3 kHz to 3.4 kHz is applied to an amplitude modulator. The USB and the LSB are both transmitted to the distant receiver, with a suppressed carrier.

Receiver: the resultant USB and LSB from the transmitter are input to an amplitude demodulator. A carrier frequency of 108 kHz is reintroduced to the demodulator.

Determine the following:

1. The LSB and USB at the output of the modulator.
2. The LSB and USB at the output of the demodulator.

Solution

1. *Transmitter*

The upper sideband:

$$\begin{aligned}\text{USB} &= f_c + f_a \\ &= 108\,\text{kHz} + (0.3\,\text{kHz} \leftrightarrow 3.4\,\text{kHz}) \\ &= 108\,\text{kHz} + 0.3\,\text{kHz} \leftrightarrow 108\,\text{kHz} + 3.4\,\text{kHz} \\ &= 108.3\,\text{kHz} \leftrightarrow 111.4\,\text{kHz}\end{aligned}$$

The lower sideband:

$$\begin{aligned}\text{LSB} &= |f_c - f_a| \\ &= |(108\,\text{kHz} - 0.3\,\text{kHz} \leftrightarrow 108\,\text{kHz} - 3.4\,\text{kHz})| \\ &= 107.7\,\text{kHz} \leftrightarrow 104.6\,\text{kHz} \\ &= 104.6\,\text{kHz} \leftrightarrow 107.7\,\text{kHz}\end{aligned}$$

This is shown in Fig. 2.3.

2. *Receiver*

Received band = 108.3 kHz to 111.4 kHz. The upper sideband:

$$\begin{aligned}\text{USB} &= f_c + f_a \\ &= 108\,\text{kHz} + (108.3\,\text{kHz} \leftrightarrow 111.4\,\text{kHz}) \\ &= 108\,\text{kHz} + 108.3\,\text{kHz} \leftrightarrow 108\,\text{kHz} + 111.4\,\text{kHz} \\ &= 216.3\,\text{kHz} \leftrightarrow 219.4\,\text{kHz}\end{aligned}$$

The lower sideband:

$$\begin{aligned}\text{LSB} &= |f_c - f_a| \\ &= |(108\,\text{kHz} - 108.3\,\text{kHz} \leftrightarrow 108\,\text{kHz} - 111.4\,\text{kHz})| \\ &= 0.3\,\text{kHz} \leftrightarrow 3.4\,\text{kHz}\end{aligned}$$

Received band = 104.6 kHz to 107.7 kHz. The upper sideband:

$$\begin{aligned}\text{USB} &= f_c + f_a \\ &= 108\,\text{kHz} + (104.6\,\text{kHz} \leftrightarrow 107.7\,\text{kHz}) \\ &= 108\,\text{kHz} + 104.6\,\text{kHz} \leftrightarrow 108\,\text{kHz} + 107.7\,\text{kHz} \\ &= 212.6\,\text{kHz} \leftrightarrow 215.7\,\text{kHz}\end{aligned}$$

The lower sideband:

$$\begin{aligned}\text{LSB} &= |f_c - f_a| \\ &= |(108\,\text{kHz} - 107.7\,\text{kHz} \leftrightarrow 108\,\text{kHz} - 104.6\,\text{kHz})| \\ &= 0.3\,\text{kHz} \leftrightarrow 3.4\,\text{kHz}\end{aligned}$$

From the above it can be seen that when the USB from the transmitter is demodulated and the LSB taken a band of 0.3 kHz to 3.4 kHz is yielded. When the LSB from the transmitter is demodulated and the LSB also taken then the band of 0.3 kHz to 3.4 kHz results.

The USB that results when the USB from the transmitter is demodulated is a frequency range beyond human hearing and of such a high frequency that it will not be able to propagate down the telephone cable pair effectively. The same applies to the USB that results when the LSB from the transmitter is demodulated.

As can be seen the USB and LSB at the transmitter carry the same information and hence it is not necessary to transmit both sidebands over the medium, such as a point-to-point circuit working over a metallic pair. Single sideband transmission results in a more efficient use of the available frequency spectrum, as only 4 kHz need be allocated to each channel instead of 8 kHz. This means that twice the number of SSB channels can be used when compared to DSB channels in the same frequency band.

2.4.4 Comparison of single sideband and double sideband transmission

A comparison of SSB and DSB transmission given in Table 2.2. This table shows the advantages of using single sideband transmission. Over poor quality media, however, it is preferable to use DSB and at the receiver select either the USB or

Table 2.2 Comparison of SSB and DSB

	Advantages of SSB
Bandwidth	Requires at most half the bandwidth of DSB for the same information. Allows better use of the available frequency spectrum.
Power	Requires less power than DSB, which leads to: Increased transmitter efficiency. Reduced amplifier distortion.

the LSB at the input to the demodulator. The selection is based on the amount of noise in the two bands as well as the signal strength in the two bands.

2.5 FREQUENCY DIVISION MULTIPLEXING

2.5.1 Introduction

Frequency division multiplexing (FDM) is used to enable the frequency spectrum of a particular medium to be efficiently used. There are a number of different systems on which FDM is used. Some of these systems are listed below:

- 3-channel open wire system.
- 12-channel low-frequency open wire system.
- 12-channel high-frequency open wire system.
- 12-channel carrier on cable system.
- 18-channel rural carrier system.
- 960-channel coaxial cable system.
- 900-channel coaxial cable system.
- 1800-channel coaxial cable system.
- 2700-channel coaxial cable system.

All the above systems are point-to-point analogue systems where each channel uses the commercial speech band.

Only the 2700-channel coaxial cable system will be discussed.

2.5.2 12 MHz coaxial cable system

Figure 2.4 shows a typical coaxial cable system which uses FDM. With FDM each channel is permanently allocated a particular frequency band for all time on the common transmission medium, irrespective of whether the channel is used to convey valid information or not.

The system is based on a basic group which comprises 12 channels. The commercial speech band (0.3 kHz to 3.4 kHz) is applied to the input of each channel. The carrier frequencies for the individual channels are given in Table 2.3.

For practical purposes the commercial speech band is considered to have a band width of 4 kHz. The actual bandwidth is 3.1 kHz. But it must be remembered that space must be left on each side of the band to ensure that the filter can select the appropriate band from the complete frequency range.

All the carrier frequencies and frequency bands are CCITT recommended. Since the time when the recommendations for FDM systems were laid down, CCITT has changed and become part of the ITU.

As can be seen in Table 2.3 the LSB is selected at the output of each channel. The carrier frequency is also suppressed, so the resultant band is 60 kHz to 108 kHz.

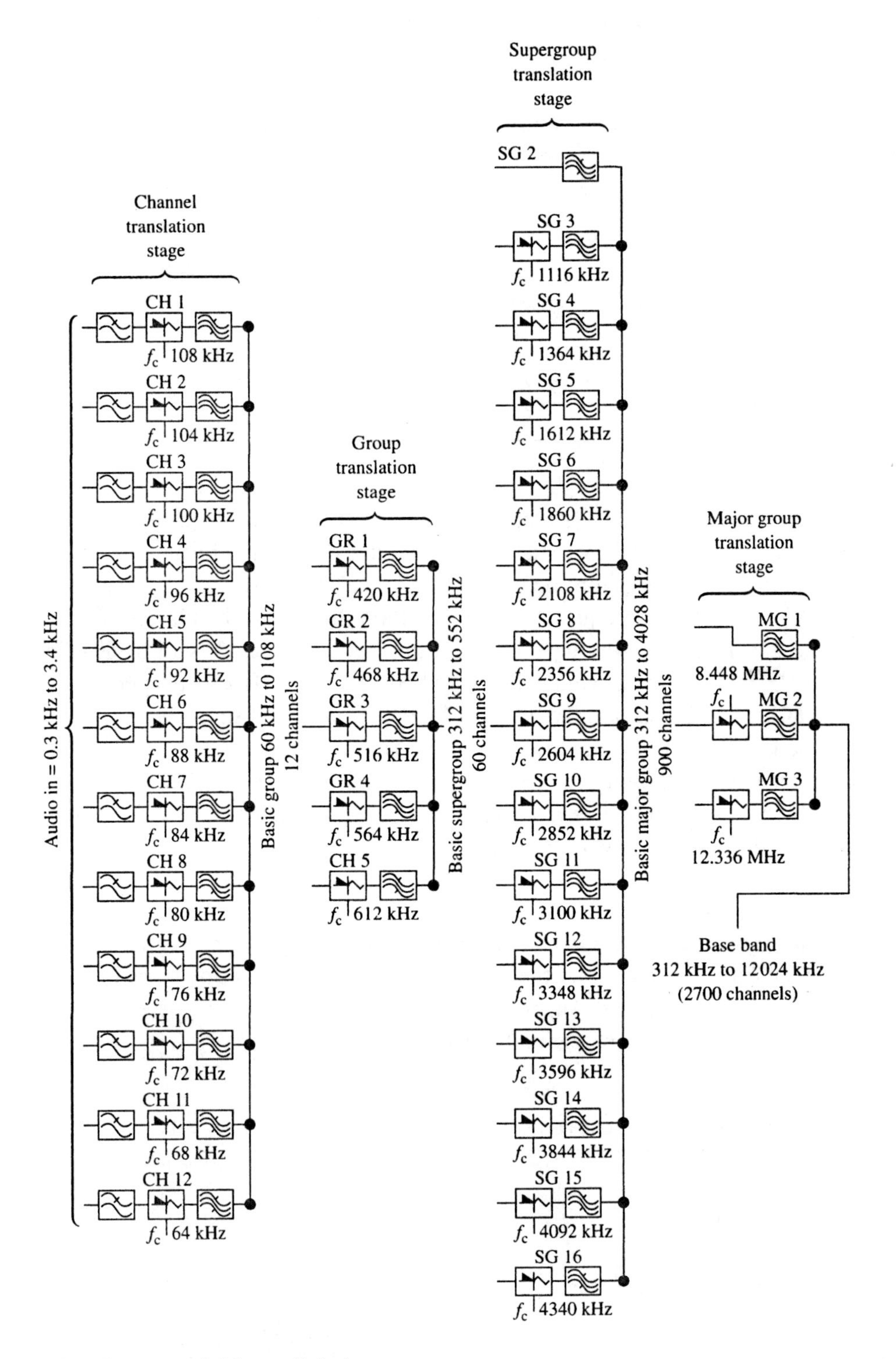

Fig. 2.4 Frequency division multiplexing

Table 2.3 Carrier frequencies for 12-channel groups

Channel number	Carrier frequency (kHz)	Output frequency band (kHz)
1	108	104–108
2	104	100–104
3	100	96–100
4	96	92–96
5	92	88–92
6	88	84–88
7	84	80–84
8	80	76–80
9	76	72–76
10	72	68–72
11	68	64–68
12	64	60–64

The output of the channel translation stage is fed into the group translation stage. The group translation stage consists of five groups and forms a basic supergroup. A basic supergroup consists of 60 channels (five groups × 12 channels per group). The carrier frequencies and resultant frequency bands are shown in Table 2.4. Note that the input modulating frequency band is from 60 kHz to 108 kHz, which gives a bandwidth of 48 kHz (12 channels × 4 kHz per channel). The 60 kHz to 108 kHz is the output of the channel translation stage.

The LSB is again taken at the output of each group card and again the carrier frequency is suppressed. The frequency band of 312 kHz to 552 kHz, which is a bandwidth of 240 kHz (five groups × 48 kHz per group), is sent to the supergroup translation equipment where a basic major group is formed. A basic major group consists of 15 supergroups, which is equivalent to 900 channels (15 supergroups × 5 groups/supergroup × 12 channels per group).

The carrier frequencies for the supergroups and the resultant frequency band is given in Table 2.5. Note that the output of the group translation stage is 312 kHz to 552 kHz. This is a basic supergroup and this is the modulating frequency band that is fed into each supergroup translation card.

Table 2.4 Group carrier frequencies and output frequency bands

Group No.	Carrier frequency (kHz)	Output frequency band (kHz)
1	420	312–360
2	468	360–408
3	516	408–456
4	564	456–504
5	612	504–552

Table 2.5 Supergroup carrier frequencies and output frequency bands

Supergroup No.	Carrier frequency (kHz)	Output frequency band (kHz)
2	—	312–552
3	1116	564–804
4	1364	812–1052
5	1612	1060–1300
6	1860	1308–1548
7	2108	1556–1796
8	2356	1804–2044
9	2604	2052–2292
10	2852	2300–2540
11	3100	2548–2788
12	3348	2796–3036
13	3596	3044–3284
14	3844	3292–3532
15	4092	3540–3780
16	4340	3788–4028

The LSB is selected at the output of each supergroup card and again the carrier is suppressed. The resultant band of 312 kHz to 4028 kHz is sent to the major group translation stage. The major group translation stage accepts three basic major groups. The carrier frequencies and resultant frequency bands are given in Table 2.6.

The LSB is selected at the output of major groups 2 and 3. Major group 1 is unmodulated and is only combined with the modulated outputs of the other two major groups. The resultant baseband is from 312 kHz to 12 024 kHz. This band is now sent to the terminal equipment where it is amplified and then transmitted down the coaxial cable.

A coaxial cable, such as that used to connect a TV aerial to a TV receiver, is made up of a thick cylindrical copper conductor in the centre which is surrounded by a copper sheath. The inner conductor is prevented from touching the outer sheath by means of spacers, i.e. insulation material placed around the inner conductor. This construction is sometimes referred to as a coaxial tube.

Table 2.6 Major group carrier frequencies and output frequency bands

Major group No.	Carrier frequency (kHz)	Output frequency band (kHz)
1	—	312–4 028
2	8 448	4 420–8 136
3	12 336	8 308–12 024

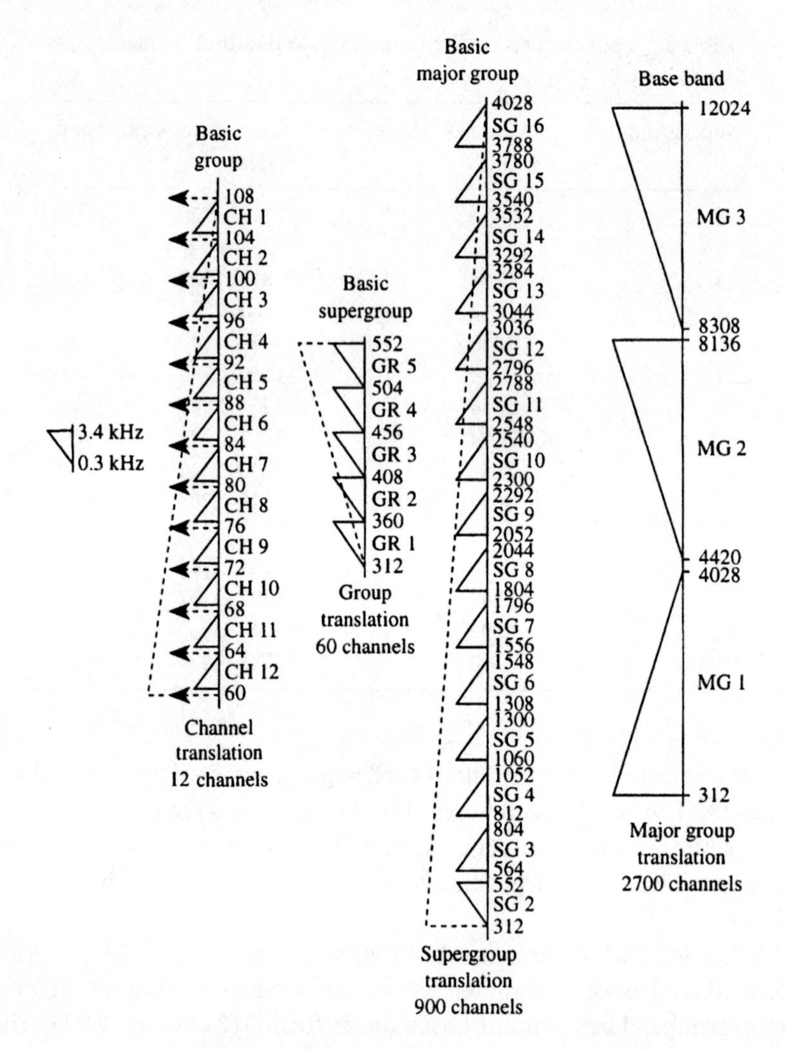

Fig. 2.5 Line graph for a 12 MHz coaxial system

For a high-capacity 12 MHz coaxial cable system two tubes are used, one for each direction of transmission. This means that each channel works on a four-wire basis. This system has a channel capacity of 2700 channels.

Figure 2.5 shows the line graph for the 12 MHz FDM system described above. Notice the sideband inversion due to the LSB being selected at the output of each stage, and being passed to the next stage.

2.6 MODULATION DEPTH

The modulation depth, sometimes referred to as the modulation index, is normally measured as a percentage. The modulation depth is a measure of the efficiency of the modulator. All modulators must be protected from overmodulation, which occurs when the input signal becomes too large. Such high-amplitude signals can damage the modulator. Overmodulation produces frequency components that can interfere with other circuits. This means that the output bandwidth is increased. Overmodulation also results in less power in the desired sidebands.

2.6.1 Mathematical treatment

The modulation depth is given by

$$m = \frac{\text{maximum amplitude} - \text{minimum amplitude}}{\text{maximum amplitude} + \text{minimum amplitude}} \times 100\%$$

$$= \frac{(V_c + V_a) - (V_c - V_a)}{(V_c + V_a) + (V_c - V_a)} \times 100\%$$

$$= \frac{(V_c - V_a - V_c + V_a)}{(V_c + V_a + V_c - V_a)} \times 100\%$$

$$= \frac{2V_a}{2V_c} \times 100\%$$

$$= \frac{V_a}{V_c} \times 100\%$$

where V_c is the peak amplitude of the carrier frequency and V_a is the peak amplitude of the modulating signal.

Example 2.4

Determine the modulation index for an amplitude modulator when a carrier frequency having peak amplitude of 2 V is modulated with a modulating frequency having a peak amplitude of 1.5 V.

Solution

$$m = \frac{V_a}{V_c} \times 100\%$$

$$= \frac{1.5\,\text{V}}{2.0\,\text{V}} \times 100\%$$

$$= 75\%$$

2.6.2 Overmodulation

Overmodulation occurs when the modulation depth is greater than 100%. Figure 2.6 shows 100% modulation. It would be reasonable to assume that all amplitude modulators should operate at this depth. This is in fact false, as the modulation index must allow for a variation in the amplitude of the input

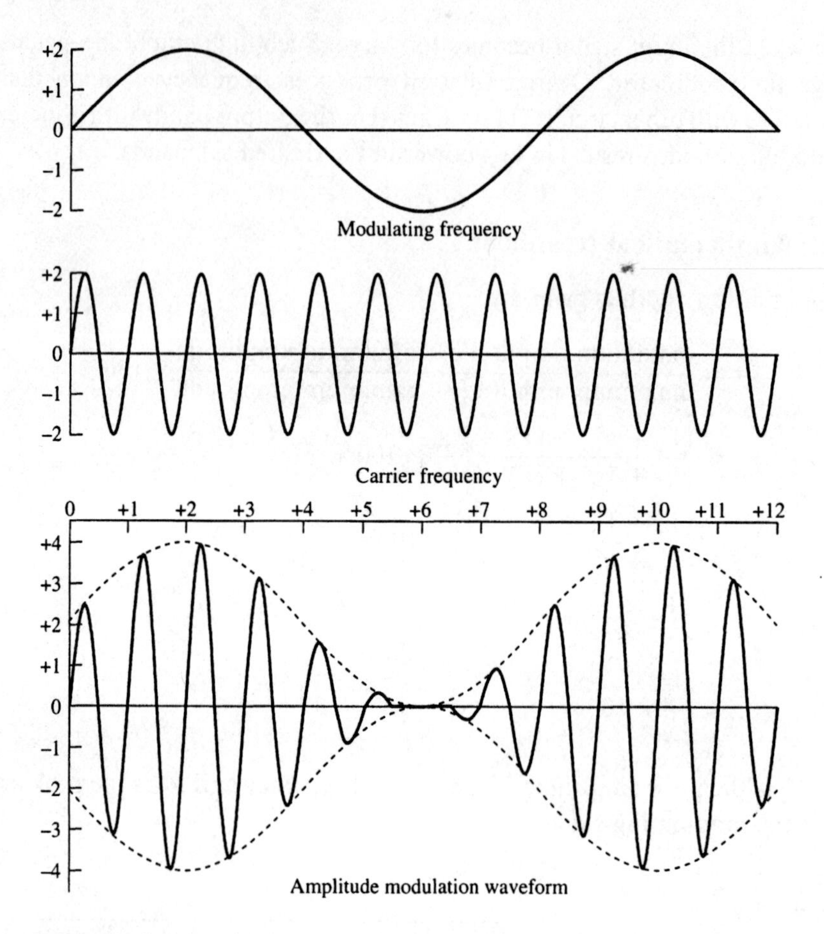

Fig. 2.6 One-hundred per cent modulation

signal and still ensure that overmodulation does not take place. Not all people talk over the telephone with the same average loudness; some talk very softly and some very loudly. The amplitude modulators must not cause overmodulation in either case. They should also not cause unnecessary distortion in the output signal relative to the input signal. As a result, practical amplitude modulators are set up to work with a modulation depth of between 70% and 80%.

Overmodulation causes frequencies to appear in the output signal other than the expected products for the modulator. It results in increased bandwidth requirements, and less power in the desired sidebands. A typical overmodulation waveform is shown in Fig. 2.7.

Example 2.5

Calculate the modulation depth for an amplitude modulator where the carrier frequency has a peak-to-peak amplitude of 500 mV and the modulating frequency has a RMS amplitude of 194.45 mV.

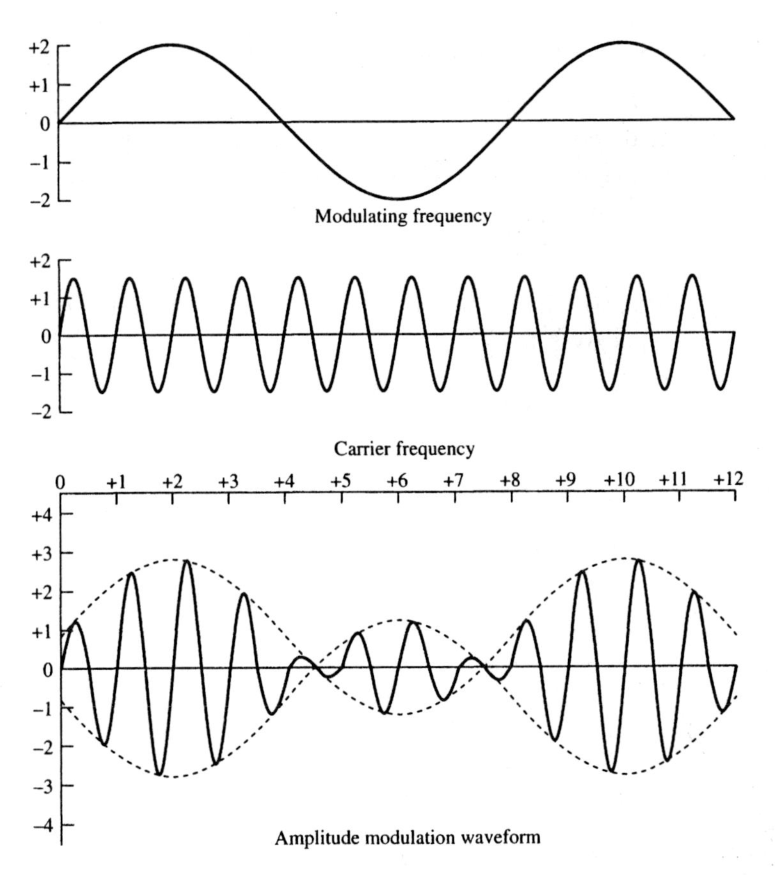

Fig. 2.7 Overmodulation

Solution

Convert to the same unit of measurement:

$$V_{a_{peak}} = V_{a_{RMS}} \times \sqrt{2}$$
$$= 194.45\,\text{mV} \times \sqrt{2}$$
$$= 274.994\,\text{mV}$$
$$= \frac{V_{c_{peak\text{-}to\text{-}peak}}}{2}$$
$$= \frac{500\,\text{mV}}{2}$$
$$= 250\,\text{mV}$$
$$m = \frac{V_a}{V_c} \times 100\%$$
$$= \frac{274.994\,\text{mV}}{250\,\text{mV}} \times 100\%$$
$$= 109.998\%$$

This example demonstrates how overmodulation can take place.

2.6.3 Power in the sidebands and carrier frequency

A second important parameter is the actual power that exists in each sideband and the amount of power that exists in the carrier. The relationship between the total power dissipated, the power in the carrier frequency and the power in the sidebands is given by

$$P_T = P_c + P_{USB} + P_{LSB}$$

The power in the carrier frequency (P_c) where the carrier amplitude is given as a peak voltage (V_c) is given by

$$P_c = \frac{(V_{c_{RMS}})^2}{\text{load impedance}}$$

But

$$V_{c_{RMS}} = \frac{V_c}{\sqrt{2}}$$

Therefore

$$P_c = \frac{\left(\frac{V_c}{\sqrt{2}}\right)^2}{R}$$

$$= \frac{(V_c)^2}{2R}$$

The power in each sideband is given by:

$$P_{USB} = P_{LSB} = \frac{\left(\frac{m}{2}\frac{V_c}{\sqrt{2}}\right)^2}{R}$$

$$= \frac{m^2 V_c^2}{8R}$$

Therefore the power in both sidebands is given by:

$$P_{SB} = \frac{2m^2 V_c^2}{8R}$$

$$= \frac{m^2 V_c^2}{4R}$$

But the total power dissipated is given by:

$$P_T = P_c + P_{USB} + P_{LSB}$$

$$= P_c + P_{SB}$$

$$= \frac{V_c^2}{2R} + \frac{m^2 V_c^2}{4R}$$

$$= \frac{V_c^2}{2R}\left(1 + \frac{m^2}{2}\right)$$

The equations given above can only be used when the voltage of the carrier is given as a peak voltage.

The efficiency η of the modulator is measured by the ratio of the power in the sidebands to the total power dissipated:

$$\eta = \frac{P_{SB}}{P_T} \times 100\%$$

Example 2.6

A sinusoidal modulating frequency is modulated with a carrier frequency having a peak voltage of 8 V. The output is connected to a 5 kΩ load. Determine each of the following:

1. The power in the carrier frequency.
2. The power in each sideband.
3. The power in both sidebands.
4. The total power.
5. The efficiency of the modulator.

For each of the following modulation indexes *m*: 25%, 50%, 75%, 100%.

Solution

$m = 25\%$:

1.
$$P_c = \frac{(V_c)^2}{2R} = \frac{(8)^2}{2 \times 5000} = \frac{64}{10\,000} = 6.4\,\text{mW}$$

2.
$$P_{USB} = P_{LSB} = \frac{m^2 V_c^2}{8R} = \frac{0.25^2 \times 8^2}{8 \times 5000} = \frac{0.0625 \times 64}{40\,000} = 0.1\,\text{mW}$$

3.
$$P_{SB} = P_{USB} + P_{LSB} = 0.1\,\text{mW} + 0.1\,\text{mW} = 0.2\,\text{mW}$$

4.
$$P_{total} = P_{SB} + P_c = 0.2\,\text{mW} + 6.4\,\text{mW} = 6.6\,\text{mW}$$

5.
$$\eta = \frac{P_{SB}}{P_T} \times 100\%$$
$$= \frac{0.2\,\text{mW}}{6.6\,\text{mW}} \times 100\% = 3\%$$

$m = 50\%$:

1.
$$P_c = \frac{V_c^2}{2R}$$
$$= 6.4\,\text{mW}$$

2.
$$P_{USB} = P_{LSB} = \frac{m^2 V_c^2}{8R}$$
$$= \frac{0.5^2 \times 8^2}{8 \times 5000}$$
$$= \frac{0.25 \times 64}{40\,000} = 0.4\,\text{mW}$$

3.
$$P_{SB} = P_{USB} + P_{LSB}$$
$$= 0.4\,\text{mW} + 0.4\,\text{mW} = 0.8\,\text{mW}$$

4.
$$P_{total} = P_{SB} + P_c$$
$$= 0.8\,\text{mW} + 6.4\,\text{mW} = 7.2\,\text{mW}$$

5.
$$\eta = \frac{P_{SB}}{P_T} \times 100\%$$
$$= \frac{0.8\,\text{mW}}{7.2\,\text{mW}} \times 100\% = 11\%$$

$m = 75\%$:

1.
$$P_c = 6.4\,\text{mW}$$

2.
$$P_{USB} = P_{LSB} = \frac{m^2 V_c^2}{8R}$$
$$= \frac{0.75^2 \times 8^2}{8 \times 5000}$$
$$= \frac{0.5625 \times 64}{40\,000} = 0.9\,\text{mW}$$

3.
$$P_{SB} = P_{USB} + P_{LSB}$$
$$= 0.9\,\text{mW} + 0.9\,\text{mW} = 1.8\,\text{mW}$$

4.
$$P_{total} = P_{SB} + P_c$$
$$= 1.8\,\text{mW} + 6.4\,\text{mW} = 8.2\,\text{mW}$$

5.
$$\eta = \frac{P_{SB}}{P_T} \times 100\%$$
$$= \frac{1.8\,\text{mW}}{8.2\,\text{mW}} \times 100\% = 22\%$$

$m = 100\%$:

1.
$$P_c = 6.4\,\text{mW}$$

2.
$$P_{USB} = P_{LSB} = \frac{m^2 V_c^2}{8R}$$
$$= \frac{1^2 \times 8^2}{8 \times 5000} = 1.6\,\text{mW}$$

3.
$$P_{SB} = P_{USB} + P_{LSB}$$
$$= 1.6\,\text{mW} + 1.6\,\text{mW} = 3.2\,\text{mW}$$

4.
$$P_{total} = P_{SB} + P_c$$
$$= 3.2\,\text{mW} + 6.4\,\text{mW} = 9.6\,\text{mW}$$

5.
$$\eta = \frac{P_{SB}}{P_T} \times 100\%$$
$$= \frac{3.2\,\text{mW}}{9.6\,\text{mW}} \times 100\% = 33\%$$

This example shows that at a modulation depth of 100% the power in both sidebands is only 33% of the total output power. This means that the power in each sideband is only 16.5%. Also shown is that the lower the modulation depth the lower the percentage power in the sidebands. These two facts indicate the disadvantages of amplitude modulation.

2.7 PRACTICAL CIRCUITS

Only diode modulators are considered here. In practice either unbalanced or balanced amplitude modulators are used. The types of circuits discussed are:

- Ring modulators, single and double balanced.
- Cowan modulator (a single balanced circuit).

2.7.1 Single balanced modulators

Ring modulator

Figure 2.8(a) shows a typical single balanced ring modulator. Note where the modulating frequency and the carrier frequency are applied. The carrier frequency is used to either forward bias or reverse bias the two diodes. When the diodes are forward biased by the carrier frequency, the modulating frequency passes through

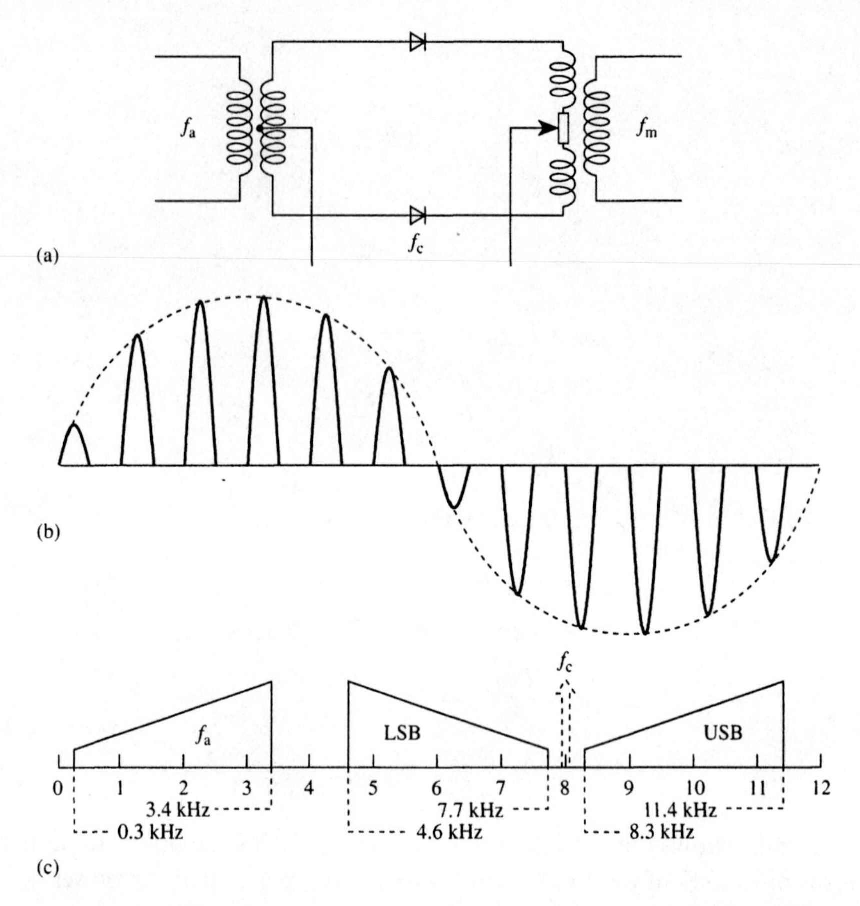

Fig. 2.8 Single balanced ring modulator: (a) circuit; (b) output waveform; (c) products of modulation

to the output. When the diodes are reverse biased by the carrier frequency, the modulating frequency is prevented from reaching the output. The resultant output waveform and line graph are shown in Fig. 2.8(b) and (c). In this figure the carrier frequency is assumed to be 8 kHz and the modulating signal is the commercial speech band.

In this case the products of modulation are the modulating frequency, the lower sideband and the upper sideband, as shown in the line graph. The carrier frequency itself is suppressed through the modulator.

Cowan modulator

Figure 2.9 shows a typical single balanced Cowan modulator. The line graph of the output products is also shown. When the diodes are forward biased by the carrier frequency, the diodes apply a short circuit to the analogue input thus preventing any signal from appearing at the output. When the diodes are reverse biased by the carrier frequency, the analogue input appears at the output.

In this case the products of modulation are the modulating frequency, the lower sideband and the upper sideband. As shown in the line graph the carrier frequency itself is suppressed through the modulator.

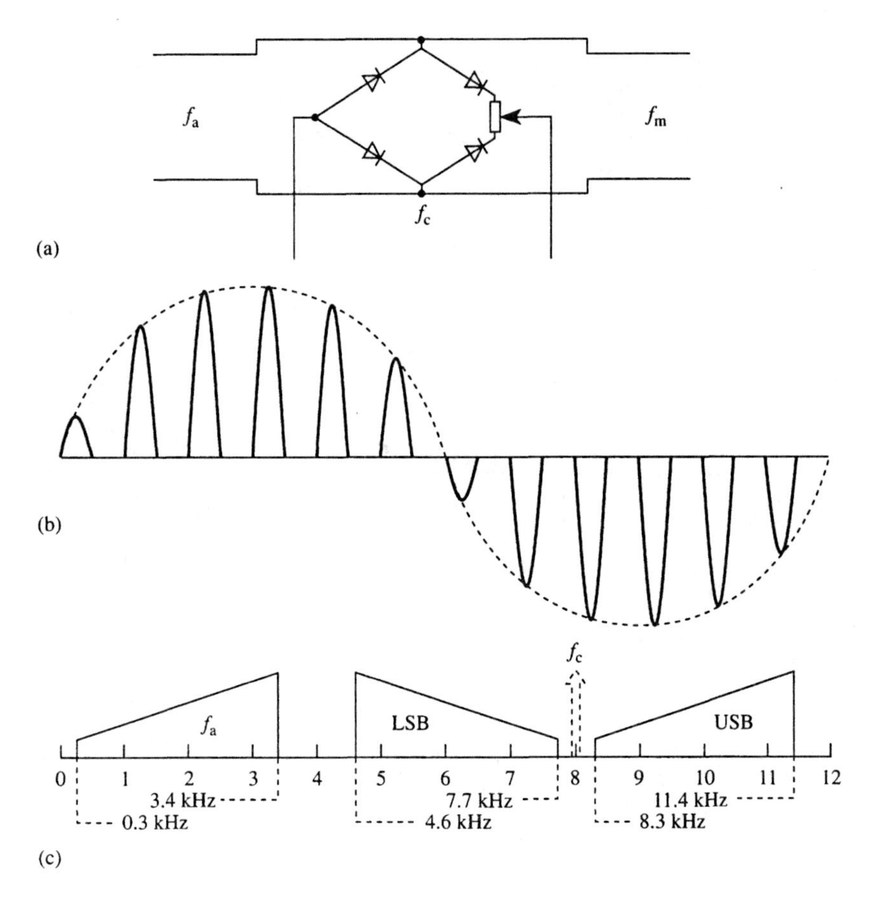

Fig. 2.9 Cowan modulator: (a) circuit; (b) output waveform; (c) products of modulation

Carrier leak

In Figs 2.8 and 2.9, a potentiometer is used to balance the modulator. This is done to ensure there is very little power in the carrier frequency at the output. In practice the carrier frequency is measured at the output of the modulator and the potentiometer is adjusted until the level of the carrier drops as low as possible. In practice the carrier leak should have a level of −38 dBmO.

2.7.2 Double balanced ring modulator

Figure 2.10(a) shows a typical double balanced ring modulator. Note where the modulating frequency and the carrier frequency are applied. The carrier frequency is used to forward bias two of the diodes at a time. The other two diodes will be reverse biased during this time. When the amplitude of the carrier frequency changes, the other two diodes are forward biased and the first two diodes are reverse biased.

This process allows the modulating frequency to always appear at the output, as shown in Fig. 2.10(b). In the figure it is assumed that the carrier frequency is

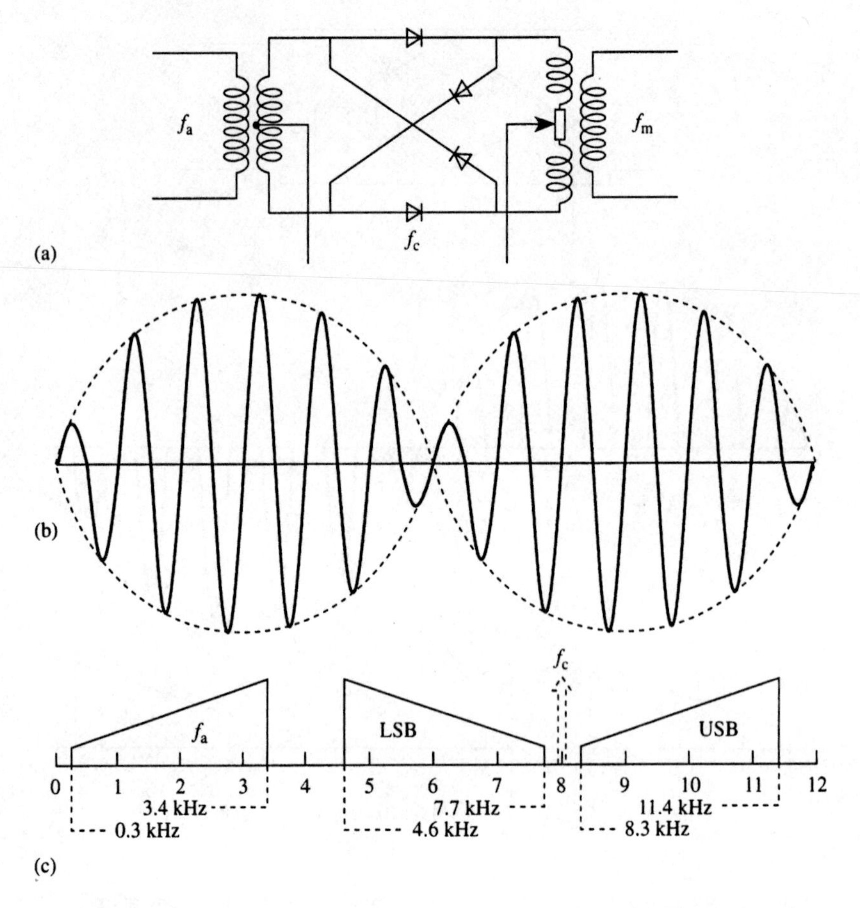

Fig. 2.10 Double balanced ring modulator: (a) circuit; (b) output waveform; (c) products of modulation

8 kHz and the modulating frequency is the commercial speech band. The line graph for the output modulation products is shown in Fig. 2.10(c). As can be seen, the output modulation products are the USB and LSB only. Again a potentiometer is incorporated for the suppression of the carrier frequency at the output and in practice the level of the carrier leak should be −38 dBmO.

2.7.3 Power in the sidebands in practical amplitude modulators

In the double balanced modulator, the carrier frequency is suppressed through the modulation stage. The output products are only the USB and LSB. As a result the output power is equally split between each sideband. The power in each sideband is thus slightly less than −3 dBmO.

In the single balanced modulator the output products of modulation are the USB, the LSB, and the modulating frequency. This means that the output power is equally split between the three products. This means that each sideband has slightly less than 33% of the total output power.

2.8 ANGLE MODULATION

This consists of:

- Frequency modulation.
- Phase modulation.

These modulation methods are very similar.

2.8.1 Frequency modulation

In frequency modulation, the carrier frequency is caused to vary by an amount proportional to the amplitude of the modulating signal, at a rate proportional to the frequency of the modulating signal.

The frequency modulation e is given by

$$e = E_0 \sin(\omega_c t + m \sin \omega_a t)$$

where E_0 = peak amplitude of the carrier frequency
$\omega_c = 2\pi f_c$
$\omega_a = 2\pi f_a$
m = modulation index

This expression can be expanded as follows:

$$\begin{aligned} e = E_0\{&J_0 m \sin \omega_c t \\ &+ J_1 m[\sin(\omega_c + \omega_a)t - \sin(\omega_c - \omega_a)t] \\ &+ J_2 m[\sin(\omega_c + 2\omega_a)t - \sin(\omega_c - 2\omega_a)t] \\ &+ J_3 m[\sin(\omega_c + 2\omega_a)t - \sin(\omega_c - 3\omega_a)t] \\ &+ \cdots\} \end{aligned}$$

where $J_n m$ = Bessel function of the first kind and nth order, with argument m

Referring to Fig. 2.11 it can be seen that the greater the amplitude of the modulating signal the greater the frequency swing. The higher the frequency of the modulating signal the faster the rate of swing.

With frequency modulation, multiple upper and lower sidebands exist in the output waveform. This is shown in Table 2.7.

The modulation index, which is sometimes known as the modulation depth, is given by

$$m = \frac{\text{deviation of carrier from mean value}}{\text{modulating frequency}}$$

$$= \frac{\Delta f_c}{f_a}$$

The frequency deviation is the difference between the carrier frequency and the lowest output frequency or the difference between the highest output frequency

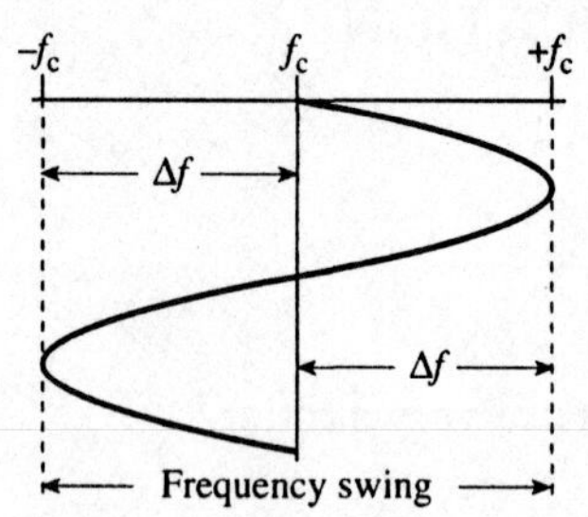

Large-amplitude low-frequency signal

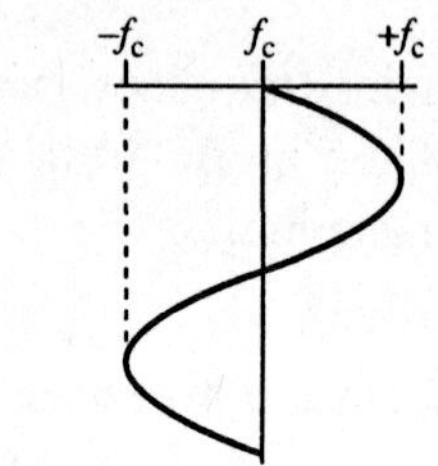

Small-amplitude low-frequency signal

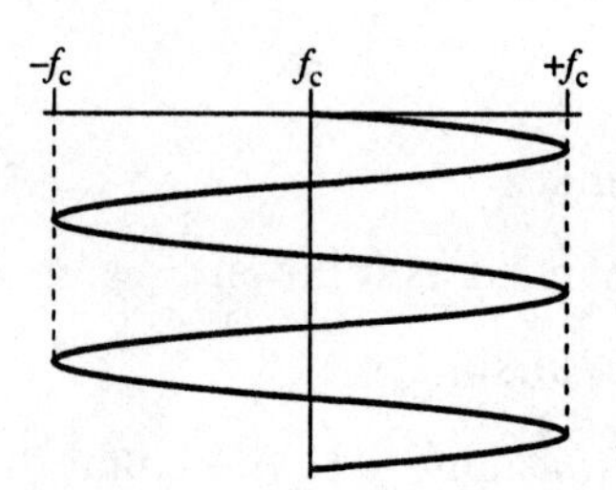

Large-amplitude high-frequency signal

Fig. 2.11 Frequency modulation

and the carrier frequency. The deviation on both sides of the carrier frequency is always the same.

The frequency swing is given as the difference between the highest output frequency and the lowest output frequency.

For modulation indices >1, the bandwidth is given by the following approximate formula:

$$B = 2(\Delta f_c + f_a)$$

Table 2.7 Frequency modulation sidebands

Order	Equation	Upper	Equation	Lower	Equation
1st	$f_c \pm f_a$	1st	$f_c + f_a$	1st	$f_c - f_a$
2nd	$f_c \pm 2f_a$	2nd	$f_c + 2f_a$	2nd	$f_c - 2f_a$
3rd	$f_c \pm 3f_a$	3rd	$f_c + 3f_a$	3rd	$f_c - 3f_a$
4th	$f_c \pm 4f_a$	4th	$f_c + 4f_a$	4th	$f_c - 4f_a$
	⋮		⋮		⋮

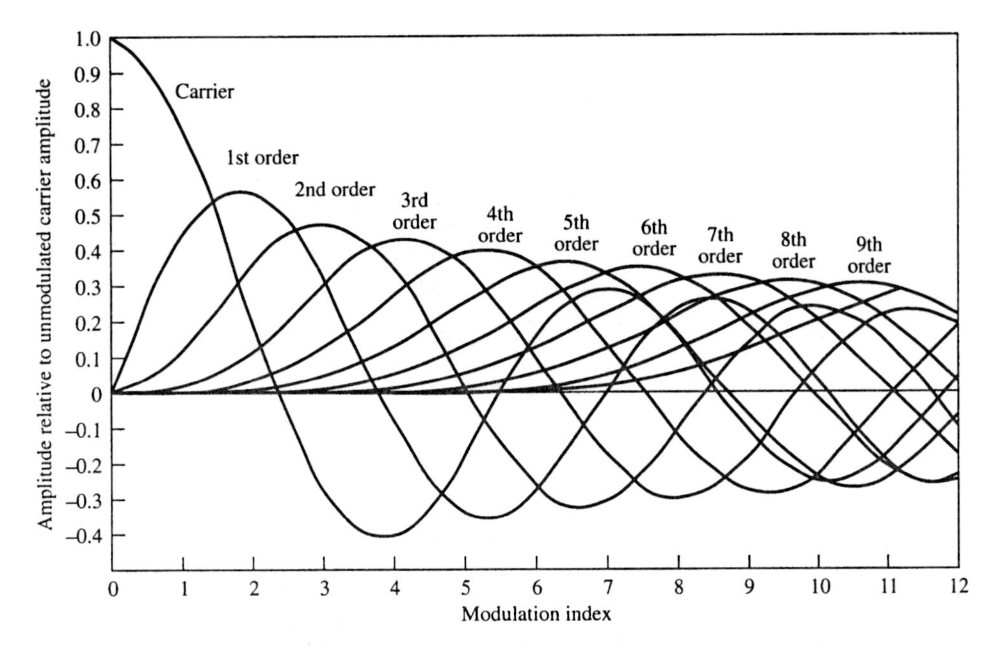

Fig. 2.12 Bessel function curves

where f_c = deviation of the carrier frequency
f_a = modulating frequency

A parameter that is more important is the maximum modulation index that is used on a system. This parameter is referred to as the deviation ratio and is given by

$$D = \frac{\Delta f_{max}}{f_{a_{max}}}$$

For a modulation depth of $m = 2.405$, 5.32, or 8.654, the carrier frequency has an amplitude of 0 V and as a result all the power is transferred to the sidebands. This can be seen when consulting the Bessel function curves shown in Fig. 2.12. This graph is a plot of the modulation index, along the x axis against the relative power in each sideband relative to the power in the unmodulated carrier frequency (y axis). To enable the graph to be used universally the relative power has been normalised.

The larger the value of modulation index the greater the number of significant sidebands. The smaller the modulation index the fewer the number of significant sidebands. A significant sideband is defined as a sideband that has at least 1% of the power in the unmodulated carrier frequency.

In high-capacity point-to-point systems a modulation index of $D < 0.5$ is used. This results in the following products of modulation, f_c, USB and LSB only. This is similar to an amplitude modulator where DSB are used. A system which has a modulation depth of $D < 0.5$ is referred to as a narrowband FM system. By consulting the Bessel function curves the above can be confirmed.

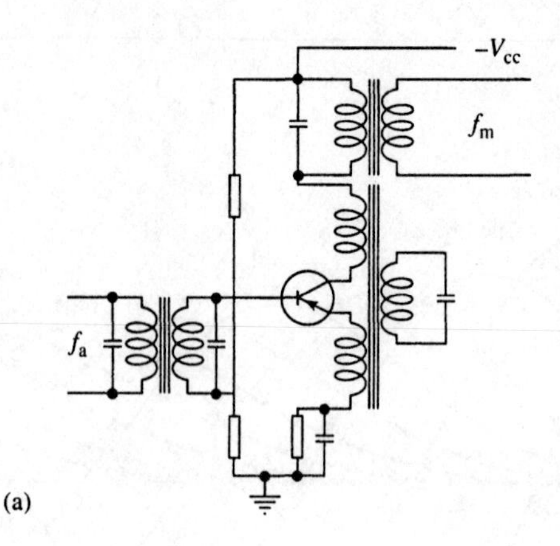

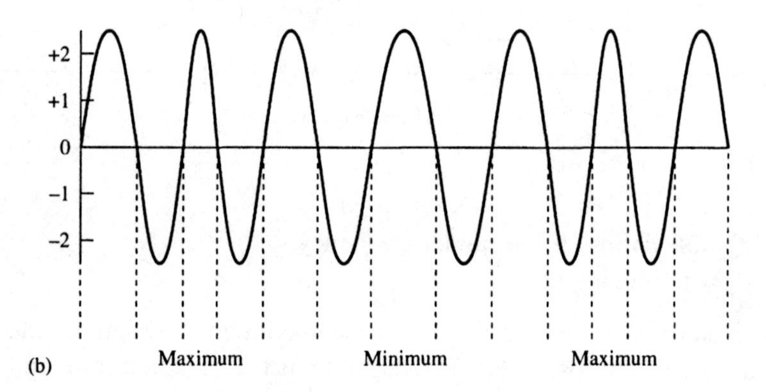

Fig. 2.13 Frequency modulator: (a) circuit; (b) output waveform

Systems which have modulation indices greater than 0.5 are referred to as wideband systems. These systems normally have a large number of significant sidebands. The larger the value of modulation index is the greater the number of significant sidebands.

A typical frequency modulator is shown in Fig. 2.13.

Advantages and disadvantages of frequency modulation

Frequency modulation has major advantages when compared with amplitude modulation in the following areas:

- *Signal-to-noise ratio*: the signal-to-noise ratio at the output of an FM receiver is greater than that at the output of an AM receiver.
- *Transmitter efficiency*: FM transmitters are more efficient as the output amplitude and hence power is constant.
- *Signal suppression*: an FM receiver can suppress the weaker of two signals simultaneously received at or near the same frequency.
- *Dynamic range*: the FM system has a larger dynamic range than the AM system.

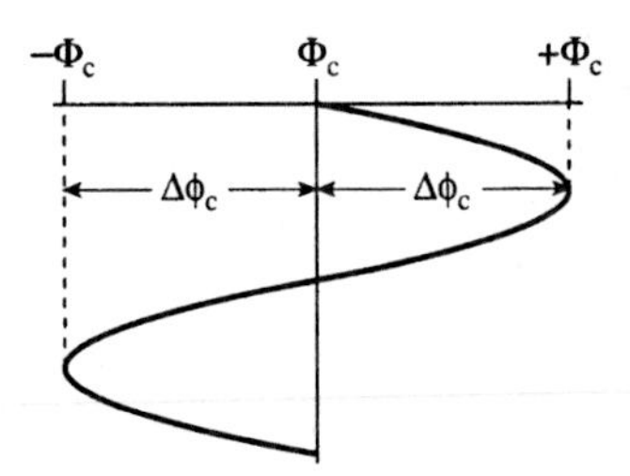

Large-amplitude low-frequency signal

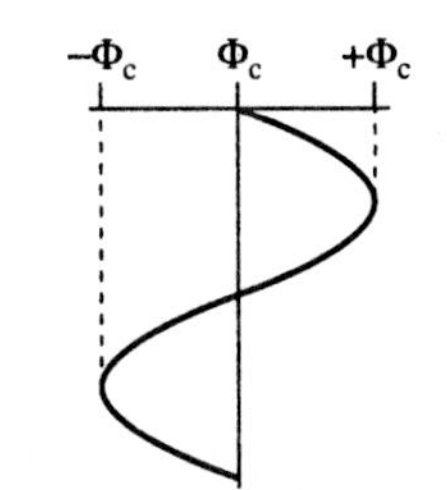

Small-amplitude low-frequency signal

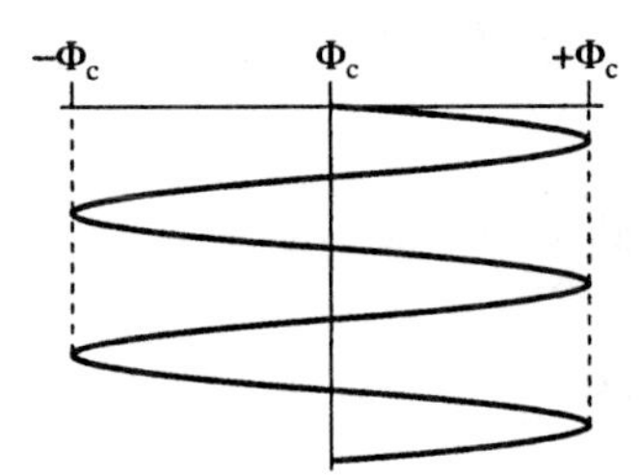

Large-amplitude high-frequency signal

Fig. 2.14 Phase modulation

However, frequency modulation also has some significant disadvantages:

- *Design*: a frequency modulator is more complex than an amplitude modulator.
- *Bandwidth*: frequency modulation uses a much larger bandwidth than amplitude modulation.

2.8.2 Phase modulation

In phase modulation, the phase of the carrier frequency is caused to vary by an amount proportional to the amplitude of the modulating signal, at a rate proportional to the frequency of the modulating signal.

Figure 2.14 shows the principle of operation of a phase modulator. Notice that the larger the amplitude of the modulating signal the greater the phase change and the higher the frequency of the modulating signal the faster the rate of change takes place.

$$\begin{aligned} e &= E_0 \sin(\omega_c t + m \sin \omega_a t) \\ &= E_0 \sin(\omega_c t + \Phi_c) \end{aligned}$$

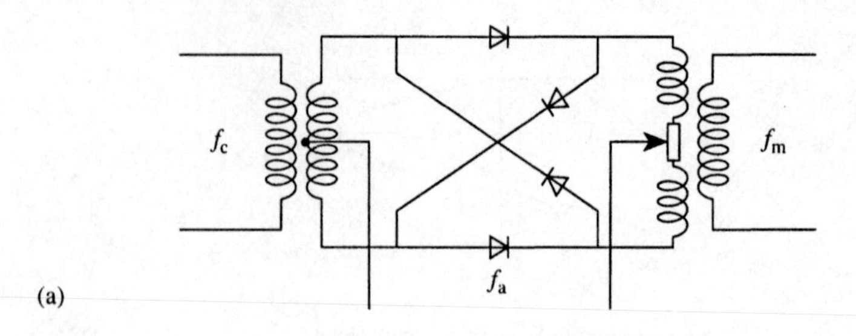

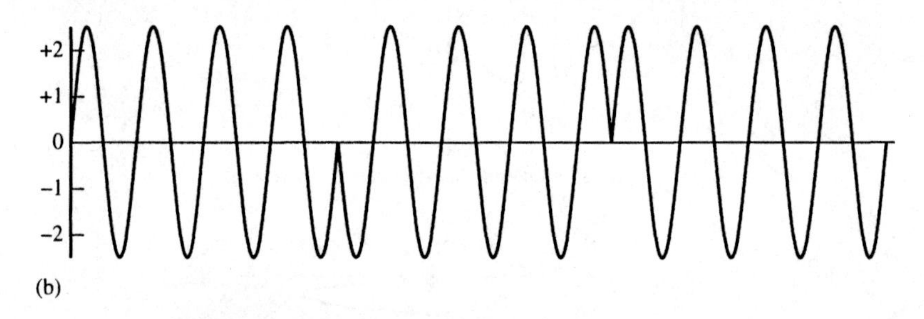

Fig. 2.15 Phase modulator: (a) circuit; (b) output waveform

where E_0 = peak amplitude of the carrier frequency
$\omega_c = 2\pi f_c$
$\omega_a = 2\pi f_a$
m = modulation index
Φ_c = phase change applied to carrier frequency

With phase modulation, multiple upper and lower sidebands exist in the output waveform which are similar to those for frequency modulation, as shown in Table 2.7.

Figure 2.15(a) shows a typical phase modulator. As can be seen the modulator is a ring modulator. The phase of the carrier frequency at the output is either the same as that of the input carrier frequency or the output carrier frequency is changed by 180° relative to the input carrier frequency. The example shown applies to a digital input waveform. The amplitude of the input digital signal is used to forward bias two of the diodes. The output phase of the carrier frequency depends on which diodes are forward biased. Note where the carrier frequency and the digital input signal are applied to the phase modulator and compare this to where the carrier frequency and analogue input signal is applied to the amplitude ring modulator.

Derivatives of phase modulation are extensively used in digital systems. Some of these modulation schemes are discussed in Chapter 4. Phase modulation is not used in analogue systems because of the complexity of the modulator circuit and the complexity of the output signal.

2.9 COMPARISON OF AMPLITUDE, PHASE AND FREQUENCY MODULATION

Table 2.8 gives a comparison between the three modulation techniques discussed.

Table 2.8 Comparison of modulation techniques

Modulation technique	Implementation	Interference	Frequency spectrum	Power	Usage
Amplitude	Easy	Badly affected	Limited	Limited power in the sidebands	Large
Phase	Difficult	Hardly affected	Very large	Dependent on the number of sidebands	Little
Frequency	Medium	Hardly affected	Very large	Dependent on the number of sidebands	Growing

2.10 REVIEW QUESTIONS

2.1 State the products of modulation at the output of an amplitude modulator, and draw the resultant output waveform for a sinusoidal input signal.

2.2 Calculate the output products of modulation for a single balanced modulator given that the input signal is from 60 kHz to 108 kHz and the carrier frequency is 340 kHz.

2.3 Given that the input frequency range to a double balanced demodulator is 312 kHz to 552 kHz and the carrier frequency applied to the demodulator is 300 kHz, determine the output products.

2.4 Calculate the modulation depth for an amplitude modulator given that the carrier frequency has a peak-to-peak voltage of 2.5 V and the RMS voltage of the modulating signal is 0.58 V.

2.5 A sinusoidal modulating frequency is modulated with a carrier frequency having a peak voltage of 3.5 V. The output is connected to a 12 kΩ load. The modulation depth is 85%. Determine each of the following:
(a) The power in the carrier frequency.
(b) The power in each sideband.
(c) The power in both sidebands.
(d) The total power.
(e) The efficiency of the modulator.

2.6 State the criteria that stipulate whether an FM system is narrowband or wideband.

2.7 State the advantages and disadvantages of frequency modulation.

3 Spread spectrum systems

Communication enables mankind to impart ideas and knowledge.

3.1 INTRODUCTION

There are numerous types of transmission systems. Spread spectrum systems are becoming more and more popular, especially as the available frequency spectrum is becoming more and more crammed. Although these systems are extremely complex and require a high cost in development, they do have numerous advantages over more conventional systems.

The major advantage of these systems is the signal security that the system presents. Unlike conventional systems it is extremely difficult to eavesdrop on a conversation that takes place over a spread spectrum system.

The system was initially developed for military applications, and some police forces use spread spectrum systems because of the signal security.

3.2. SPREAD SPECTRUM SYSTEMS

Conventional systems try to cram as much information into as small a bandwidth as possible. For police usage this means that the bands can be easily monitored. For military use, these systems can easily be jammed by high-power jamming frequencies that cover the frequency band of the particular system. Conventional systems also normally output a relatively high power to the antenna.

Spread spectrum systems spread the signal over as wide bandwidth as possible. Also they try to hide the transmitted signal as close to the background noise as possible. This makes the communication very difficult to find in the frequency spectrum, which means that they cannot be easily tracked. Also they are more difficult to jam.

There are three different types of spread spectrum techniques that can be used. These are:

- Time hopping.

- Direct sequence.
- Frequency hopping.

3.3 SPREAD SPECTRUM SYSTEM CRITERIA

For a system to be described as a spread spectrum system the following two criteria must be satisfied:

- The transmitted signal must occupy a bandwidth much greater than the bandwidth of the modulating signal (i.e. the input signal to the system).
- The bandwidth occupied by the transmitted signal must be determined by a prescribed waveform and not by the modulating frequency (i.e. carrier frequency).

3.4 REASONS FOR USE OF SPREAD SPECTRUM SYSTEMS

There are three major reasons for the use of spread spectrum techniques in communication systems today.

- They aid privacy of the transmission, since the spectral density of the spread spectrum may be less than the noise spectral density of the receiver.
- The despreading process in the receiver will spread the spectra of unwanted narrowband signals, thus improving interference rejection.
- The effect on a spread spectrum receiver, that receives a spread spectrum from a different spread spectrum system using the same frequency bands but implementing a different spreading pattern, approximates to noise in the receiver.

3.5 PSEUDORANDOM CODE GENERATORS, SCRAMBLERS AND DESCRAMBLERS

3.5.1 Pseudorandom noise code generators

All spread spectrum systems make use of pseudorandom code generators. This code is referred to as pseudorandom noise code (PN code). This code is used to determine the frequency spectrum that the output signal will occupy, i.e. the output bandwidth. In other words these generators are used to control the spreading pattern of the system. A simple 3-bit pseudorandom coder is shown in Fig. 3.1. Take note of the truth table that results from this circuit.

The truth table output produces the following output states:

$$000_2,\ 110_2,\ 101_2,\ 010_2,\ 111_2,\ 011_2,\ 001_2,\ 000_2$$

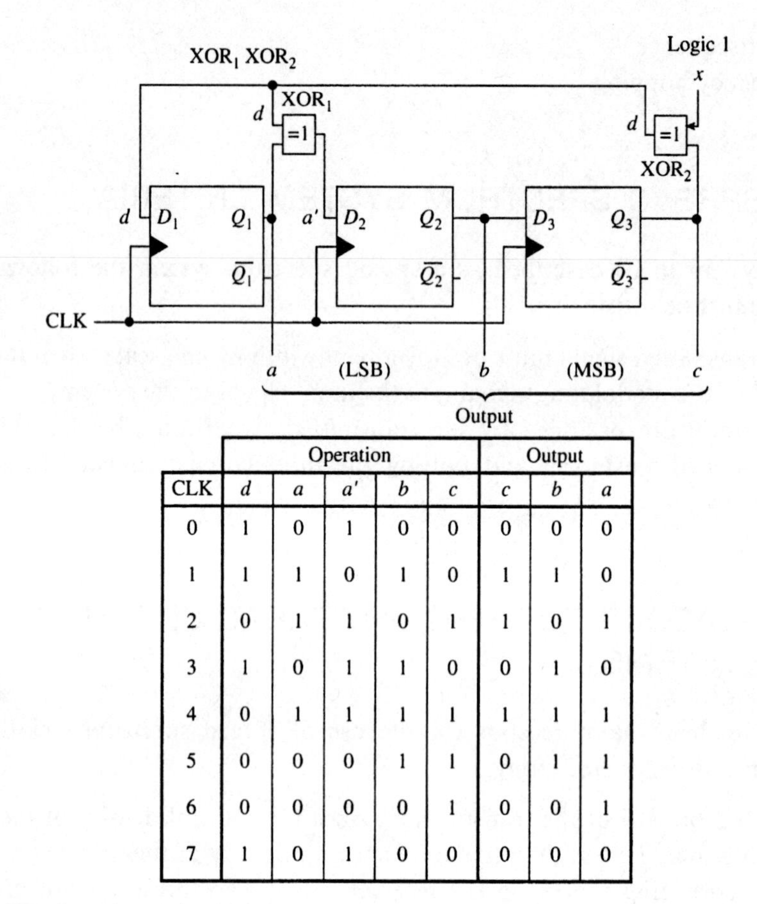

	Operation					Output		
CLK	d	a	a'	b	c	c	b	a
0	1	0	1	0	0	0	0	0
1	1	1	0	1	0	1	1	0
2	0	1	1	0	1	1	0	1
3	1	0	1	1	0	0	1	0
4	0	1	1	1	1	1	1	1
5	0	0	0	1	1	0	1	1
6	0	0	0	0	1	0	0	1
7	1	0	1	0	0	0	0	0

Fig. 3.1 Pseudorandom code generator

As can be seen, the output logic states do not follow a logical sequence. The output sequence is determined by the outputs of the two exclusive-OR (XOR) gates that are incorporated in the circuit.

In this example the sequence will repeat itself after six clock pulses. In practice very large generators are used, which means that the sequence will only repeat itself after a large number of clock pulses.

In Fig. 3.1 two XOR gates are used. The truth table for an XOR gate is given in Table 3.1. Notice that the output is a logic high only when there are opposite logic levels on the inputs.

The input x is permanently connected to a logic high. Initially the three flipflops are all reset, which means that Q_1, Q_2 and Q_3 are all logic low. Point d is a logic high, as a logic low XORed with a logic high results in a logic high. Point a' goes to a logic high. Now D_1 is logic high, D_2 is logic high and D_3 is logic low. After the first clock pulse, Q_1 goes high, Q_2 high and Q_3 goes low.

The logic low on Q_3 is fed back to the one input to XOR_2. This results in d remaining at a logic high. Point a' goes to a logic low. Thus D_1 is now a logic

Table 3.1 Truth table for XOR gate

b	a	Output
0	0	0
0	1	1
1	0	1
1	1	0

high, D_2 is a logic low and D_3 is a logic high. After the second clock pulse, Q_1 goes to a logic high, Q_2 goes to a logic low and Q_3 goes to a logic high.

Q_1 is connected to output a, which is the least significant bit (LSB), Q_2 is connected to b and Q_3 is connected to c, which is the most significant bit (MSB).

3.5.2 Scramblers and descramblers

In order to make the information even more difficult to be decoded by unauthorised receivers the information is scrambled. Figure 3.2 shows the circuits for a simple 3-bit scrambler and a 3-bit descrambler. This process is also referred to as encryption and decryption. In this example the input data is a continuous string of logic 1s. In practice the data that is applied to the circuit is digitised speech, digitised video, or data. As a result the data that is transmitted is random due to the scrambler.

The operation of this circuit is the same as that of the 3-bit pseudorandom noise generator. The only difference is that the output of the PN generator is XORed with the data input. As a result the data stream of a continuous string of logic 1s is changed into a sequence of

$$1101000 \quad 1101000_2$$

As can be seen the output repeats itself. In practice, however, the data itself is random and the scrambler consists of many more stages, and hence the output is a pseudorandom sequence.

The descrambler circuit is identical to that of the scrambler and works in exactly the same way. In this case the output of the PN generator is XORed with the input pseudorandom data stream and as shown in this example a data stream consisting of a sequence of logic 1s results. What is important here is to remember that the descrambler and the scrambler must be properly synchronised or else a totally corrupted data output will result. The descrambler must produce the identical PN code to that of the scrambler, and the descrambler must start to descramble the data at the same point that the scrambler scrambled the data.

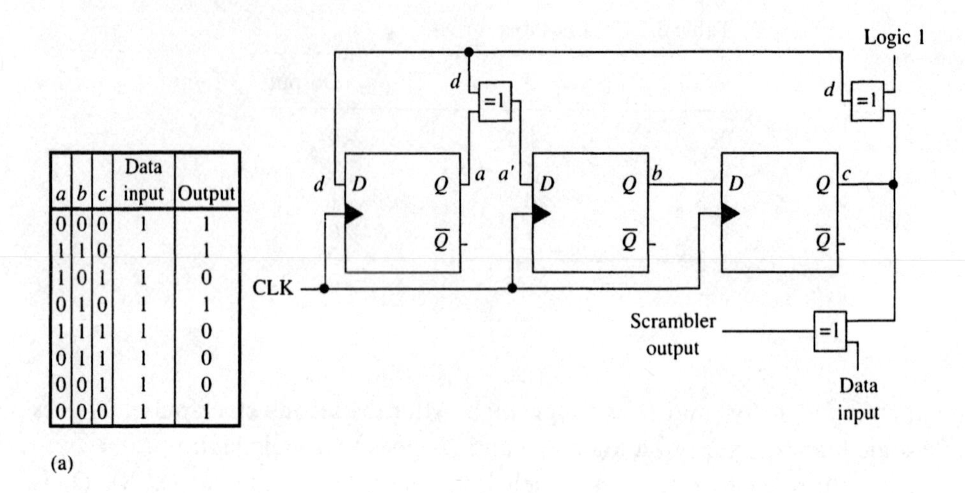

a	b	c	Data input	Output
0	0	0	1	1
1	1	0	1	1
1	0	1	1	0
0	1	0	1	1
1	1	1	1	0
0	1	1	1	0
0	0	1	1	0
0	0	0	1	1

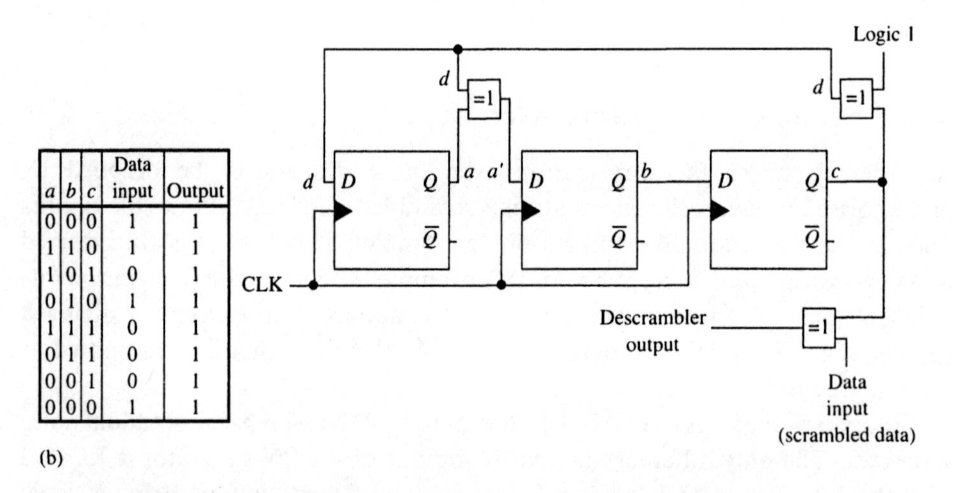

a	b	c	Data input	Output
0	0	0	1	1
1	1	0	1	1
1	0	1	0	1
0	1	0	1	1
1	1	1	0	1
0	1	1	0	1
0	0	1	0	1
0	0	0	1	1

Fig. 3.2 (a) Scrambler; (b) descrambler

3.6 TYPES OF SPREAD SPECTRUM TECHNIQUES

3.6.1 Direct sequence spread spectrum

In this method, phase modulation or a derivative of phase modulation is used. The data is first scrambled. The scrambling process is achieved by mixing the actual data with the output of a PN coder as described in Section 3.5.2.

The resultant scrambled data is then modulated in a binary phase shift key (BPSK) or quadrature phase shift key (QPSK) modulator. The output of the BPSK modulator is then transmitted. A basic block diagram of the system using a BPSK modulator is shown in Fig. 3.3.

In this example the PN code generator is running at a higher clock frequency than that used for the data. The resultant scrambled data is now transmitted at

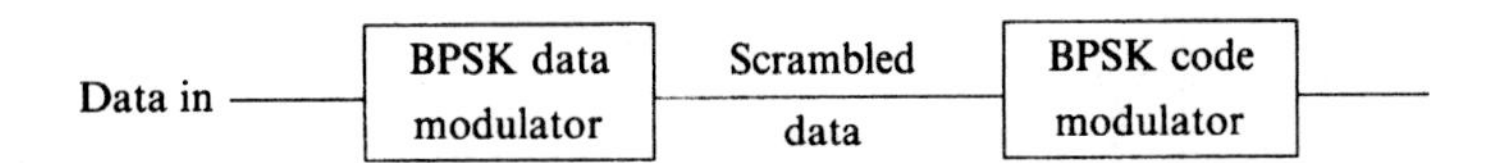

Fig. 3.3 Scrambler system using BPSK modulation

a much higher rate than that of the data. When a logic high is applied to the BPSK modulator the carrier undergoes a phase change of 180°. When a logic low is applied to the BPSK modulator the carrier undergoes a 0° phase change. This is shown in Fig. 3.4.

The carrier frequency containing the phase change information is received by the distant receiver. This signal is then descrambled by a descrambler that relies on a PN code generator producing PN code identical to that at the transmitter. The signal is then converted back into logic levels to produce the original data.

A spread spectrum system that does not have a coder identical to that at the transmitter will not be able to decode the data correctly.

Figure 3.4 shows the processes that take place in a direct sequence spread spectrum system.

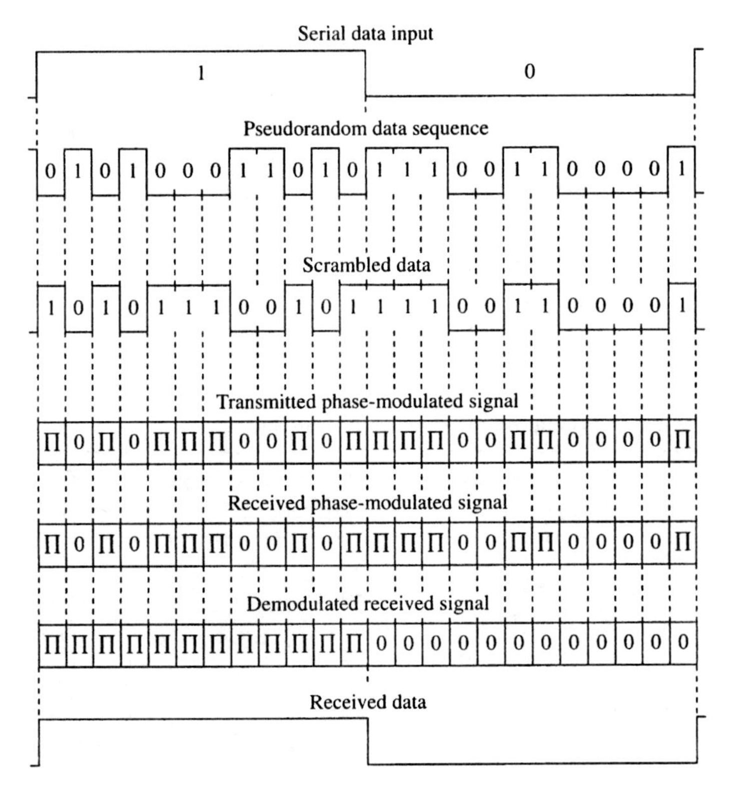

Fig. 3.4 Direct sequence spread spectrum

3.6.2 Frequency hopping

In this method frequency shift key (FSK) modulation is used. Before describing how the system works it is necessary to define the terms that are used with respect to the system:

- *Hop set*: this is the number of channels that are used by the system (i.e. the number of different frequencies utilised).
- *Dwell time*: this is the length of time that the system transmits on an individual channel (i.e. the length of time spent on one frequency).
- *Hop rate*: this is the rate at which the hopping takes place (i.e. how fast the system changes from one channel to another or from one frequency to another).

Frequency allocation

A theoretical model using the 3-bit PN coder described in Section 3.5.1 will be used to explain the principles behind the frequency-hopping spread spectrum system. Consider the FSK modulator where the carrier frequency is controlled by a 3-bit PN code generator and produces the carrier frequencies shown in Table 3.2.

In Table 3.2 the hop set is equal to 8. This means that eight different channels or frequencies are used. These different channels are indicated in Fig. 3.5.

It is imperative that the receiver has an identical PN code generator and that the PN coder at the receiver is synchronised with that at the transmitter.

Pseudorandom coder

A 3-bit PN coder produces the binary codes and the associated output frequencies shown in Table 3.3. The hop set equals 8 according to Table 3.2.

Assume that the dwell time is 1 μs. The hop rate is determined as follows:

$$\text{hop rate} = \frac{1}{\text{dwell time} \times \text{hop set}}$$

$$= \frac{1}{1\,\mu\text{s} \times 8} = 125\,\text{kHz}$$

Table 3.2 Control codes and frequencies for an FSK modulator

Control codes	Carrier frequency (kHz)	Derivation of output frequency (kHz)	Output frequency (kHz)
000	8	8 + 1.75	9.75
001	8	8 + 1.25	9.25
010	8	8 + 0.75	8.75
011	8	8 + 0.25	8.25
100	8	8 − 0.25	7.75
101	8	8 − 0.75	7.25
110	8	8 − 1.25	6.75
111	8	8 − 1.75	6.25

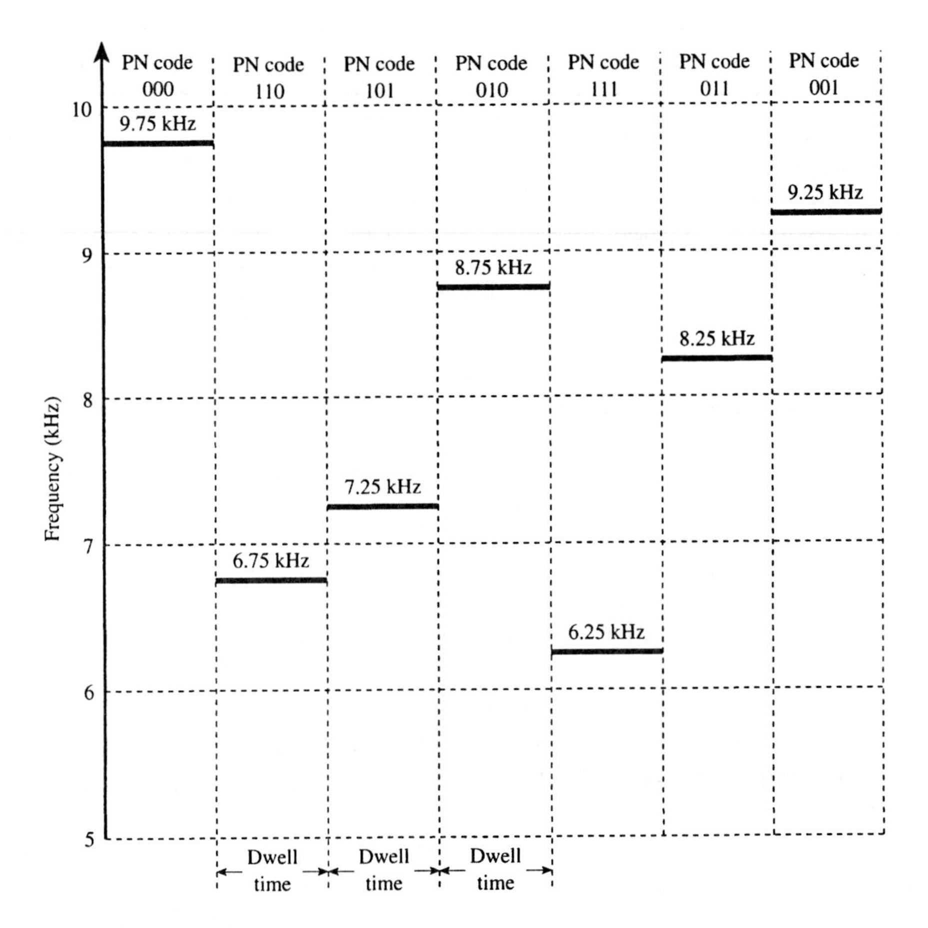

Fig. 3.5 Frequency hopping

Assume the PN code that is applied to the frequency synthesiser is

$$000 \quad 110 \quad 101 \quad 010 \quad 111 \quad 011 \quad 001_2$$

The frequency synthesiser produces the carrier frequencies as shown in Table 3.3 which are in accordance with the frequency allocation for the different PN codes as shown in Table 3.2.

Table 3.3 Codes and frequencies for a 3-bit PN decoder

a	*b*	*c*	Output frequency (kHz)
0	0	0	9.75
1	1	0	6.75
1	0	1	7.25
0	1	0	8.75
1	1	1	6.25
0	1	1	8.25
0	0	1	9.25

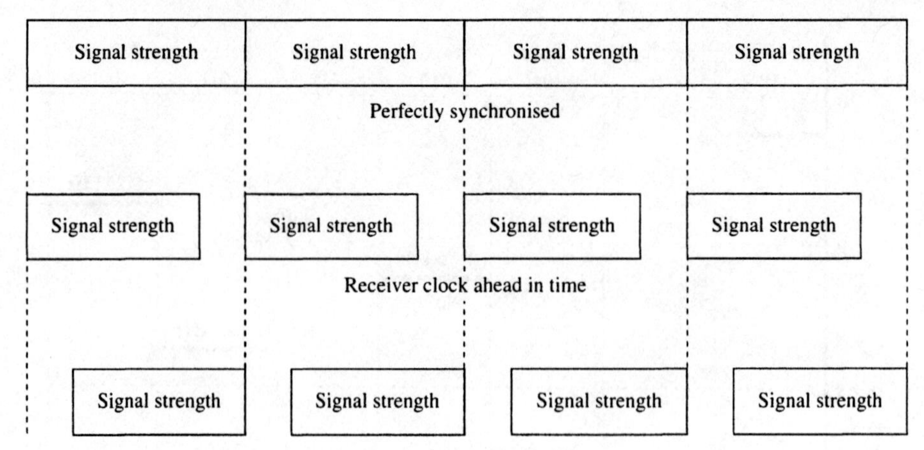

Fig. 3.6 Synchronisation

The digitised speech is then modulated with each of these carrier frequencies for a dwell time of 1 μs. It is then spread over a hop set of eight channels, in this example, in a pseudorandom fashion. The overall bandwidth of the system is $6.25\,\text{kHz} - f_a$ to $9.75\,\text{kHz} + f_a$. For a speech band $f_a = 3.4\,\text{kHz}$, the output frequency would be from 2.85 kHz to 13.15 kHz, which gives a bandwidth of 10 kHz.

For a conventional system using frequency modulation the output bandwidth would be 6.8 kHz and as can be seen the spread spectrum system produces a bigger output bandwidth. In practice the carrier frequency would be much higher than 8 kHz and the carrier frequency deviation would be much greater that shown in Table 3.2. Also the PN code would be much larger. This would result in a much greater output bandwidth than given in the example.

Synchronisation

One of the most important aspects of a spread spectrum system is the synchronisation of the receiver with the transmitter. A tracking circuit in the receiver is used to synchronise the receiver with the transmitter.

The tracking circuit compares the signal strength of the despread signal over the dwell time. If the receiver is exactly synchronised then the signal strength is constant over the whole dwell time, as shown in Fig. 3.6. If the receiver clock is ahead of the transmitter clock then the signal strength will move from high to low during the dwell time. This means that the receiver is hopping to the next frequency prior to the transmitter, as shown in Fig. 3.6.

If the receiver clock is behind the transmitter clock then the signal strength will move from low to high during the dwell time. This indicates that the transmitter has hopped to the next channel prior to the receiver, as shown in Fig. 3.6.

A control loop in the receiver is used to either advance or retard the PN code generator count in the receiver.

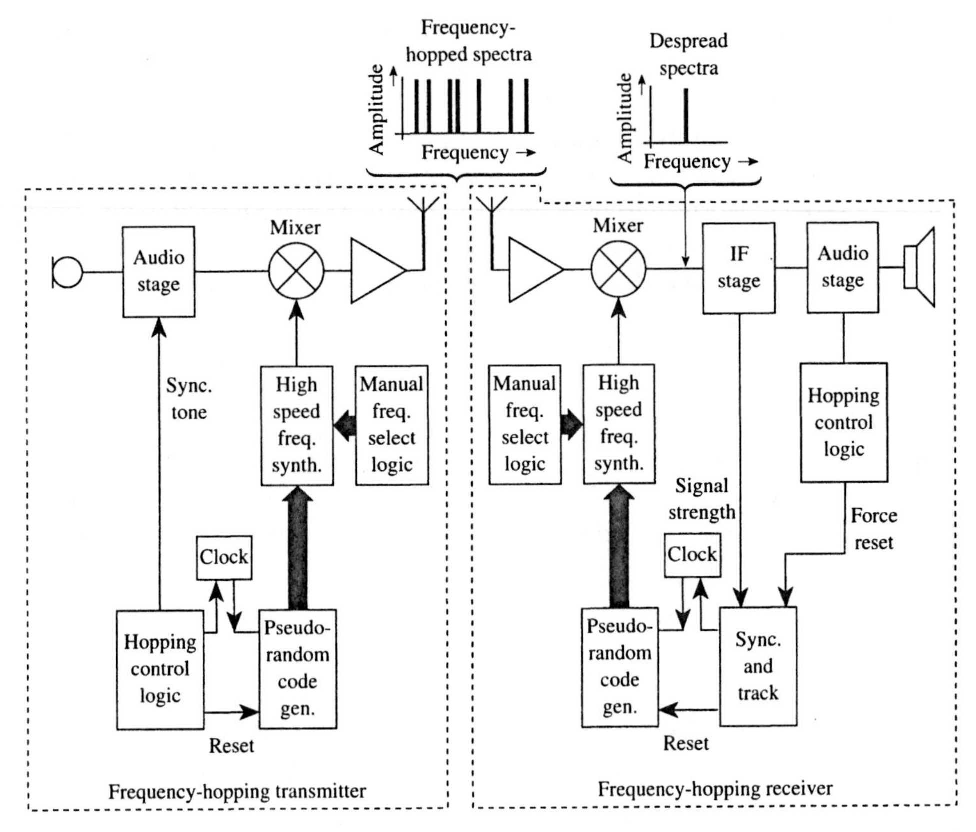

Fig. 3.7 Frequency-hopping spread spectrum system

System operation

Figure 3.7 shows a typical block diagram of a frequency-hopping spread spectrum system.

The transmitter

At the transmitter the hopping control circuit controls the whole system. This is the circuit into which the user will enter his/her personal identification number (PIN). The PIN will then enable the PN code generator. The user group number is entered into the manual frequency select logic, which allocates the channels for that user to the system. The PN code generator output drives a high-speed frequency synthesiser, which then allocates the correct channels in a pseudorandom manner.

The output frequency from the frequency synthesiser is then fed as the carrier frequency to the mixer where the input audio is modulated. The resultant modulated frequency is then amplified and transmitted.

The receiver

The received signal is first amplified by a wideband amplifier. The signal is then demodulated with the carrier frequency from the high-speed synthesiser. The resultant intermediate frequency (IF) is applied to a tracking circuit. The user must input into the manual frequency select logic the same user group code as was used at the transmitter. The user must also enter his/her PIN into the hopping control logic circuit.

The sync and tracking circuit compares the signal strength over the dwell time and develops control signals that either advance or retard the PN code generator. Once the system is correctly aligned then the IF will be demodulated to the audio range and communication can then take place.

The block diagram in Fig. 3.7 only shows one direction. Obviously this circuitry is repeated for the other direction for full communication to take place.

3.6.3 Time hopping

Time hopping is where a short pulse, called a chirp, is transmitted having either a pseudorandom pulse duration or transmitted in a random position relative to the input bit period. Thus time hopping can be implemented in two different ways.

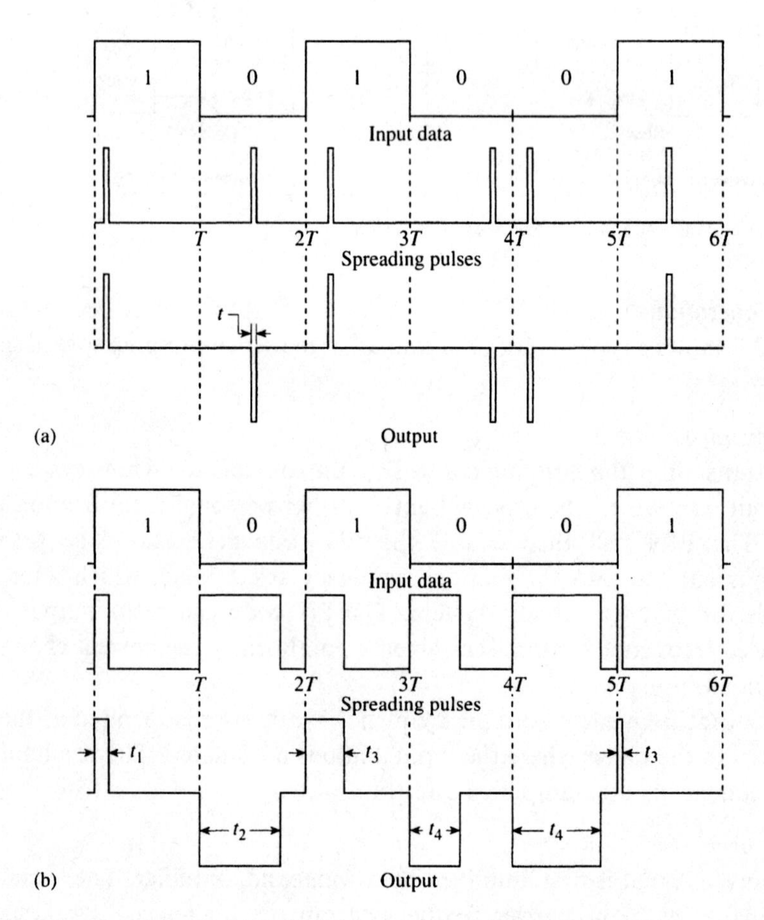

Fig. 3.8 Time hopping methods: (a) method 1, equal chirp durations; (b) method 2, varying chirp durations

Method 1

With this technique each binary bit is transmitted as a short pulse. The actual interval in which it is transmitted is determined by a PN code generator.

This technique seeks to accomplish uncertainty in the communication channel for any unwanted receiver by varying the time intervals between transmitted pulses. By referring to Fig. 3.8(a) it can be seen that the spreading pulses are very short pulses that appear in a pseudorandom position during the bit period. The polarity of the output signal indicates whether the input binary was a logic high or a logic low.

During each bit period a chirp (subdivision) is selected at random by the PN code generator. Each chirp has the same duration as shown in Fig. 3.8(a).

Method 2

With this technique each chirp has a different duration. The start of each chirp takes place at the same point during the bit period. The PN code generator causes the chirp duration to be altered. Referring to Fig. 3.8(b), the chirp duration for the spreading pulse is different. Again the polarity of the output signal indicates whether the input binary was a logic high or a logic low.

3.7 ADVANTAGES AND DISADVANTAGES OF SPREAD SPECTRUM TECHNIQUES

The advantages of spread spectrum techniques as applied to communication systems can be summarised as follows:

- *All types:*
  Multiple usage by different user groups, as each user group can be allocated a different PN code.

  This code is the key necessary to unlock the message from the system. Without this key it is very difficult, almost impossible to extract the information.

  A spread spectrum will see another spread spectrum signal as interference and reject it as it would for a narrowband signal.
- *Frequency hopping:*
  A narrow band will only cause minimal interference on the wideband signal.

  Frequency jamming is extremely difficult and can only be effectively achieved if the jamming receiver has the same PN code and channel allocation.

  The greater the hop set, the smaller the dwell time and the greater the bandwidth the smaller the interference from narrowband signals.
- *Frequency hopping and direct sequence:*
  The output power of the spread spectrum is spread over a large bandwidth. This means that the spectrum has a very low spectral density; 1 W output power over an 8 MHz band gives a spectral density of 125 nW/Hz.

  These systems are particularly useful to the military and police. The low spectral density may not even be recognised as valid communication, thus leading to low probability of interception and recognition.

Disadvantages of spread spectrum techniques are:

- *All types:*
  Complex circuitry.
  Expensive to develop.
  Very large bandwidths.
- *Time hopping:*
  Easily jammed and hence is not generally used in its true form.

3.8 REVIEW QUESTIONS

3.1 State the criteria that specify a system as a spread spectrum system.
3.2 Discuss the reasons for using spread spectrum systems.
3.3 Briefly discuss the direct sequence spread spectrum system.
3.4 Describe the terms hop rate, dwell time and hop set with respect to a frequency-hopping spread spectrum system.
3.5 Describe the basic operation of a frequency-hopping spread spectrum system.
3.6 Describe the synchronisation process used on a frequency-hopping spread spectrum system.
3.7 Briefly discuss the two different methods that can be used in a time-hopping spread spectrum system.
3.8 State the advantages and disadvantages of spread spectrum systems.

4 Digital modulation techniques

Knowledge empowers man to use technology wisely.

4.1 INTRODUCTION

The techniques used to modulate digital information so that it can be transmitted via microwave, satellite or down a cable pair are different to that of analogue transmission. However, it must be remembered that when the data is transmitted via satellite or microwave, the radio frequency (RF) stage is an analogue stage and that the data is transmitted as an analogue signal.

This means that the techniques used to transmit analogue signals are used to transmit digital signals. The problem is to convert the digital signal to a form that can be treated as an analogue signal. This chapter deals with the different modulation techniques that are used to convert a digital signal into an analogue signal that is then in the appropriate form to either be transmitted down a twisted cable pair or to be applied to the RF stage where it is modulated to a frequency that can be transmitted via microwave or satellite.

Although digital transmission is on the increase worldwide, many of the existing circuits are still analogue. This means that the digital signals must be changed into an analogue format to enable them to be transmitted over the existing metallic pairs that form the link. The equipment that implements this conversion for a personal computer (PC) that is used to communicate with other PCs via the public switched telephone network (PSTN) is a modem. The term modem is made up from the words 'modulator' and 'demodulator'.

A modem accepts a serial data stream from the PC and converts it into an analogue format that matches the transmission medium. A typical system arrangement is as shown in Fig. 4.1. As can be seen in the figure the data terminal equipment (DTE) is the point at which the data originates and also the end point for received data. The data circuit equipment (DCE) comprises the modem. The connection between the DTE and DCE is via an interface such as the RS232C. The modem is connected to the PSTN by means of a suitable circuit such as a cable pair.

There are many different modulation techniques that can be utilised in a modem. Some of these techniques are:

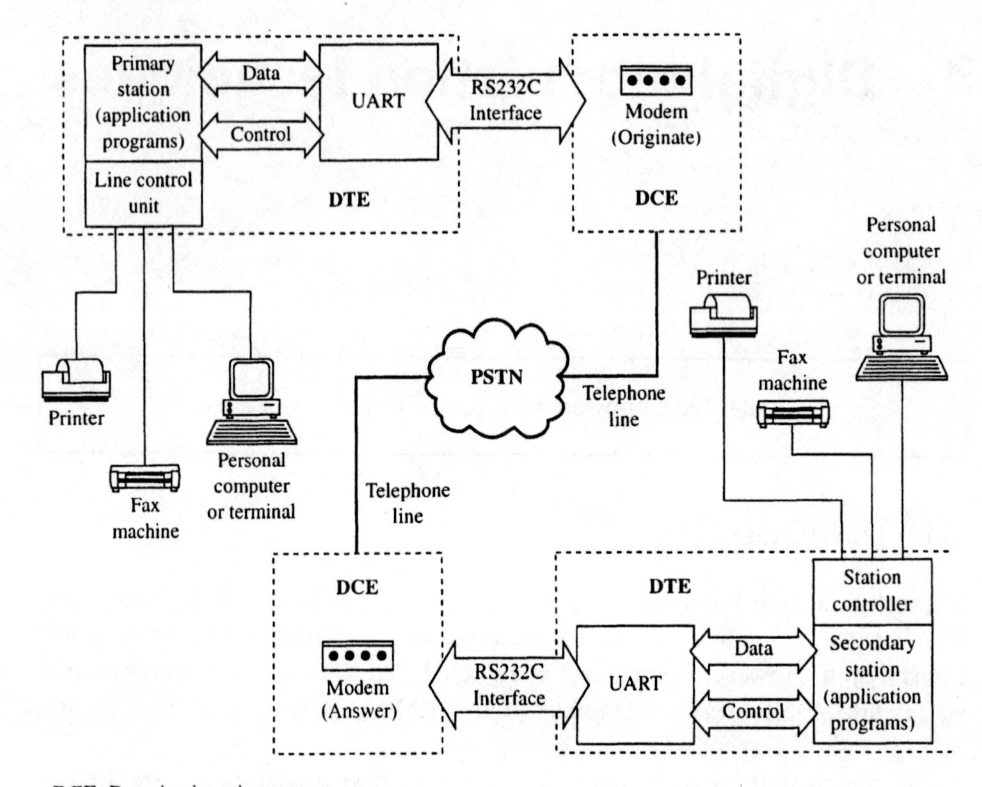

Fig. 4.1 Low-speed data link

- Amplitude shift key modulation (ASK).
- Frequency shift key modulation (FSK).
- Binary-phase shift key modulation (BPSK).
- Quadrature-phase shift key modulation (QPSK).
- Quadrature amplitude modulation (QAM).

4.2 AMPLITUDE SHIFT KEY MODULATION

In this method the amplitude of the carrier assumes one of two amplitudes dependent on the logic states of the input bit stream. Unlike an amplitude modulator used for modulating analogue signals, the carrier frequency need not be totally suppressed. The typical output waveform of a ring modulator is shown in Fig. 4.2. The frequency components are the USB and LSB with a small residual carrier frequency. The carrier suppression resistor is omitted from this modulator so that when a logic 0 is sent, a low-amplitude carrier frequency is allowed to be transmitted. This ensures that at the receiver the logic 1 and logic 0 conditions can be recognised uniquely.

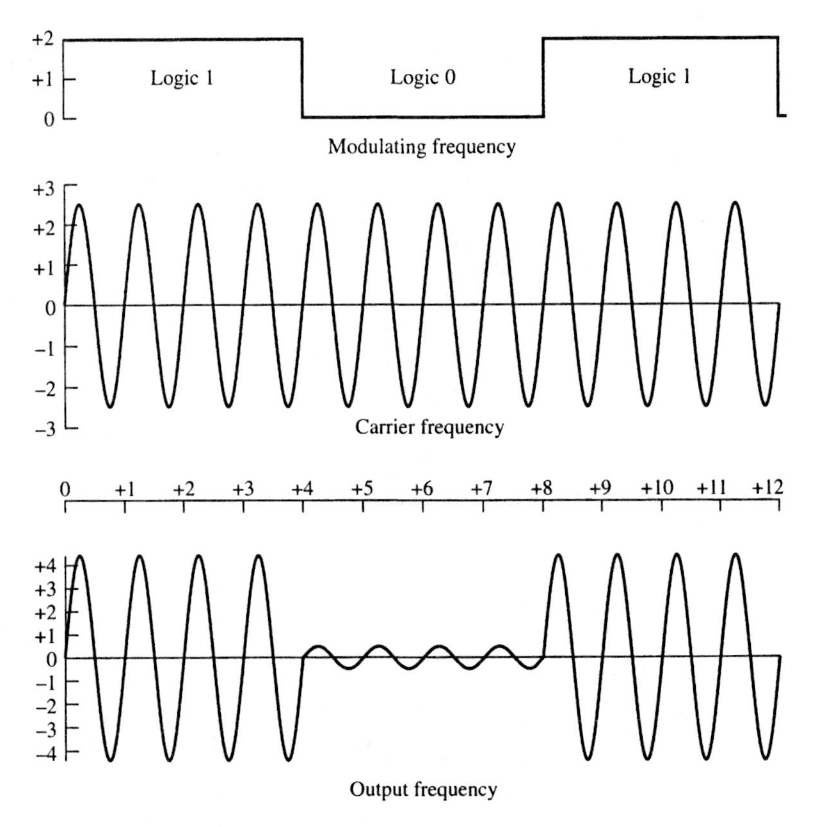

Fig. 4.2 Amplitude shift key modulation

A typical single sideband modulator is shown in Fig. 4.3(a). The carrier frequency is used to either forward bias or reverse bias the two diodes. When the diodes are forward biased then the digital signals are passed to the output. When the diodes are reverse biased then the digital signal is prevented from reaching the output and only a residual carrier frequency appears at the output.

A typical Cowan modulator is shown in Fig. 4.3(b). In this case, when the diodes are forward biased by the carrier frequency the input is short circuited and the input digital signal is prevented from reaching the output. When the carrier frequency reverse biases the diodes then the digital signal is passed to the output.

Figure 4.3(c) shows a ring modulator. The phase of the carrier frequency is used to forward bias two of the diodes and reverse bias the other two diodes.

4.3 FREQUENCY SHIFT KEY MODULATION

In this method the frequency of the carrier is changed to two different frequencies depending on the logic state of the input bit stream. The typical output waveform

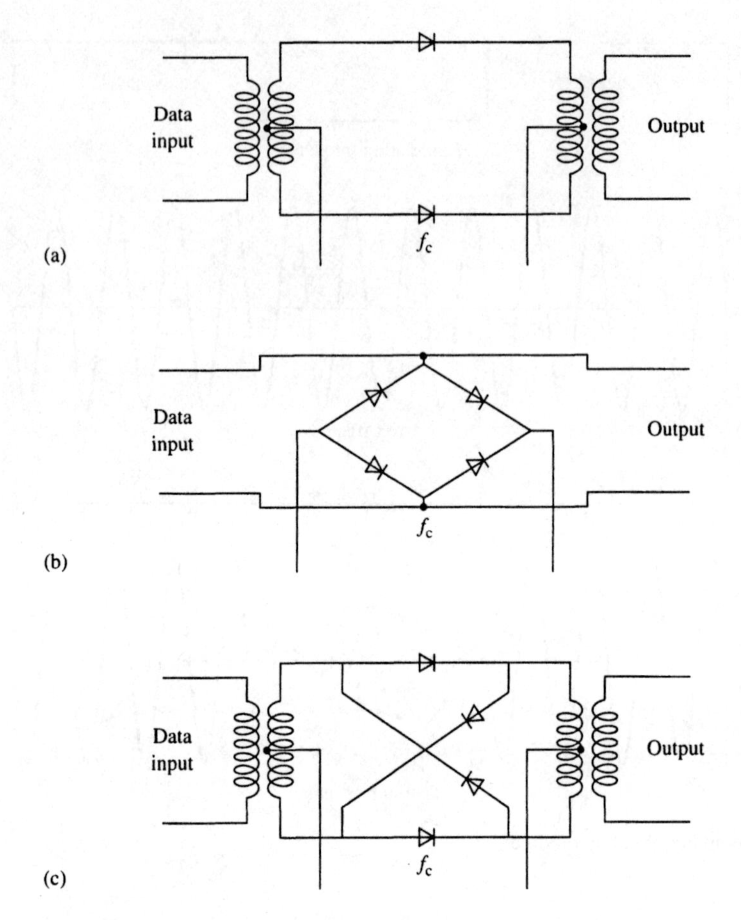

Fig. 4.3 (a) Single balanced modulator; (b) Cowan modulator; (c) ring modulator

is shown in Fig. 4.4. The carrier frequency is more conveniently referred to as the centre frequency. Notice that a logic high causes the centre frequency to increase to a maximum and a logic low causes the centre frequency to decrease to a minimum.

A typical FSK transmitter and receiver is shown in Fig. 4.5. The circuits shown use phase-locked loop circuits.

The transmitter

The input digital signal is applied to a digital-to-analogue (D/A) converter. The resultant analogue signal is summed with the control voltage produced by the phase comparator. The resultant control voltage is used to change the frequency of the voltage-controlled oscillator (VCO). The output of the VCO is the FM output which is applied to the input of the next stage.

Some of the output signal from the VCO is divided and this frequency is phase-compared with a frequency produced by the reference oscillator. The resultant output of the phase comparator produces a control voltage. The greater the

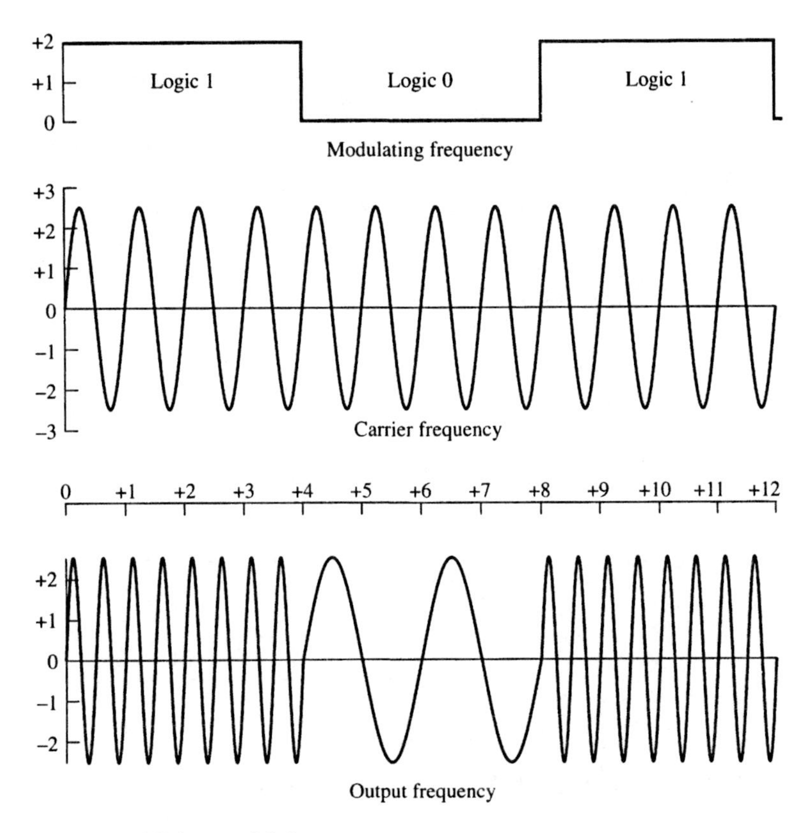

Fig. 4.4 Frequency shift key modulation

difference between the two input frequencies the larger the control voltage. The control voltage will be either positive or negative, depending on whether the output frequency of the VCO is higher or lower than the reference frequency at the input to the phase comparator. The divider incorporated into the circuit enables the output frequency of the VCO to be much larger than the reference frequency.

If there is no digital input then the output of the VCO is the carrier frequency or centre frequency. Hence by changing the control voltage the centre frequency is either increased to a maximum or decreased to a minimum.

The receiver

The FM signal is applied to the receiver input. This input signal is phase-compared with the output of the VCO. The output of the phase comparator produces a d.c. control voltage, which contains the information relating to the original digital signal. The control voltage is used to change the frequency of the VCO. In this way the control voltage is changing in accordance with the original digital signal.

The control voltage is then applied to an analogue-to-digital (A/D) converter, which produces the original digital signal.

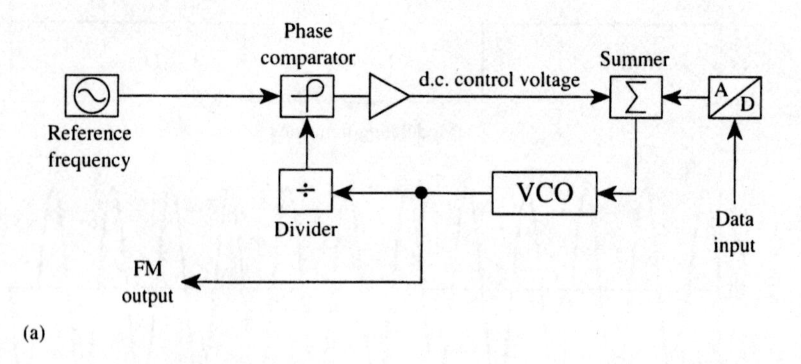

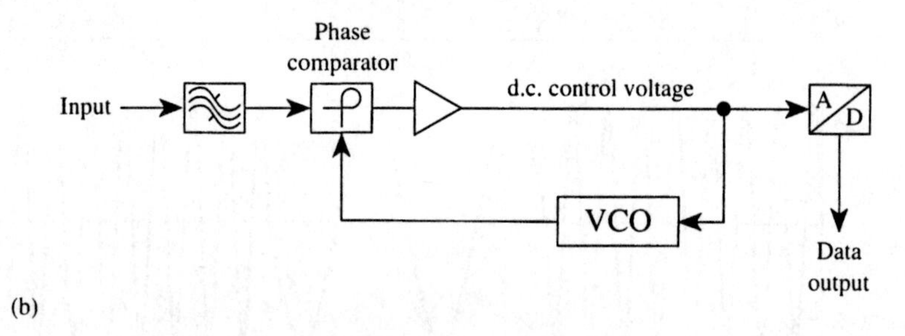

Fig. 4.5 (a) FSK transmitter using a VCO; (b) FSK receiver using a VCO

4.3.1 Frequency modulation generalities

The modulation index is given by

$$m = \frac{\Delta f}{f_a}$$

When the modulating frequency f_a is at a maximum and the frequency deviation Δf is at a maximum then the modulation index is referred to as the deviation ratio (D). The deviation ratio is determined by

$$D = \frac{\max \Delta f}{\max f_a}$$

A generalised formula to determine the required bandwidth is

$$B = 2(\Delta f + f_a)$$

Typical frequency allocation for low-speed modems is given in Table 4.1.

For an FSK modulator the deviation ratio is determined by:

$$D = \frac{\text{tone separation}}{\text{bit rate}}$$

The bandwidth is given by

$$B = \text{tone separation} + \text{bit rate}$$

Table 4.1 Frequency allocation for a low-speed modem

Bit rate (b/s)	Direction	Frequencies (Hz)		Nominal frequency (Hz)
		Logic 0	Logic 1	
≤300	A–B	1180	980	1080
	B–A	1850	1650	1710
600	A–B B–A	1700	1300	1500
1200	A–B B–A	2100	1300	1700

Example 4.1

Calculate the deviation ratio and bandwidth required to transmit a 1200 b/s digital signal.

$$D = \frac{\text{tone separation}}{\text{bit rate}}$$

Tone separation: from Table 4.1,

$$\begin{aligned}\text{tone separation} &= 2100 - 1300 \\ &= 800\,\text{Hz}\end{aligned}$$

$$\begin{aligned}D &= \frac{\text{tone separation}}{\text{bit rate}} \\ &= \frac{800}{1200} \\ &= 0.67\end{aligned}$$

The bandwidth is given by

$$\begin{aligned}B &= \text{tone separation} + \text{bit rate} \\ &= 800 + 1200 \\ &= 2000\,\text{kHz}\end{aligned}$$

If the bandwidth is limited to the bit rate then the receiver will still be able to recover the original data. The tone difference is 800 Hz which is smaller than the 2 kHz given above.

As the deviation ratio is approximately equal to the energy concentrated in the carrier and the first-order sidebands it is not necessary to transmit all the sidebands. The receiver need only recover the original logic states from the received signal.

4.4 PHASE SHIFT KEY MODULATION

With this method the phase of the carrier changes between different phases determined by the logic states of the input bit stream.

There are several different types of phase shift key (PSK) modulators:

- Two-phase (2 PSK).
- Four-phase (4 PSK).
- Eight-phase (8 PSK).
- Sixteen-phase (16 PSK).
- Sixteen-quadrature amplitude (16 QAM).

The 16 QAM is a composite modulator consisting of amplitude modulation and phase modulation. However, it is more closely associated with phase modulation than amplitude modulation.

The 2 PSK, 4 PSK, 8 PSK and 16 PSK modulators are generally referred to as binary phase shift key (BPSK) modulators, whereas the QAM modulators are generally referred to as quadrature phase shift key (QPSK) modulators.

4.4.1 Some phase shift key generalities

The symbol rate S of a binary PSK modulator can be determined by

$$S = \frac{D}{n} \text{ symbols/s}$$

where D = data rate (b/s)
n = number of bits per symbol

The total number of possible symbols at the output is given by

$$M = 2^n \text{ symbols}$$

where M = number of possible symbols at the output
n = number of bits per symbol

The phase difference between each symbol is given by

$$P = \frac{360°}{M} = \frac{2\pi}{M} \text{ rad}$$

The maximum input bit rate C can be determined by

$$C = B \log_2 M$$

where B = bandwidth
M = maximum possible number of symbols at the output

4.4.2 Two-phase shift key modulation

In this modulator the carrier assumes one of two phases. A logic 1 produces no phase change and a logic 0 produces a 180° phase change. The output waveform for this modulator is shown in Fig. 4.6.

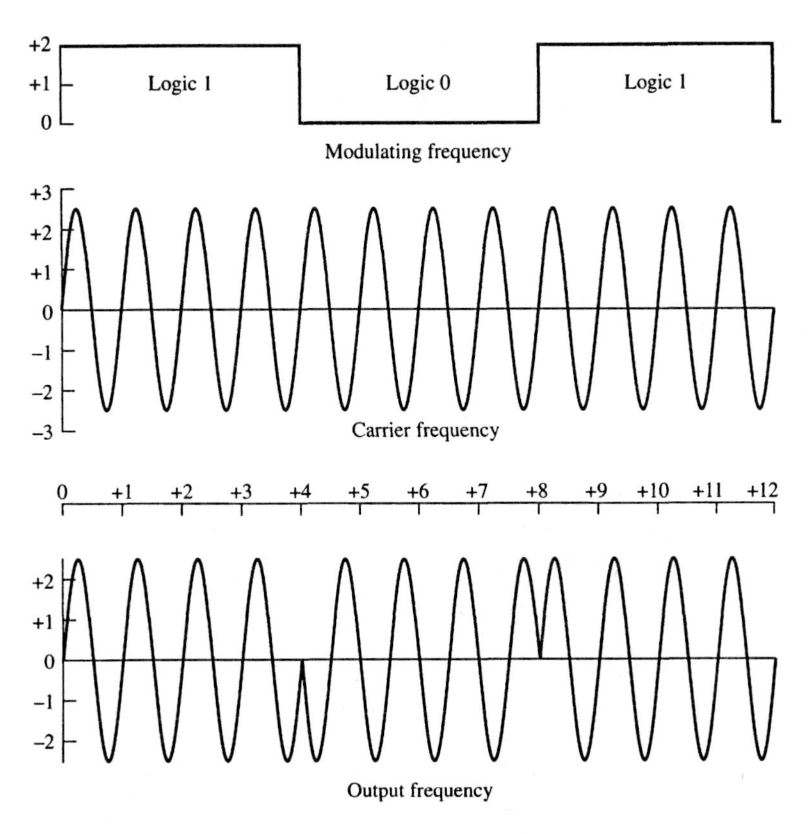

Fig. 4.6 Phase shift key modulation

A ring modulator (Fig. 4.3(c)) can be used for this purpose. The only difference between the ASK modulator and the 2 PSK modulator is where the carrier frequency and the digital modulating signal are applied. The carrier frequency and digital modulating signal are swapped around for the 2 PSK modulator so that the digital signal is used to forward bias and reverse bias the diodes.

Example 4.2

A 2 PSK modulator has an input bit rate of 2400 b/s and works into a commercial speech band circuit.

Determine:

1. The number of possible symbols at the output.
2. The symbol rate.
3. The phase difference between the symbols.
4. The maximum bit rate.

Solution

1. Each input bit causes a phase change of either 0° or 180°. Hence $n = 1$ and

$$M = 2^n = 2^1 = 2$$

2.
$$S = \frac{D}{n} \text{ symbols/s}$$
$$= 2400 \text{ symbols/s}$$

i.e. the symbol rate is the same as the bit rate.

3.
$$P = \frac{360°}{M}$$
$$= \frac{360°}{2} = 180°$$

This shows that there are only two output vectors for the output data.

4. The bandwidth is 300 Hz to 3400 Hz ($B = 3100$ Hz). Hence

$$C = B \log_2 M$$
$$= 3100 \log_2 2$$
$$= 3100 \text{ b/s}$$

This means that the maximum input bit rate that can be applied to the circuit is 3100 b/s. If a higher bit rate was applied to the input then the received data would be errored.

4.4.3 Four-phase shift key modulation

With 4 PSK, 2 bits are processed to produce a single phase change. In this case each symbol consists of 2 bits, which are referred to as a dibit.

Example 4.3

A 4 PSK modulator has an input bit rate of 2400 b/s and works into a commercial speech band circuit.

Determine:

1. The number of possible symbols at the output.
2. The symbol rate.
3. The phase difference between the symbols.
4. The maximum bit rate.

Solution

1. Two bits produce a single phase change. Hence $n = 2$ and

$$M = 2^n = 2^2 = 4$$

2.
$$S = \frac{D}{n} \text{ symbols/s}$$
$$= \frac{2400}{2} = 1200 \text{ symbols/s}$$

In this case the output symbol rate is half of the input bit rate.

3.
$$P = \frac{360°}{M}$$
$$= \frac{360°}{4} = 90°$$

This shows that there are four output phases for the output data.

4. The bandwidth is 300 Hz to 3400 Hz ($B = 3100$ Hz). Hence

$$C = B \log_2 M$$
$$= 3100 \log_2 4$$
$$= 3100 \frac{\log_{10} 4}{\log_{10} 2}$$
$$= 6200 \text{ b/s}$$

This means that the maximum input bit rate that can be input into the circuit is 6200 b/s. If a higher bit rate was applied to the input then the received data would be errored.

The actual phases that are produced by the 4 PSK modulator are shown in Table 4.2.

Because the output bit rate is less than the input bit rate, this results in a smaller bandwidth.

A typical 4 PSK circuit is shown in Fig. 4.7(a) and the constellation is shown in Fig. 4.7(b). The input signal is applied to a 2-bit splitter or a 1-to-2 demultiplexer. The output bit rate on each of the two outputs is half of that at the input to the splitter.

The two outputs are each connected to the digital inputs to the two ring modulators. The carrier frequency is applied directly to the carrier input to the top ring modulator. The carrier input to the bottom ring modulator undergoes a 90° phase shift. The phasor output of the top ring modulator is along the

Table 4.2 Phases produced by a 4 PSK modulator

Dibit	Phase change
00	+225°/−135°
01	+135°/−225°
10	+315°/−45°
11	+45°/−315°

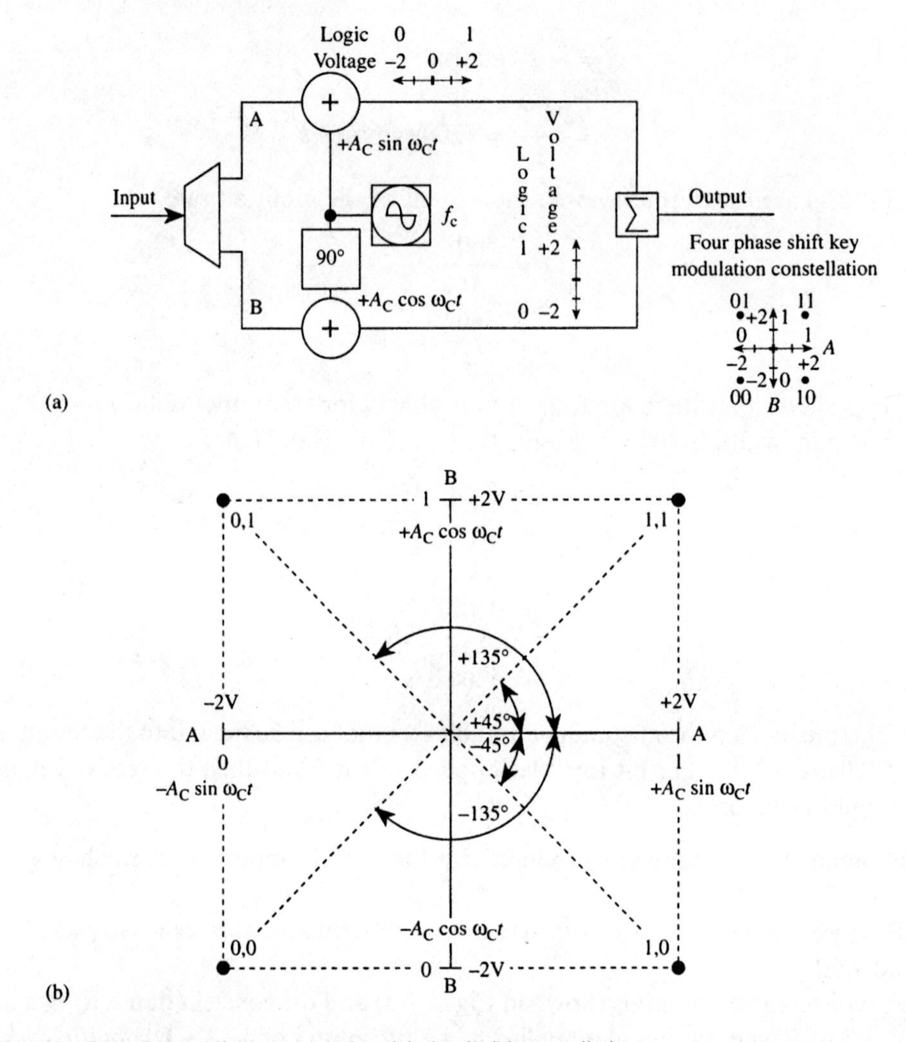

Fig. 4.7 Four-phase shift key modulation: (a) circuit; (b) constellation

horizontal axis and the phasor output from the bottom ring modulator is along the vertical axis.

A logic high at the input to each ring modulator produces a voltage of +2 V at the output and a logic low at the input produces an output voltage of −2 V.

The outputs from the two ring modulators are then summed together in a linear summer. The output of the linear summer produces the constellation shown in Fig. 4.7(b).

4.4.4 Eight-phase shift key modulation

With this modulator 3 bits are processed to produce a single phase change. This means that each symbol consists of 3 bits.

Example 4.4

An 8 PSK modulator has an input bit rate of 2400 b/s and works into a commercial speech band circuit.

Determine:

1. The number of possible symbols at the output.
2. The symbol rate.
3. The phase difference between the symbols.
4. The maximum bit rate.

Solution

1. Three bits produce a single phase change. Hence $n = 3$ and

$$M = 2^n = 2^3 = 8$$

2.

$$S = \frac{D}{n} \text{ symbols/s}$$

$$= \frac{2400}{3} = 800 \text{ symbols/s}$$

In this case the output symbol rate is a third of the input bit rate.

3.

$$P = \frac{360°}{M}$$

$$= \frac{360°}{8} = 45°$$

This shows that there are eight output vectors for the output data.

4. The bandwidth is 300 Hz to 3400 Hz ($B = 3100$ Hz). Hence

$$C = B \log_2 M$$

$$= 3100 \log_2 8$$

$$= 3100 \frac{\log_{10} 8}{\log_{10} 2}$$

$$= 9300 \text{ b/s}$$

This means that the maximum input bit rate that can be input into the circuit is 9300 b/s. If a higher bit rate was applied to the input then the received data would be errored.

A theoretical 8 PSK modulator

Figure 4.8(a) shows a typical circuit for this modulator. With this modulator bit A controls the output polarity of the first digital-to-analogue converter (DAC 1). Bit B is used to control the output polarity of the second digital-to-analogue converter (DAC 2) and bit C is used to control the output amplitude of both the digital-to-analogue converters (DAC 1 and DAC 2).

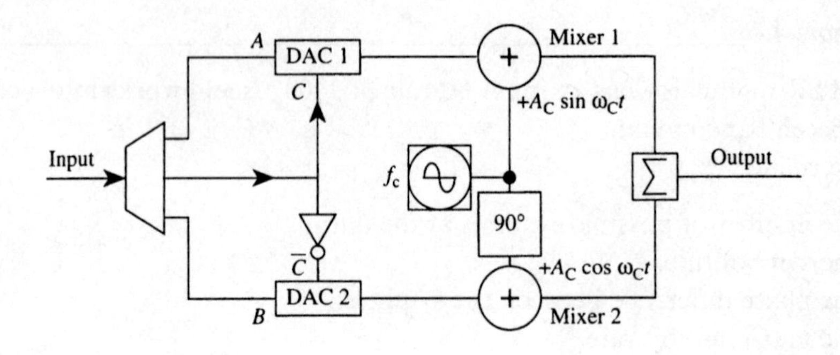

(a)

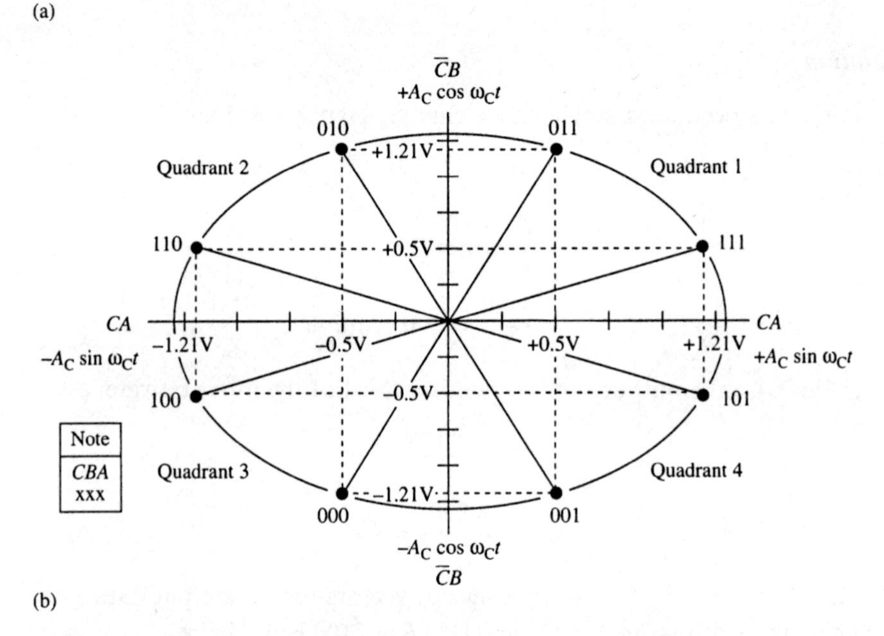

(b)

Fig. 4.8 Theoretical eight-phase shift key modulator: (a) circuit; (b) constellation

The input digital signal is applied to a 3-bit splitter or a 1-to-3 demultiplexer. This means that the output bit rate on each of the three outputs is a third of that at the input to the splitter. The data on lead A and lead B are applied to the inputs of the two digital-to-analogue converters (DAC 1 and DAC 2). The output on lead C is applied to the second input to DAC 1 directly but is negated before being applied to the second input to DAC 2.

The carrier frequency from the oscillator is connected directly to the carrier input on mixer 1 (ring modulator) and it passes through a 90° phase shifter before being applied to the carrier input for mixer 2 (ring modulator). The two mixers are identical circuits. The resultant carrier frequency phases are as shown in the circuit diagram indicated in Fig. 4.8(b). As can be seen the phase of the carrier for quadrants 1 and 2 is positive, i.e. axis $\bar{C}B$, and the phase of the carrier for quadrants 4 and 5 is negative, i.e. axis $\bar{C}B$. The digital-to-analogue conversion follows the conditions given in Table 4.3.

Table 4.3 Digital-to-analogue conversion conditions for 8 PSK modulator

A	Polarity	B	Polarity	C	Amplitude	$\bar{C}$	Amplitude
0	–	0	–	0	0.5	1	1.21
1	+	1	+	1	1.21	0	0.5

The conditions shown in Table 4.3 produce the positions as shown in Table 4.4 for all the different input permutations.

The constellation diagram can now be drawn according to this table and is shown in Fig. 4.8(b). The operation of each mixer is the same as described in Section 4.5.2. By referring to Table 4.4 it will be seen that in each quadrant there are two phases.

The outputs of the two modulators (mixer 1 and mixer 2) are applied to a linear summer. The output of the linear summer then produces the constellation as shown in Fig. 4.8(b).

A practical 8 PSK modulator

In practice the phase of the carrier frequencies is as shown in Fig. 4.9(a). The digital-to-analogue conversion follows the conditions given in Table 4.3.

The circuit is identical to that of the theoretical 8 PSK modulator except for the phasing of the carrier frequency. On the $\bar{C}B$ axis the polarity must be multiplied by a negative due to the phasing of the carrier frequency. This produces the positions shown in Table 4.5. In this case the phase of the carrier is negative for the $\bar{C}B$ axis for quadrants 1 and 2 and the phase of the carrier is positive for the $\bar{C}B$ axis for quadrants 3 and 4.

The constellation diagram can now be drawn according to this table and is shown in Fig. 4.9(b). As can be seen, again there are two phases in each quadrant. The operation of the circuit is identical to that of the theoretical 8 PSK circuit.

Table 4.4 Input permutations and positions (theoretical)

$\bar{C}$	C	B	A	C	A	Polarity	Amplitude	$\bar{C}$	B	Polarity	Amplitude	Quad
1	0	0	0	0	0	–	0.5	1	0	–	1.21	3
1	0	0	1	0	1	+	0.5	1	0	–	1.21	4
1	0	1	0	0	0	–	0.5	1	1	+	1.21	2
1	0	1	1	0	1	+	0.5	1	1	+	1.21	1
0	1	0	0	1	0	–	1.21	0	0	–	0.5	3
0	1	0	1	1	1	+	1.21	0	0	–	0.5	4
0	1	1	0	1	0	–	1.21	0	1	+	0.5	2
0	1	1	1	1	1	+	1.21	0	1	+	0.5	1

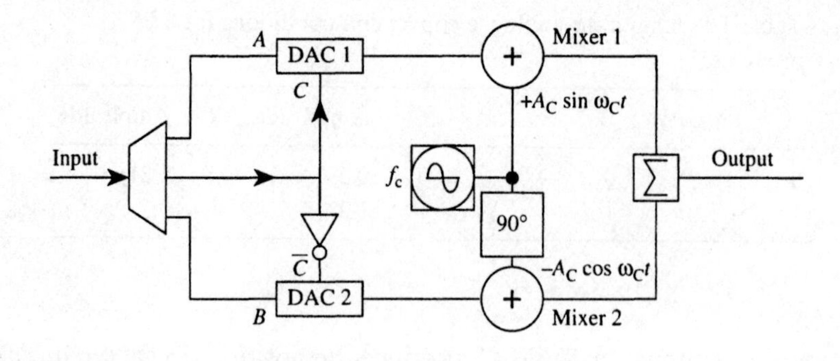

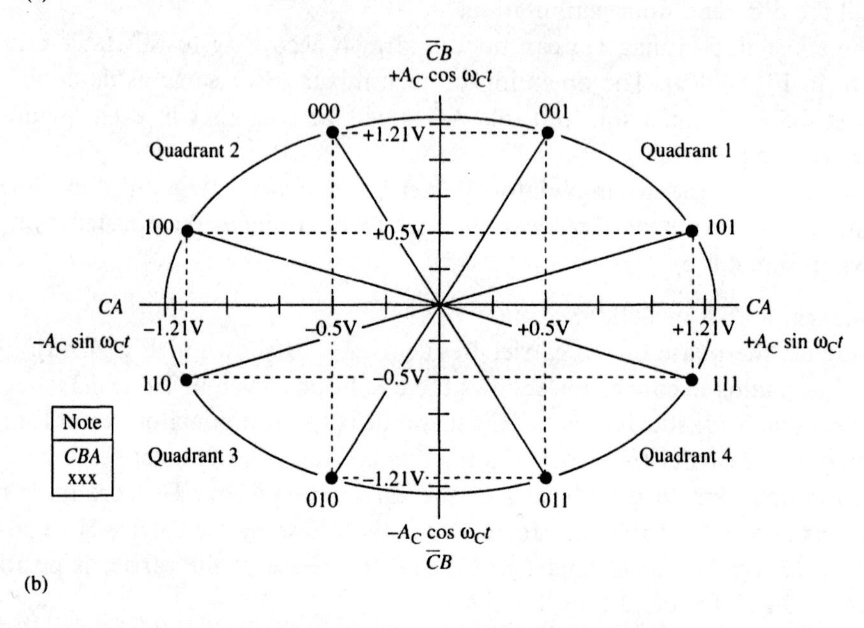

Fig. 4.9 Practical eight-phase shift key modulator: (a) circuit; (b) constellation

Table 4.5 Input permutations and positions (*practical*)

$\bar{C}$	C	B	A	C	A	Polarity	Amplitude	$\bar{C}$	B	Polarity	Amplitude	Quad
1	0	0	0	0	0	−	0.5	1	0	$(-\times-)=+$	1.21	2
1	0	0	1	0	1	+	0.5	1	0	$(-\times-)=+$	1.21	1
1	0	1	0	0	0	−	0.5	1	1	$(-\times+)=-$	1.21	3
1	0	1	1	0	1	+	0.5	1	1	$(-\times+)=-$	1.21	4
0	1	0	0	1	0	−	1.21	0	0	$(-\times-)=+$	0.5	2
0	1	0	1	1	1	+	1.21	0	0	$(-\times-)=+$	0.5	1
0	1	1	0	1	0	−	1.21	0	1	$(-\times+)=-$	0.5	3
0	1	1	1	1	1	+	1.21	0	1	$(-\times+)=-$	0.5	4

4.4.5 Sixteen-phase shift key modulation

With this modulator 4 bits are processed to produce a single phase change. This means that each symbol consists of 4 bits.

Example 4.5

A 16 PSK modulator has an input bit rate of 2400 b/s and works into a commercial speech band circuit.

Determine:

1. The number of possible symbols at the output.
2. The symbol rate.
3. The phase difference between the symbols.
4. The maximum bit rate.

Solution

1. Four bits produce a phase change. Hence $n = 4$ and

$$M = 2^n = 2^4 = 16$$

2.

$$S = \frac{D}{n} \text{ symbols/s}$$

$$= \frac{2400}{4} = 600 \text{ symbols/s}$$

In this case the output symbol rate is a quarter of the input bit rate.

3.

$$P = \frac{360°}{M}$$

$$= \frac{360°}{16} = 22.5°$$

This shows that there are 16 output vectors for the output data.

4. The bandwidth is 300 Hz to 3400 Hz ($B = 3100$ Hz). Hence

$$C = B \log_2 M$$

$$= 3100 \log_2 16$$

$$= 3100 \frac{\log_{10} 16}{\log_{10} 2}$$

$$= 12\,400 \text{ b/s}$$

This means that the maximum input bit rate that can be input into the circuit is 12 400 b/s. If a higher bit rate was applied to the input then the received data would be errored.

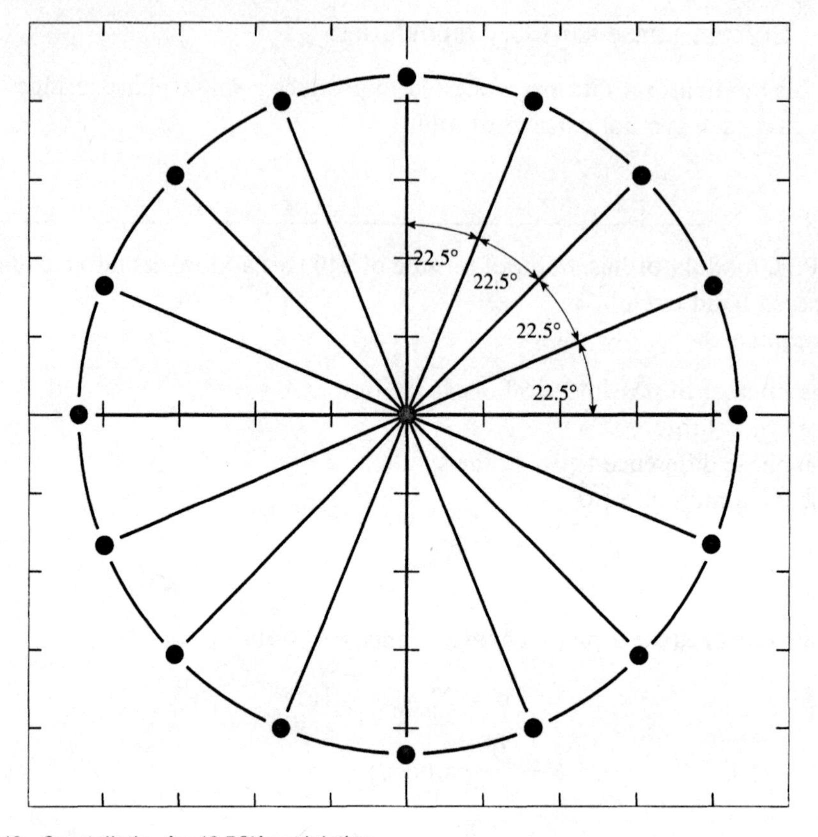

Fig. 4.10 Constellation for 16 PSK modulation

The constellation of a 16 PSK modulator is shown in Fig. 4.10. Only the constellation for this modulation scheme is shown. A better modulator is the 16 QAM, which is described in more detail below. Notice that in the constellation for the 16 PSK, each phasor is separated from the adjacent phases by 22.5°.

4.5 SIXTEEN-QUADRATURE AMPLITUDE MODULATION

With this modulator, 4 bits are processed to produce a single vector. Bits A and C determine the quadrant in which the phasor will lie and bits B and D determine the position within the quadrant that the phasor will assume. Bit A is the MSB and bit D is the LSB. This is a better method than 16 PSK modulation. The complete constellation is shown in Fig. 4.11.

A typical circuit is shown in Fig. 4.12. The circuit consists of four identical ring modulators. The carrier to the bottom two modulators undergoes a 90° phase shift.

The top two modulator outputs are summed together to produce phases on the horizontal axis through a linear summer. The outputs of the bottom two

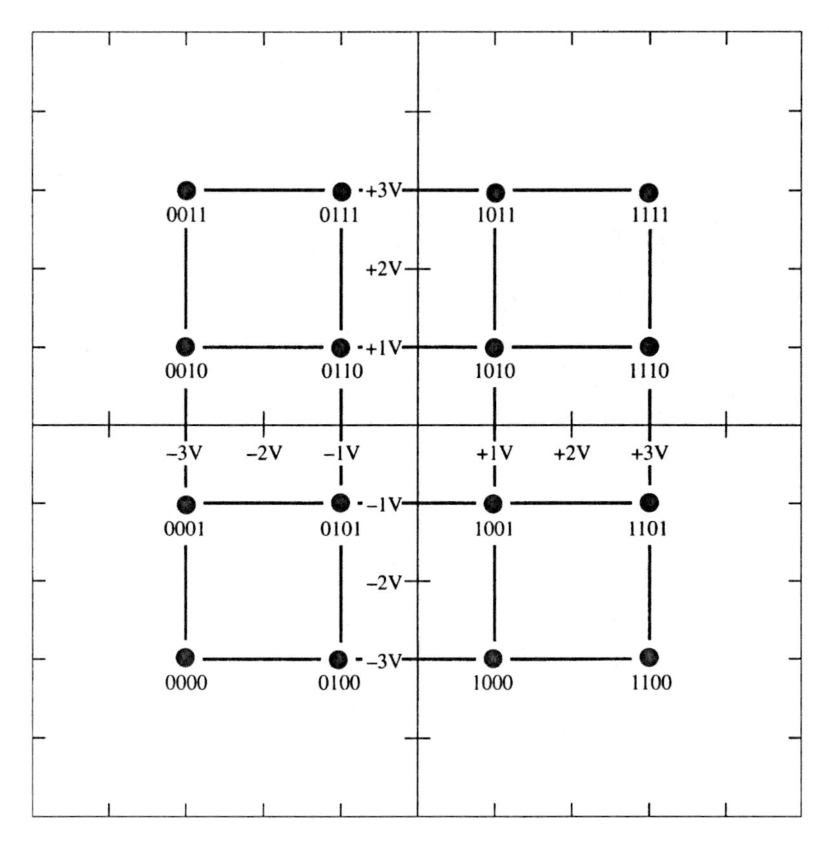

Fig. 4.11 Constellation for 16 QAM showing the 4-bit binary codes

modulators are summed together to produce the resultant phases on the vertical axis through a linear summer. The vertical and horizontal vectors are summed together to produce the 16 QAM constellation through a linear summer.

The two 6 dB attenuators, or pads, are used to ensure that the output voltage is half of the input voltage. The 6 dB pads are connected to the outputs of the ring modulators for bits B and D.

The following mathematical manipulations show that the 6 dB pads cause the output voltage to be half of the input voltage:

$$N = 20 \log_{10} \frac{V_{\text{out}}}{V_{\text{in}}} \text{ dB}$$

$$-6 \text{ dB} = 20 \log_{10} \frac{V_{\text{out}}}{V_{\text{in}}}$$

$$-0.3 = \log_{10} \frac{V_{\text{out}}}{V_{\text{in}}}$$

$$\frac{V_{\text{out}}}{V_{\text{in}}} = 10^{-0.3} = 0.5$$

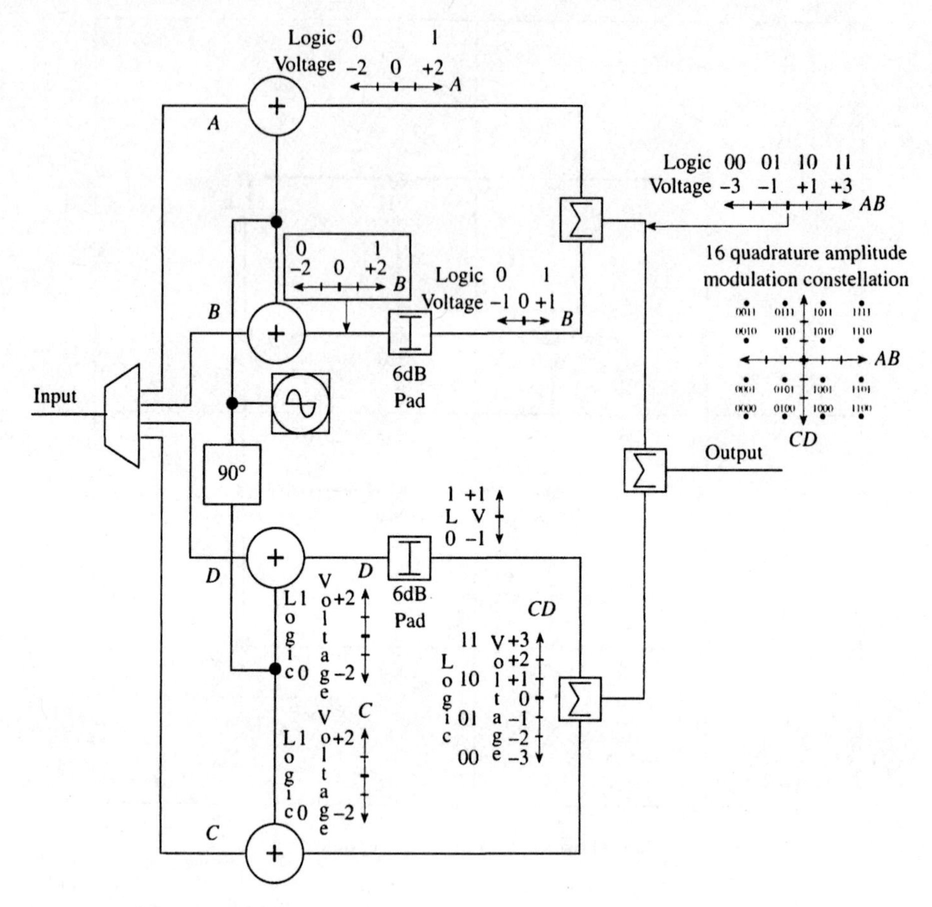

Fig. 4.12 Modulator circuit for 16 QAM

The circuit produces the phase angles in each quadrant as shown in Fig. 4.13. This figure, together with Fig. 4.14, clearly indicate the three different amplitudes. Figure 4.13 together with Fig. 4.15 indicate the 12 different phases that result. The three different amplitudes and 12 different phases together with the binary codes for each phasor are shown in Fig. 4.16. This figure shows more clearly the 4-bit binary codes for each phasor.

The modulator works according to the data given in Table 4.6. From this the positions in the constellation can be plotted as shown in Fig. 4.11.

When the logic condition in lead *A* is a logic high, the output of the ring modulator produces a phase having an amplitude of +2 V. When the logic condition on lead *A* is a logic low then the output of the ring modulator produces a phase having an amplitude of −2 V. The same applies to the logic conditions on leads *B*, *C* and *D* and the outputs from the other ring modulators. The 6 dB pad that is associated with the output of the ring modulator for lead *B* produces an output voltage of either +1 V for an input voltage of +2 V or −1 V for an input voltage of −2 V. The same applies to the 6 dB pad associated with the ring modulator for lead *D*.

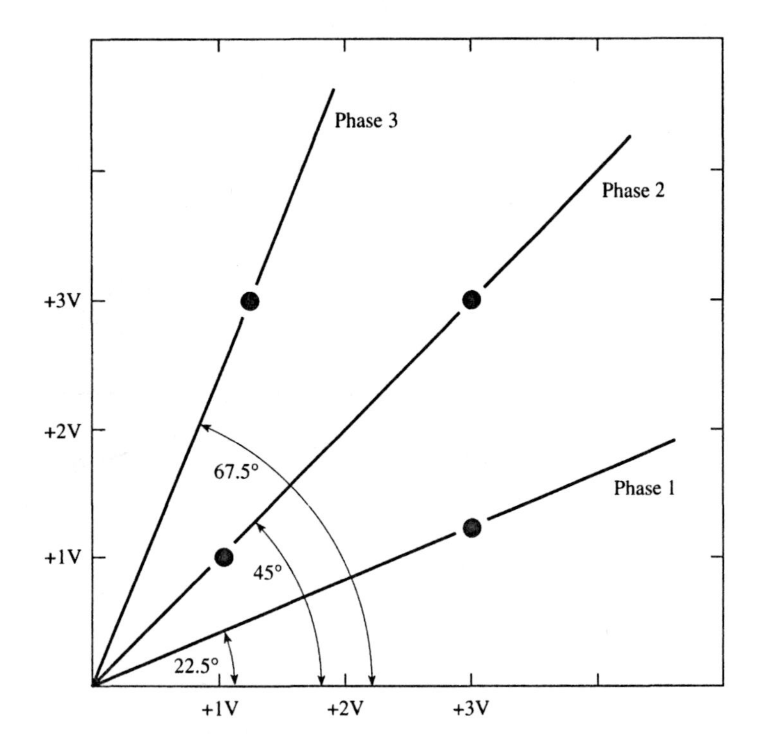

Fig. 4.13 Phase angles in the first quadrant for 16 QAM

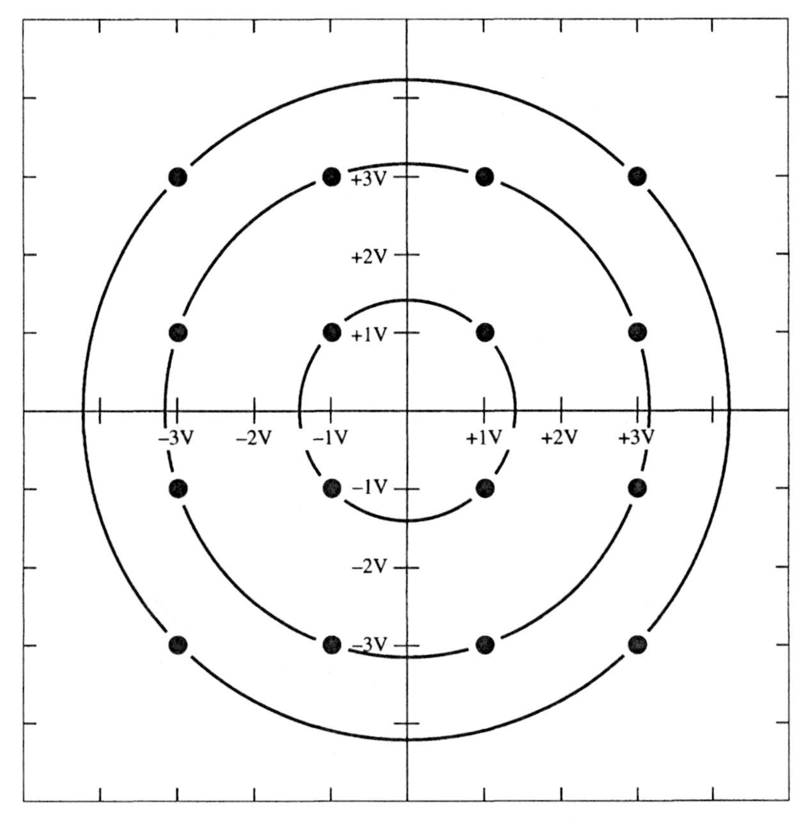

Fig. 4.14 Constellation for 16 QAM showing three amplitudes

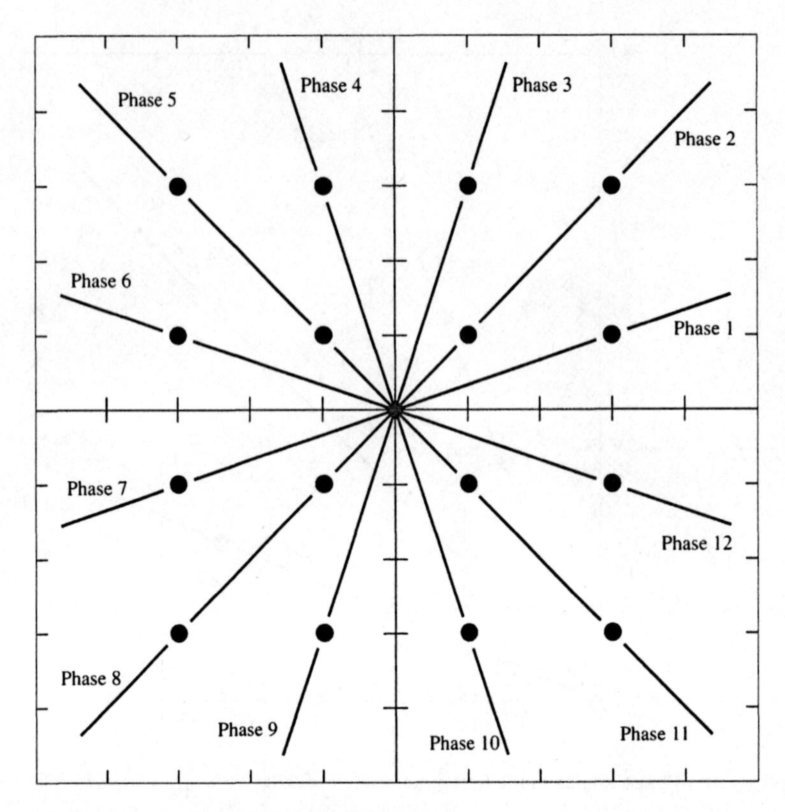

Fig. 4.15 Constellation for 16 QAM showing 12 phases

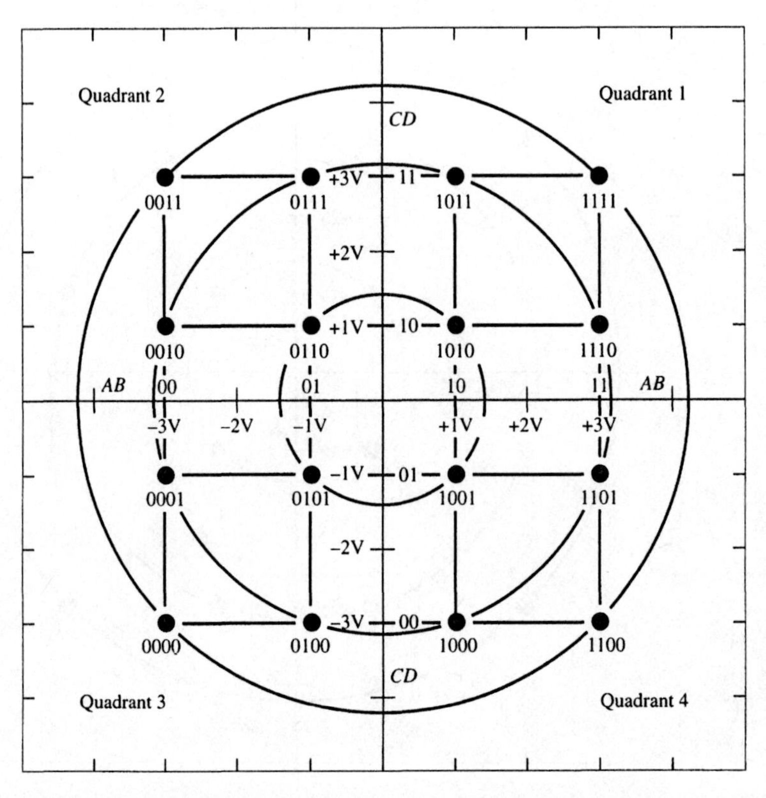

Fig. 4.16 Constellation for 16 QAM showing amplitudes, phases and codes

Table 4.6 Input permutations and positions (16 QAM)

A	*B*	*C*	*D*	*A*	*B*	Polarity	Amplitude	*C*	*D*	Polarity	Amplitude	Quad
0	0	0	0	0	0	–	3	0	0	–	3	3
0	0	0	1	0	0	–	3	0	1	–	1	3
0	0	1	0	0	0	–	3	1	0	+	1	2
0	0	1	1	0	0	–	3	1	1	+	3	2
0	1	0	0	0	1	–	1	0	0	–	3	3
0	1	0	1	0	1	–	1	0	1	–	1	3
0	1	1	0	0	1	–	1	1	0	+	1	2
0	1	1	1	0	1	–	1	1	1	+	3	2
1	0	0	0	1	0	+	1	0	0	–	3	4
1	0	0	1	1	0	+	1	0	1	–	1	4
1	0	1	0	1	0	+	1	1	0	+	1	1
1	0	1	1	1	0	+	1	1	1	+	3	1
1	1	0	0	1	1	+	3	0	0	–	3	4
1	1	0	1	1	1	+	3	0	1	–	1	4
1	1	1	0	1	1	+	3	1	0	+	1	1
1	1	1	1	1	1	+	3	1	1	+	3	1

The resultant constellation consists of three different amplitudes distributed in 12 different phases as shown in Figs 4.11 and 4.14–4.16.

4.6 BANDWIDTHS

An important consideration in any communication circuit is bandwidth. Today the usable frequency spectrum is becoming increasingly crowded, with less and less available. This means that bandwidths must be limited more and more.

4.6.1 Nyquist bandwidth

This is the minimum bandwidth required to transmit and receive a digital signal to ensure error-free communication. The Nyquist bandwidth is much smaller than the complete signal bandwidth. The receiver need only be able to identify the different logic states.

4.6.2 Spectrum analyser comparison

Consider a modulator using a carrier frequency of 140 MHz when a 143.36 Mb/s bit stream is applied to it. Figure 4.17 shows these comparisons.

2 PSK modulator (Fig. 4.17(a))

The entire band stretches from −3.36 Hz to 283.36 MHz. This is derived from

$$140\,\text{MHz} - 143.36\,\text{MHz to } 140\,\text{MHz} + 143.36\,\text{MHz}$$

In practice the band is from 0 Hz to 143.35 MHz as there are no negative frequencies.

The Nyquist bandwidth is 68.32 MHz to 211.68 MHz. The frequency deviation is thus 71.68 MHz.

4 PSK modulator (Fig. 4.17(b))

The entire band stretches from 68.32 MHz to 211.68 MHz. This is derived from

$$143.36/2 = 71.68\,\text{Mb/s}$$

$$140\,\text{MHz} - 71.68\,\text{MHz to } 140\,\text{MHz} + 71.68\,\text{MHz}$$

The Nyquist bandwidth is 104.16 MHz to 175.84 MHz. The frequency deviation is thus 35.84 MHz. This is a smaller bandwidth than the 2 PSK.

8 PSK modulator (Fig. 4.17(c))

The entire band stretches from 104.16 MHz to 175.84 MHz. This is derived from

$$143.36/4 = 35.84\,\text{Mb/s}$$

$$140\,\text{MHz} - 35.84\,\text{MHz to } 140\,\text{MHz} + 35.84\,\text{MHz}$$

The Nyquist bandwidth is 122.08 MHz to 157.92 MHz. The frequency deviation is thus 17.92 MHz. This is a smaller bandwidth than the 4 PSK.

16 PSK modulator (Fig. 4.17(d))

The entire band stretches from 122.08 MHz to 157.92 MHz. This is arrived at by

$$143.36/8 = 17.92\,\text{Mb/s}$$

$$140\,\text{MHz} - 17.92\,\text{MHz to } 140\,\text{MHz} + 17.92\,\text{MHz}$$

The Nyquist bandwidth is 131.04 MHz to 148.96 MHz. The frequency deviation is thus 8.96 MHz. This is a smaller bandwidth than the 8 PSK.

64 QAM

From the above it is obvious that if 64 QAM was implemented then the Nyquist bandwidth would be

$$146.36\,\text{MHz}/16 = 8.96\,\text{MHz}$$

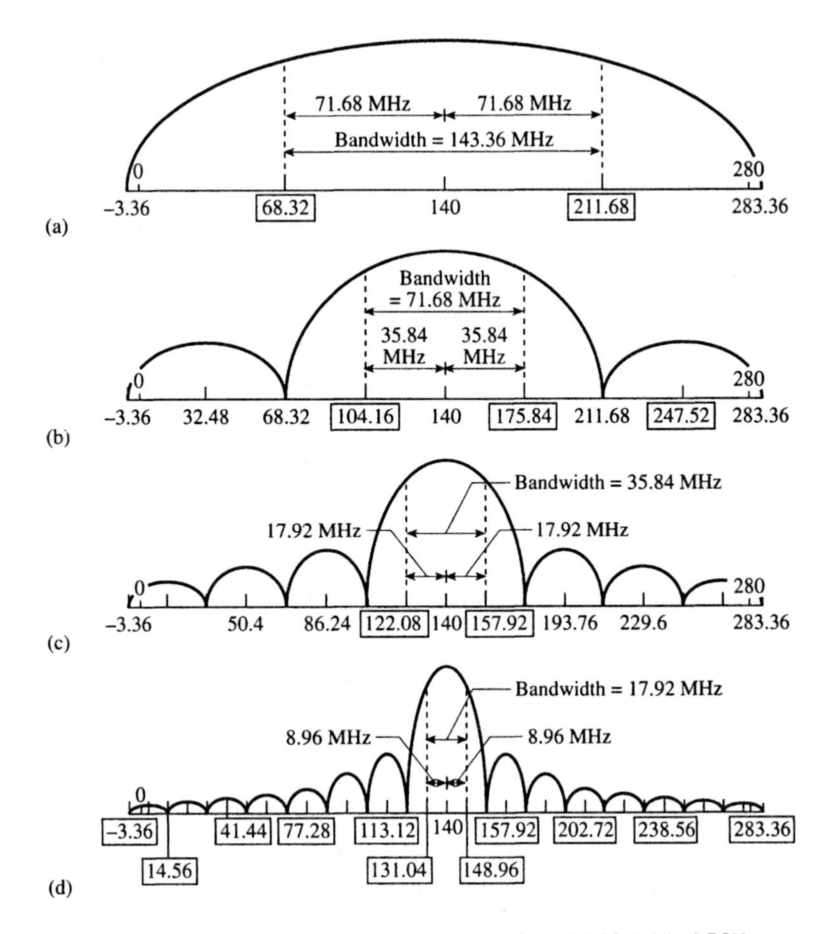

Fig. 4.17 Spectrum analyser comparison: (a) 2 PSK; (b) 4 PSK; (c) 8 PSK; (d) 16 PSK

with a deviation of 4.48 MHz. The advantages of QAM modulation are clearly seen. However, there is a limit, as the higher the modulation scheme becomes the closer the phasors are to one another. When the signal is transmitted and noise is introduced it is possible for phasors to cross over and the result is corrupted data.

4.7 DIFFERENTIAL PHASE MODULATION

Differential phase modulation is used on radio and microwave systems which employ BPSK, QPSK or QAM modulators. On these systems the data is transmitted in different phasors and/or amplitudes.

If differential modulation techniques are not used then the received signal will be decoded incorrectly as the constellation can and does rotate as it propagates through the air. Differential encoding enables the relative phase and relative amplitude information to be transmitted instead of the actual phase and actual amplitude information. This ensures the correct detection and restoration of the digital information.

4.7.1 PSK differential modulation

A typical 8 PSK differential modulator circuit is shown in Fig. 4.18(a). Bits C, B and A, together with bits c, b and a, form the address for the EPROM. Bits C, B and A are the input bits, with bit C being the MSB and bit A the LSB. The output of the EPROM produces bits c, b and a, with bit c being the MSB and bit a being the LSB. These bits are then applied to the 8 PSK modulator.

By considering the constellation diagram of a practical 8 PSK modulator, as shown in Fig. 4.19, bits B and A determine the quadrant and bit C determines the position within the quadrant.

It is important to be able to position the phasors in the quadrant. Figure 4.19(a) shows the axis for a theoretical 8 PSK modulator. As can be seen, the x–x axis contains the binary codes for CA which are 10_2, 00_2, 01_2, 11_2 from left to right.

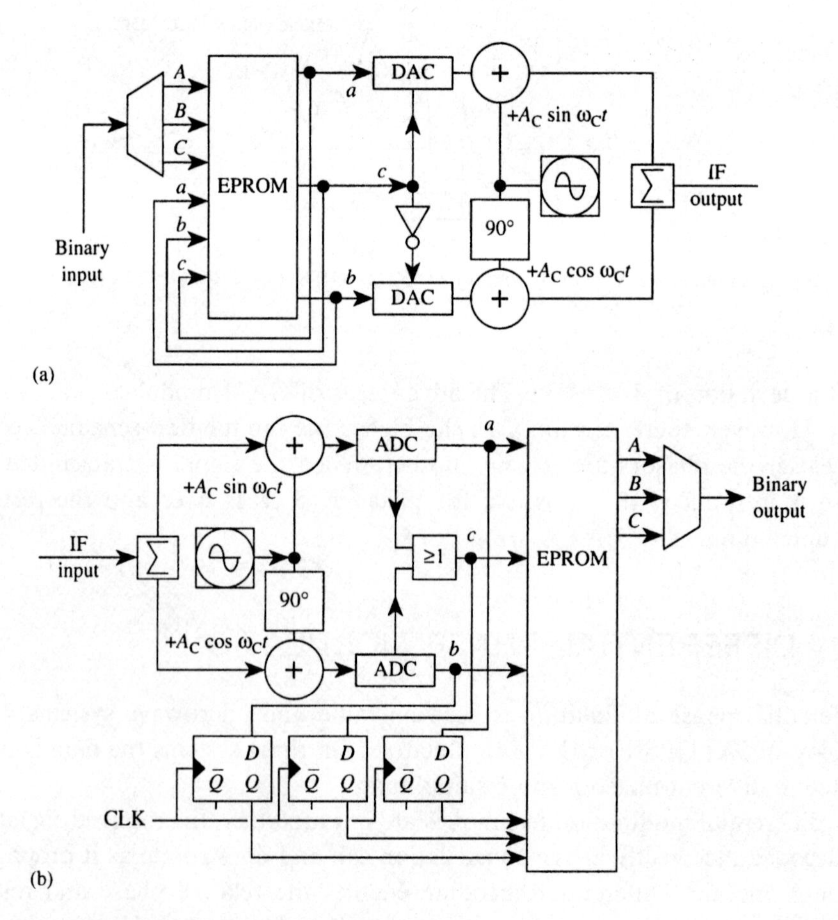

Fig. 4.18 (a) 8 PSK modulator (b) 8 PSK demodulator

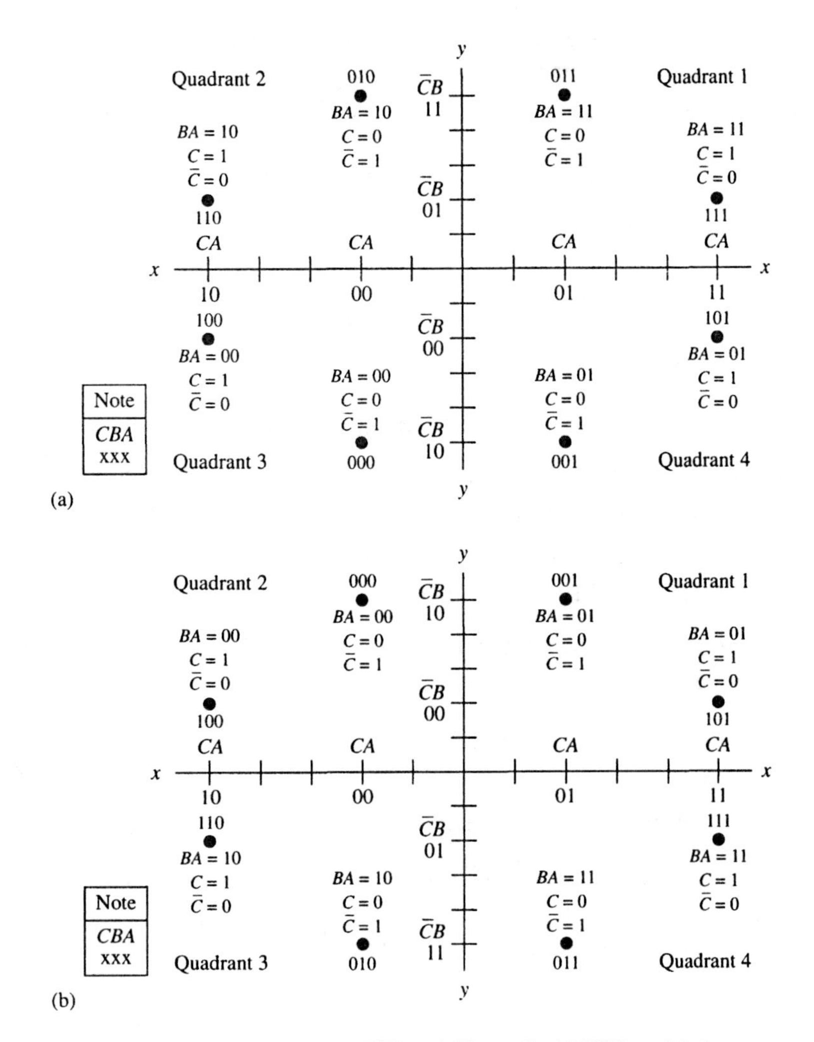

Fig. 4.19 Constellations for (a) theoretical 8 PSK and (b) practical 8 PSK modulation

Likewise the y–y axis contains the binary codes for $\bar{C}B$, which are 10_2, 00_2, 01_2, 11_2 from bottom to top.

Figure 4.19(b) shows the axis for the constellation for the practical 8 PSK modulator. In this case the x–x axis contains the binary codes for CA which are 10_2, 00_2, 01_2 and 11_2. The y–y axis contains the binary codes for $\bar{C}B$, which are 11_2, 01_2, 00_2 and 10_2.

As an example, the binary code 001_2 produces a phasor in quadrant 4 for the theoretical 8 PSK modulator and quadrant 1 for the practical 8 PSK modulator as $CA = 01_2$ and $\bar{C}B = 10_2$.

The differential encoding takes place according to the data given in Table 4.7. In this example the logic state of bit C remains unchanged through the differential encoder.

Table 4.7 Differential encoding data

Bin code		Phase angle
B/*b*	*A*/*a*	
0	0	0°
0	1	+90°
1	0	−90°
1	1	±180°

Example 4.6

A practical 8 PSK differential modulator produces the constellation as shown in Fig. 4.9. The following bit stream is applied to the input of the modulator. The differential encoding takes place according to the data given in Table 4.7.

$$\text{Data} = (110)\quad 101\quad 000\quad 010\quad 100\quad 011\quad 101_2$$

Determine the codes for the encoded data at the output of the EPROM. Assume the default code to be 110_2.

Solution

The encoding is done as shown in Table 4.8.

1. Find the position on the constellation diagram for $110_2 \rightarrow$ 3rd quadrant.
   Find the position on the constellation diagram for $101_2 \rightarrow$ 1st quadrant.
   Relative to 110_2, 101_2 is found by moving 110_2 by $\pm 180° \rightarrow$ bits $ba = 11$, bit $c =$ bit $C = 1$.
   The data is $cba = 111_2$.
2. Feed 111_2 back to the input.
   Find the position on the constellation diagram for $111_2 \rightarrow$ 4th quadrant.
   Find the position on the constellation diagram for $000_2 \rightarrow$ 2nd quadrant.

Table 4.8 Differential encoding for 8 PSK modulator

Present	Quad	Next	Quad	Phase change	Output code
C B A		*C B A*			*c b a*
1 1 0	3rd	1 0 1	1st	±180°	1 1 1
1 1 1	4th	0 0 0	2nd	±180°	0 1 1
0 1 1	4th	0 1 0	3rd	−90°	0 1 0
0 1 0	3rd	1 0 0	2nd	−90°	1 1 0
1 1 0	3rd	0 1 1	4th	+90°	0 0 1
0 0 1	1st	1 0 1	1st	0°	1 0 0

Relative to 111_2, 000_2 is found by moving 111_2 by $\pm 180° \rightarrow$ bits $ba = 11$, bit $c =$ bit $C = 0$.
The data is 011_2.

3. Feed 011_2 back to the input.
   Find the position on the constellation diagram for $011_2 \rightarrow$ 4th quadrant.
   Find the position on the constellation diagram for $010_2 \rightarrow$ 3rd quadrant.
   Relative to 011_2, 010_2 is found by moving 011_2 by $-90° \rightarrow$ bits $ba = 10$, bit $c =$ bit $C = 0$.
   The data is 010_2.
4. Feed 010_2 back to the input.
   Find the position on the constellation diagram for $010_2 \rightarrow$ 3rd quadrant.
   Find the position on the constellation diagram for $100_2 \rightarrow$ 2nd quadrant.
   Relative to 010_2, 100_2 is found by moving 010_2 by $-90° \rightarrow$ bits $ba = 10$, bit $c =$ bit $C = 1$.
   The data is 110_2.
5. Feed 110_2 back to the input.
   Find the position on the constellation diagram for $110_2 \rightarrow$ 3rd quadrant.
   Find the position on the constellation diagram for $011_2 \rightarrow$ 4th quadrant.
   Relative to 110_2, 011_2 is found by moving 110_2 by $+90° \rightarrow$ bits $ba = 01$, bit $c =$ bit $C = 0$.
   The data is 001_2.
6. Feed 001_2 back to the input.
   Find the position on the constellation diagram for $001_2 \rightarrow$ 1st quadrant.
   Find the position on the constellation diagram for $101_2 \rightarrow$ 1st quadrant.
   Relative to 001_2, 101_2 is found by moving 110_2 by $0° \rightarrow$ bits $ba = 00$, bit $c =$ bit $C = 1$.
   The data is 100_2.

This gives the following data streams:

		CBA	*CBA*	*CBA*	*CBA*	*CBA*	*CBA*	*CBA*
Input data	=	(110)	101	000	010	100	011	101_2

		cba	*cba*	*cba*	*cba*	*cba*	*cba*	*cba*
Diff. encoded data	=	(110)	111	011	010	110	001	100_2

The differentially encoded data is eventually transmitted to the distant receiver. The receiver must correctly decode the information to obtain the original data stream.

A typical circuit for the differential decoder is shown in Fig. 4.18(b). In this case the intermediate frequency (IF) input is equally split and applied to the two ring demodulators. The IF is demodulated and the outputs are connected to analogue-to-digital converters (ADCs). The resultant digital signals on leads b and a are used as part of the address for the EPROM. The ADC also produce the logic condition for lead c which is also used as part of the address for the EPROM. The leads c, b, and a are then connected to the inputs to three D flipflops. The Q

outputs of these flipflops produce the same logic conditions one clock pulse later. These outputs are used as part of the address for the EPROM.

The output of the EPROM produce the logic conditions for outputs C, B and A. These three outputs are multiplexed together to produce a serial output bit stream.

Example 4.7

A practical 8 PSK differential demodulator produces the constellation shown in Fig. 4.9. The following bit stream is received from the distant transmitter. The differential decoding takes place according to the data given in Table 4.7.

$$\text{DATA} = (110)\quad 111\quad 011\quad 010\quad 110\quad 001\quad 100_2$$

Determine the output data of the decoder. Assume the default code to be 110_2.

Solution

To enable the decoding to be done refer to the constellation diagram for the practical 8 PSK modulator shown in Fig. 4.9(b). The decoding is done as shown in Table 4.9.

1. Find 110_2 on the constellation → 3rd quadrant.
   Determine where 110_2 is moved to.
   Next input data is $111_2 \rightarrow C = 1$, $ba = 11$.
   $11_2 = \pm 180°$.
   110_2 is moved to the 1st quadrant → $BA = 01$.
   Data = 101_2.
2. Feed 111_2 back to the input.
   Find 111_2 on the constellation → 4th quadrant.
   Determine where 111_2 is moved to.
   Next input data is $011_2 \rightarrow C = 0$, $ba = 11$.
   $11_2 = \pm 180°$.

Table 4.9 Differential decoding for 8 PSK demodulator

Present	Quad	Next	Phase change	Output code
c b a		*c b a*		*C B A*
1 1 0	3rd	1 1 1	±180°	1 0 1
1 1 1	4th	0 1 1	±180°	0 0 0
0 1 1	4th	0 1 0	−90°	0 1 0
0 1 0	3rd	1 1 0	−90°	1 0 0
1 1 0	3rd	0 0 1	+90°	0 1 1
0 0 1	1st	1 0 0	0°	1 0 1

111_2 is moved to the 2nd quadrant $\rightarrow BA = 00$.
Data $= 000_2$.

3. Feed 011_2 back to the input.
Find 011_2 on the constellation $\rightarrow$ 4th quadrant.
Determine where 011_2 is moved to.
Next input data is $010_2 \rightarrow C = 0, ba = 10$.
$10_2 = -90°$.
011_2 is moved to the 3rd quadrant $\rightarrow BA = 10$.
Data $= 010_2$.
4. Feed 010_2 back to the input.
Find 010_2 on the constellation $\rightarrow$ 3rd quadrant.
Determine where 010_2 is moved to.
Next input data is $110_2 \rightarrow C = 1, ba = 10$.
$10_2 = -90°$.
010_2 is moved to the 2nd quadrant $\rightarrow BA = 00$.
Data $= 100_2$
5. Feed 110_2 back to the input.
Find 110_2 on the constellation $\rightarrow$ 3rd quadrant.
Determine where 110_2 is moved to.
Next input data is $001_2 \rightarrow C = 0, ba = 01$.
$01_2 = +90°$.
110_2 is moved to the 4th quadrant $\rightarrow BA = 11$.
Data $= 011_2$.
6. Feed 001_2 back to the input.
Find 001_2 on the constellation $\rightarrow$ 1st quadrant.
Determine where 001_2 is moved to.
Next input data is $100_2 \rightarrow C = 1, ba = 00$.
$00_2 = 0°$.
001_2 remains in the 1st quadrant $\rightarrow BA = 01$.
Data $= 101_2$.
This gives the following data stream:

$$\begin{array}{lcccccc} & CBA & CBA & CBA & CBA & CBA & CBA \\ \text{Output data} = & 101 & 000 & 010 & 100 & 011 & 101_2 \end{array}$$

This is the same as the original data at the transmitter prior to the differential encoding.

4.7.2 Sixteen-quadrature amplitude differential modulation

A typical circuit for a 16 QAM differential modulator is shown in Fig. 4.20. Bits A, B, C and D together with bits a, b, c and d form the address for the EPROM. Bits A, B, C and D are the input bits with bit A being the MSB and bit D the LSB. The output of the EPROM produces bits a (MSB), b, c and d (LSB). These bits are then applied to the 16 QAM modulator.

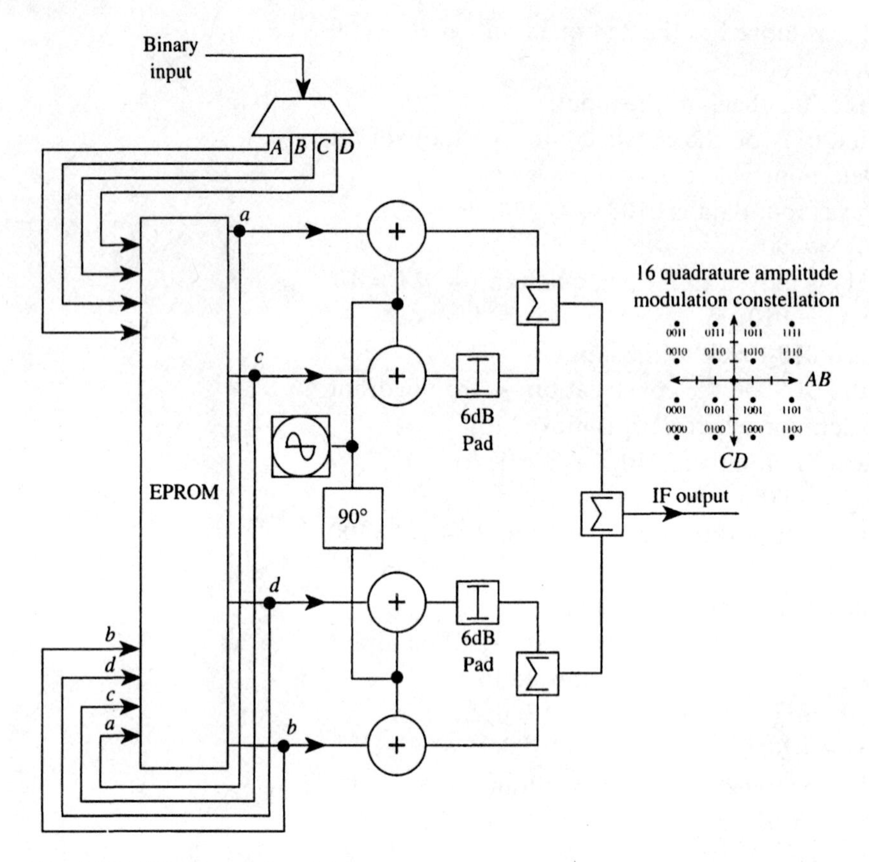

Fig. 4.20 A 16 QAM differential modulator

The differential modulation method given in the solution to Example 4.8 relies on being able to position the phasors in the correct quadrant in the constellation. By referring to Fig. 4.21 it can be seen that the x–x axis contains the binary codes for AB, which are 00_2, 01_2, 10_2, 11_2 from left to right. Likewise the y–y axis contains the binary codes for CD which are 00_2, 01_2, 10_2, 11_2 from bottom to top.

Using this information it is a fairly simple matter to position each phasor in the correct quadrant. As shown, $AC = 11_2$ for quadrant 1, $AC = 01_2$ for quadrant 2, $AC = 00_2$ for quadrant 3 and $AC = 10_2$ for quadrant 4. Hence the logic conditions for AC determine the quadrant in which the phasor lies. The logic conditions for BD determine the position within the quadrant as in each quadrant the logic states of BD assume values of 00_2, 01_2, 10_2 or 11_2.

As an example the phasor having the code 0010_2 means that $AC = 01_2$ and $BD = 00_2$. The code 1000_2 causes the phasor to lie in quadrant 4 as $AC = 10$ and $BD = 00_2$.

The differential encoding takes place according to the data given in Table 4.10.

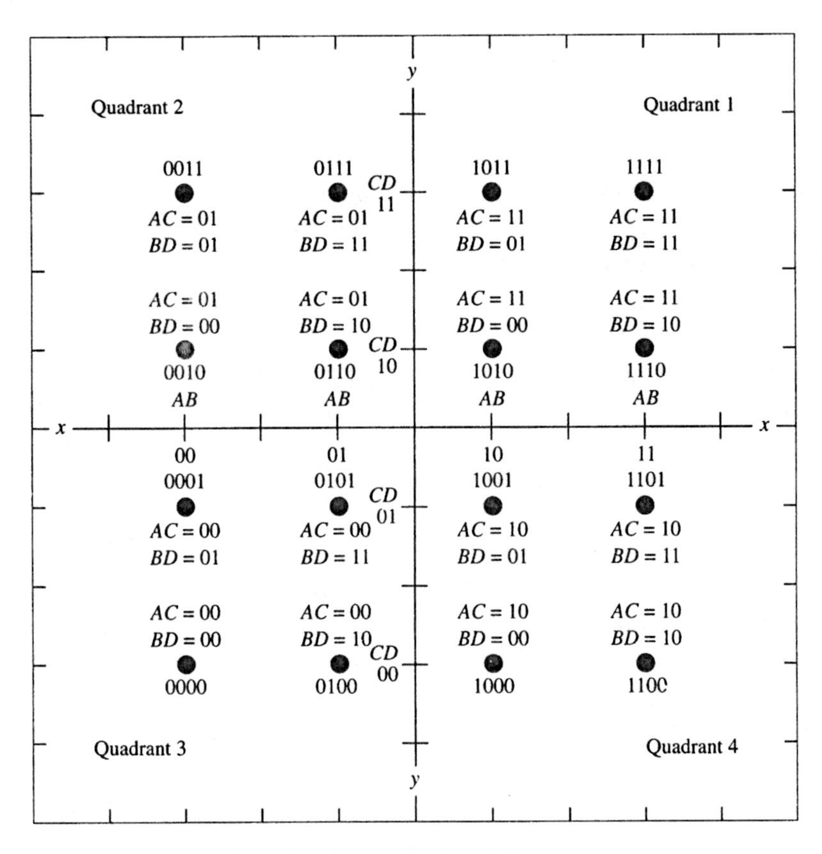

Fig. 4.21 Constellation for 16 QAM showing positioning of phasors

Example 4.8

A 16 QAM differential modulator produces the constellation as shown in Fig. 4.11. The following bit stream is applied to the input of the modulator. The differential encoding takes place according to Table 4.10.

$$\text{Data} = (0110)\quad 1101\quad 1000\quad 0001\quad 1100\quad 1010_2$$

Determine the codes for the encoded data at the output of the EPROM. Assume the default code to be 0110_2.

Table 4.10 Differential encoding data

Bin code		Phase angle	Bin code		Amplitude	
A/a	*C/c*		*B/b*	*D/d*	*b/B*	*d/D*
0	0	0°	0	0	1	1
0	1	−90°	0	1	1	0
1	0	+90°	1	0	0	1
1	1	±180°	1	1	0	0

Solution

The differential coding is done according to the data given in Table 4.11.

1. Find the position on the constellation diagram for $0110_2 \rightarrow$ 2nd quadrant.
   Find the position on the constellation diagram for $1101_2 \rightarrow$ 4th quadrant.
   Relative to 0110_2, 1101_2 is found by moving 0110_2 by $\pm 180° \rightarrow$ bits $ac = 11$.
   Bits $BD = 11$, therefore bits $bd = 00$.
   The data is 1010_2.
2. Feed 1010_2 back to the input.
   Find the position on the constellation diagram for $1010_2 \rightarrow$ 1st quadrant.
   Find the position on the constellation diagram for $1000_2 \rightarrow$ 4th quadrant.
   Relative to 1010_2, 1000_2 is found by moving 1010_2 by $-90° \rightarrow$ bits $ac = 01$.
   Bits $BD = 00$, therefore bits $bd = 11$.
   The data is 0111_2.
3. Feed 0111_2 back to the input.
   Find the position on the constellation diagram for $0111_2 \rightarrow$ 2nd quadrant.
   Find the position on the constellation diagram for $0001_2 \rightarrow$ 3rd quadrant.
   Relative to 0111_2, 0001_2 is found by moving 0111_2 by $+90° \rightarrow$ bits $ac = 10$.
   Bits $BD = 01$, therefore bits $bd = 10$.
   The data is 1100_2.
4. Feed 1100_2 back to the input.
   Find the position on the constellation diagram for $1100_2 \rightarrow$ 4th quadrant.
   Find the position on the constellation diagram for $1100_2 \rightarrow$ 4th quadrant.
   Relative to 1100_2, 1100_2 is found by moving 1100_2 by $0° \rightarrow$ bits $ac = 00$.
   Bits $BD = 10$, therefore bits $bd = 01$.
   The data is 0001_2.
5. Feed 0001_2 back to the input.
   Find the position on the constellation diagram for $0001_2 \rightarrow$ 3rd quadrant.
   Find the position on the constellation diagram for $1010_2 \rightarrow$ 1st quadrant.
   Relative to 0001_2, 1010_2 is found by moving 0001_2 by $\pm 180° \rightarrow$ bits $ac = 11$.
   Bits $BD = 00$, therefore bits $bd = 11$.
   The data is 1111_2.

Table 4.11 Differential encoding for 16 QAM modulator

Present				Quad	Next				Quad	Phase			Amplitude				Output data			
A	*B*	*C*	*D*		*A*	*B*	*C*	*D*		Change	*a*	*c*	*B*	*D*	*b*	*d*	*a*	*b*	*c*	*d*
0	1	1	0	2nd	1	1	0	1	4th	180°	1	1	1	1	0	0	1	0	1	0
1	0	1	0	1st	1	0	0	0	4th	−90°	0	1	0	0	1	1	0	1	1	1
0	1	1	1	2nd	0	0	0	1	3rd	+90°	1	0	0	1	1	0	1	1	0	0
1	1	0	0	4th	1	1	0	0	4th	0°	0	0	1	0	0	1	0	0	0	1
0	0	0	1	3rd	1	0	1	0	1st	180°	1	1	0	0	1	1	1	1	1	1

This gives the following data streams:

	ABCD	*ABCD*	*ABCD*	*ABCD*	*ABCD*	
Input data =	1101	1000	0001	1100	1010_2	
	abcd	*abcd*	*abcd*	*abcd*	*abcd*	*abcd*
Diff. encoded data =	(0110)	1010	0111	1100	0001	1111_2

Once the data has been differentially encoded it will eventually be transmitted to the distant receiver. The receiver must be able to correctly decode the data to obtain the original data.

A typical circuit is shown in Fig. 4.22. The IF input is equally split to the four mixers where each input is demodulated with the same carrier frequency. The carrier oscillator is connected directly to the top two mixers and the carrier frequency is shifted by 90° before being applied to the bottom two mixers. The top two mixers produce bits a and c while the bottom two mixers produce bits b and d. These four bit streams are then applied to the EPROM as part of the address. They are also applied to the D inputs of four flipflops and one

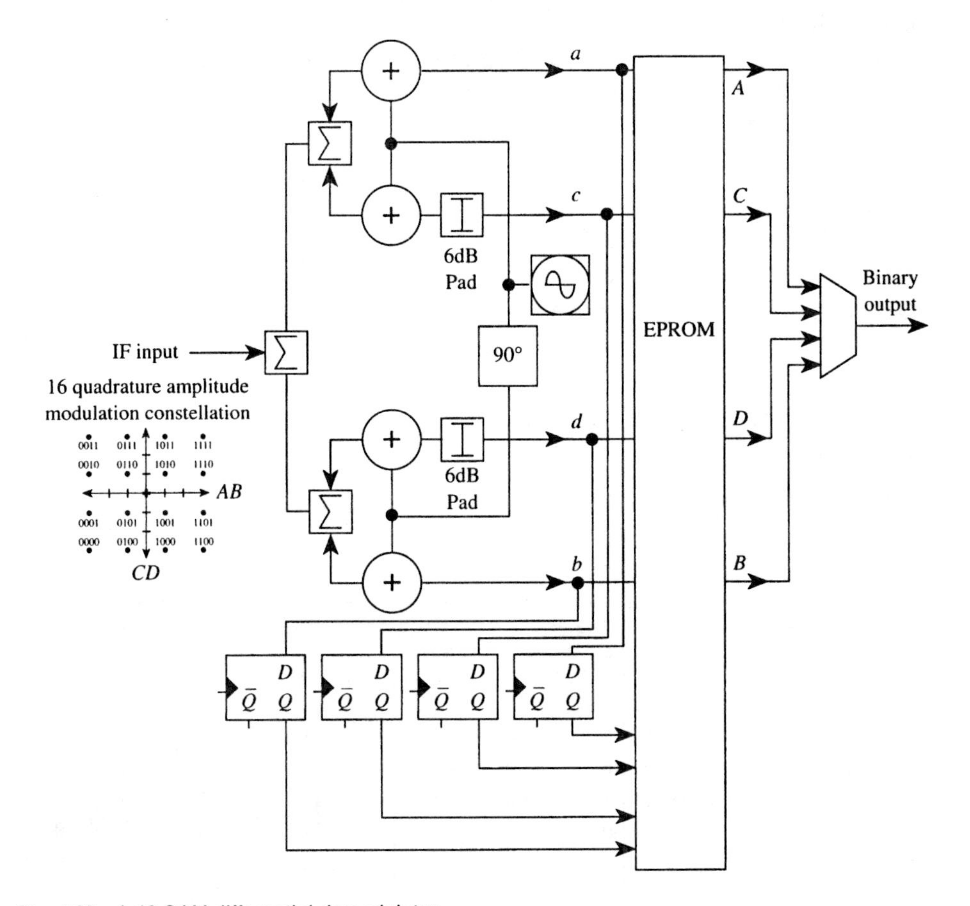

Fig. 4.22 A 16 QAM differential demodulator

clock cycle later the same logic conditions appear on the four Q outputs and form the other part of the EPROM address. The output of the EPROM produces the bit streams on leads A, B, C and D. These four bit streams are then multiplexed together to form a serial output bit stream.

Example 4.9

A 16 QAM differential demodulator produces the constellation as shown in Fig. 4.11. The following bit stream is received from the distant transmitter. The differential decoding takes place according to the data given in Table 4.10.

$$\text{Data} = (0110) \quad 1010 \quad 0111 \quad 1100 \quad 0001 \quad 1111_2$$

Determine the output data of the decoder. Assume the default code to be 0110_2.

Solution

The decoding is done as shown in Table 4.12.

1. Find 0110_2 on the constellation → 2nd quadrant.
   Determine where 0110_2 is moved to.
   Next input data is $1010_2 \rightarrow bd = 00$ and $BD = 11 \rightarrow ac = 11$.
   $11_2 = \pm 180°$.
   0110_2 is moved to the 4th quadrant.
   $AC = 10$.
   Data $= 1101_2$.
2. Feed 1010_2 back to the input.
   Find 1010_2 on the constellation → 1st quadrant.
   Determine where 1010_2 is moved to.
   Next input data is $0111_2 \rightarrow bd = 11$ and $BD = 00 \rightarrow ac = 01$.
   $01_2 = -90°$.
   1010_2 is moved to the 4th quadrant.
   $AC = 10$.
   Data $= 1000_2$.

Table 4.12 Differential decoding for 16 QAM demodulator

Present	Quad	Next	Phase					Amplitude		Output data
a b c d		*a b c d*	*a c*	Change	Quad	*A C*		*b d*	*B D*	*A B C D*
0 1 1 0	2nd	1 0 1 0	1 1	180°	4th	1 0		0 0	1 1	1 1 0 1
1 0 1 0	1st	0 1 1 1	0 1	−90°	4th	1 0		1 1	0 0	1 0 0 0
0 1 1 1	2nd	1 1 0 0	1 0	+90°	3rd	0 0		1 0	0 1	0 0 0 1
1 1 0 0	4th	0 0 0 1	0 0	0°	4th	1 0		0 1	1 0	1 1 0 0
0 0 0 1	3rd	1 1 1 1	1 1	180°	1st	1 1		1 1	0 0	1 0 1 0

3. Feed 0111_2 back to the input.
   Find 0111_2 on the constellation $\rightarrow$ 2nd quadrant.
   Determine where 0111_2 is moved to.
   Next input data is $1100_2 \rightarrow bd = 10$ and $BD = 01 \rightarrow ac = 10$.
   $10_2 = +90°$.
   0111_2 is moved to the 3rd quadrant.
   $AC = 00$.
   Data $= 0001_2$.
4. Feed 1100_2 back to the input.
   Find 1100_2 on the constellation $\rightarrow$ 4th quadrant.
   Determine where 1100_2 is moved to.
   Next input data is $0001_2 \rightarrow bd = 01$ and $BD = 10 \rightarrow ac = 00$.
   $00_2 = 0°$.
   1100_2 remains in the 4th quadrant.
   $AC = 10$.
   Data $= 1100_2$.
5. Feed 0001_2 back to the input.
   Find 0001_2 on the constellation $\rightarrow$ 3rd quadrant.
   Determine where 0001_2 is moved to.
   Next input data is $1111_2 \rightarrow bd = 11$ and $BD = 00 \rightarrow ac = 11$.
   $11_2 = \pm180°$.
   0001_2 is moved to the 1st quadrant.
   $AC = 11$.
   Data $= 1010_2$.

This gives the following data streams:

	ABCD	*ABCD*	*ABCD*	*ABCD*	*ABCD*
Output data =	1101	1000	0001	1100	1010_2

This is the same as the original data that was input into the transmitter.

4.8 REVIEW QUESTIONS

4.1 Describe the operation of the ASK modulators shown in Fig. 4.3.
4.2 Explain with the aid of a diagram how a VCO can be used as an FSK modulator.
4.3 Explain with the aid of a diagram how a VCO can be used as an FSK demodulator.
4.4 Draw a typical circuit diagram of a PSK modulator.
4.5 Draw the constellation of a typical theoretical 8 PSK modulator.
4.6 Show the positions on the constellation of a practical 8 PSK modulator for the following data stream.

$$\text{Data} = 011 \quad 101 \quad 110 \quad 001_2$$

4.7 Determine the output data for a 16 QAM differential modulator given that the default code is 1001_2. The modulator codes the data according to the information given in Table 4.10. The input data stream is as follows:

$$\text{Data} = 0111 \quad 0011 \quad 0000 \quad 0110 \quad 1111_2$$

4.8 Decode the bit stream obtained in Question 4.7 using a 16 QAM demodulator which uses a default code of 1001_2. The demodulator decodes the data according to the information given in Table 4.10.

5 Pulse code modulation

> The human being is a community animal and cannot survive without communication with other human beings.

5.1 INTRODUCTION

The method of converting analogue signals to digital format for telephones is of prime importance, as most businesses and most homes still have a cable pair linking the telephone or telephone system to the exchange. There are presently two standards, the North American standard and the European standard. The UK has followed the European standard and hence this chapter deals mainly with the pulse code modulation (PCM) system that is based on the European standard.

The PCM system relies on many different processes and this chapter is devoted to explaining these different concepts. The concepts are then placed in the full context of a typical PCM system.

5.2 TIME DIVISION MULTIPLEXING

Time division multiplexing is a method of allocating each channel to a single line in a specific time slot. This means that each channel has full access to the transmission line during its particular time slot only. All the channels use exactly the same frequency band. This is unlike an FDM system (discussed in Chapter 2). Each channel on the FDM system is connected to the line for the whole time but each channel is allocated a different frequency band.

Figure 5.1 shows how an analogue signal applied to one channel, on a PCM system, is sampled and the resultant pulse amplitude modulated (PAM) signal that is obtained. The PAM signal is then transmitted. Notice that the PAM signal consists of a number of samples. Observe that the dotted line on the PAM signal touches the peaks of each sample and that the shape of this dotted line is the same as that of the original modulating frequency.

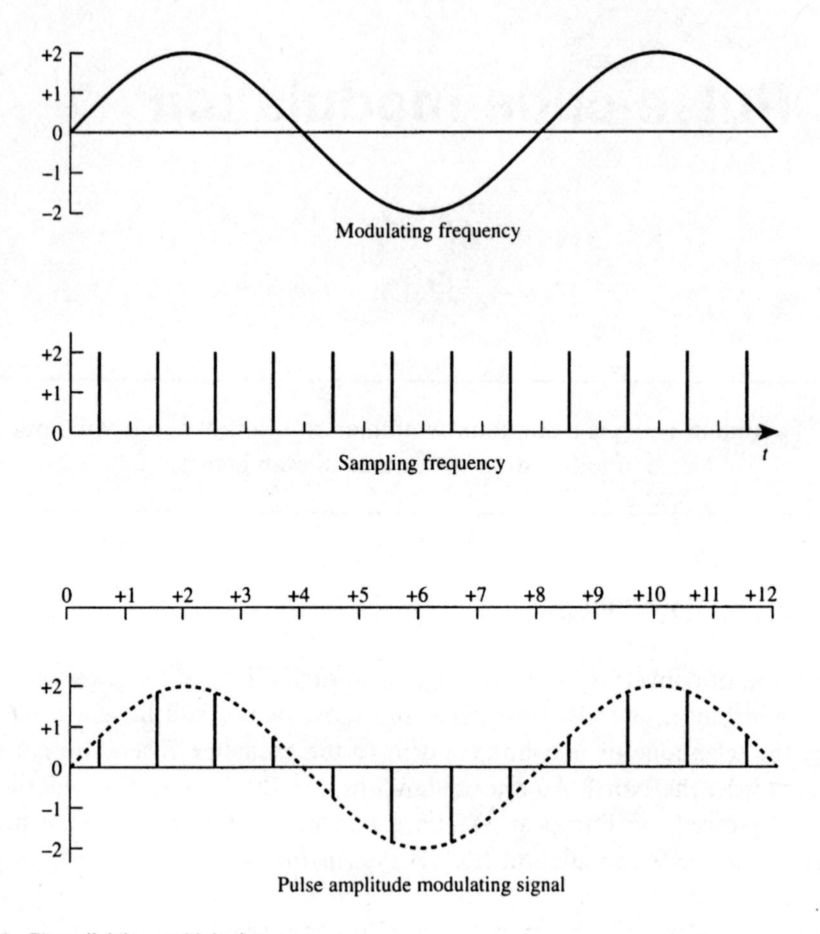

Fig. 5.1 Time division multiplexing

5.3 PRINCIPLE OF OPERATION

For the principle to be effectively used, a number of analogue signals are sampled to produce a time division multiplexed (TDM) signal. The principle of TDM is shown in Fig. 5.2. As shown in this figure, each sample in the TDM signal is quantised and encoded into a binary signal. The binary signal is then further processed and then transmitted down the transmission medium.

Figure 5.2(a) shows two mechanical switches. One mechanical switch has multiple inlets and the second mechanical switch has multiple outlets. The wiper arms of both switches start their rotation at the same position. The two motors driving both switches must be synchronised so that the correct inlet is connected to the correct outlet at the same time. This means that when the wiper on switch 1 is connected to inlet 1, the wiper arm of switch 2 must be connected to outlet 1. The wipers only stay in this position for a short duration and the wiper of switch 1 rotates to inlet 2. At the same time the wiper of switch 2 rotates to outlet 2. This process then continues until, as in this case, inlet 3 at the transmitter is connected to outlet 3 at the receiver. Both wiper

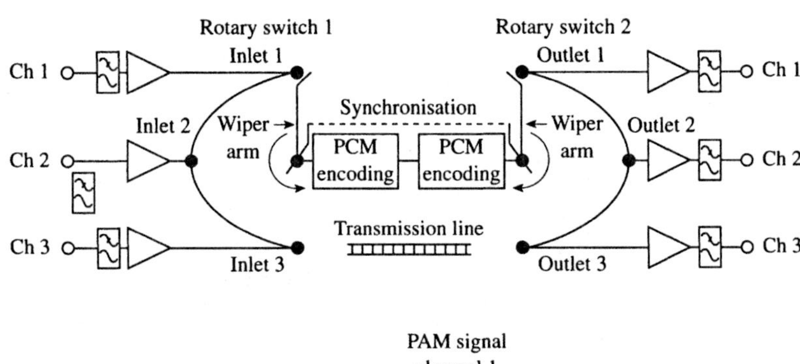

(a)

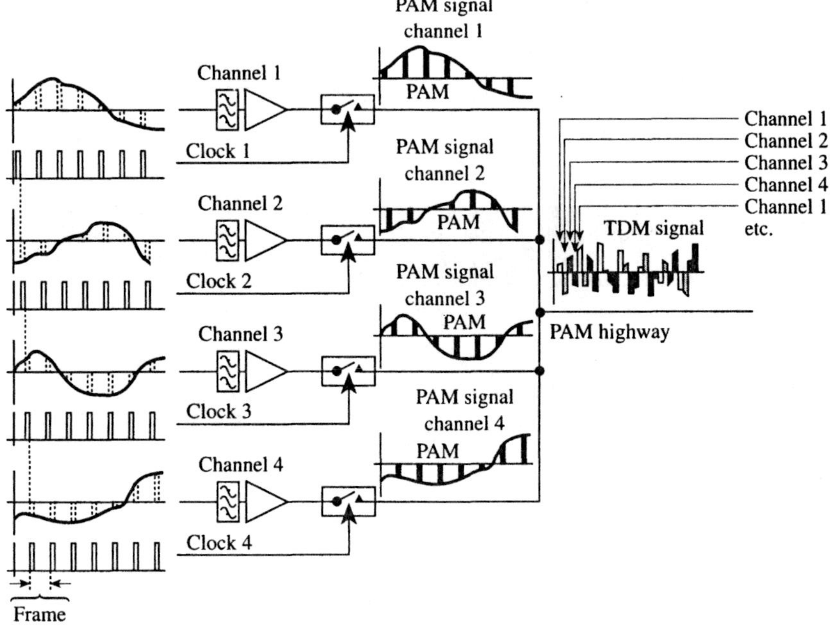

(b)

Fig. 5.2 Principle of TDM: (a) mechanical switches; (b) simple system

arms then rotate to connect channel 1 at the transmitter to channel 1 at the receiver.

In practice the mechanical switches are too slow and the principle is implemented using electronic gates instead.

As can be seen, the signal on each channel is connected to the line for only a short time. The signal that is developed for each channel is a PAM signal and because there are multiple switches the resultant signal is a TDM signal that is transmitted down the line.

Referring to Fig. 5.2(b), the mechanical switches have been replaced by electronic switches or analogue gates. Each analogue gate is controlled by means of a sampling frequency which is a rectangular waveform. Each time the rectangular pulse appears on the control lead of that particular gate the gate closes for the duration of the rectangular pulse. Each pulse has an extremely short width. Thus the gate only remains closed for a very short period of time.

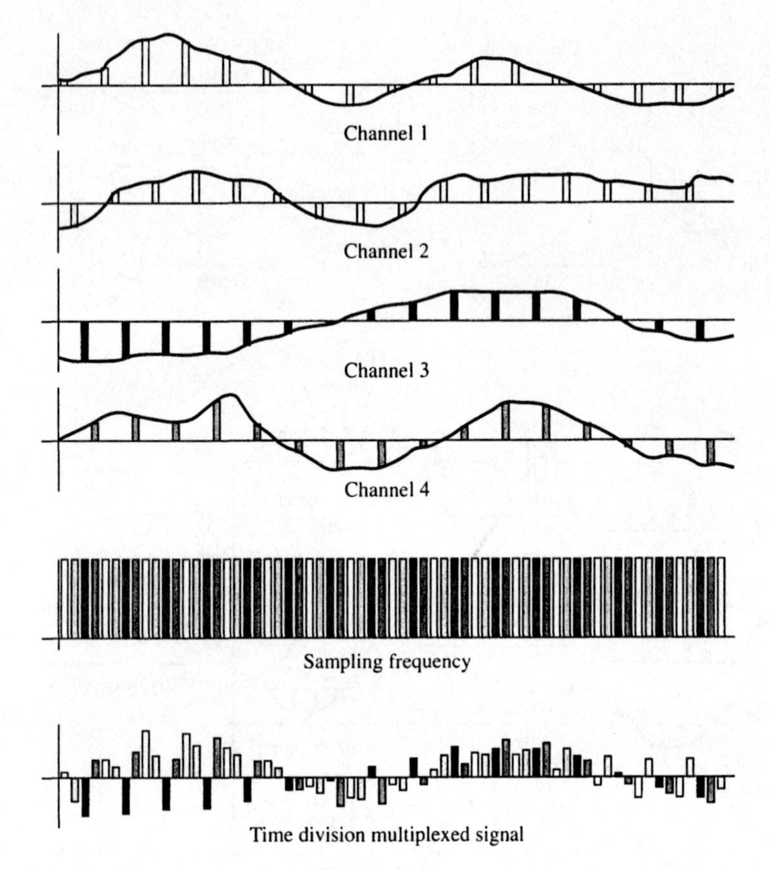

Fig. 5.3 Signal sampling for a four-channel TDM system

The output of each channel gate produces a PAM signal. However, as can be seen, the outputs of all the gates are commoned together and hence a TDM signal is obtained on the PAM highway.

5.3.1 The time division multiplexed signal

Figure 5.3 shows how the TDM signal is developed on a four-channel system. As shown, each channel is sampled using a sampling frequency. The sampling frequency for each individual channel is displaced in time relative to the preceding and succeeding channel. This means that the sampling takes place successively, so that channel 1 is sampled first, followed by channels 2, 3 and 4, then channel 1 is sampled again and so on.

5.4 RECOMMENDED STANDARDS

The International Telecommunication Union (ITU), which is a subcommittee of the United Nations (UN) has recommended two standards. These are:

- North American standard.
- European standard.

The North American standard is based on a 24-channel PCM system which uses the μ-law (or mu-law) for quantising. The system produces a gross line bit rate of 1.55 Mb/s. The system was designed for channel-associated signalling.

In 1959 the European countries formed a non-political organisation 'CEPT'. The English translation of this mnemonic is: 'The European Committee of Postal and Telecommunication Administrations'. This committee recommended standards for compatibility of the telecommunications systems employed throughout Europe.

From this committee the European standard based on the 30/32 channel PCM system was recommended. This system contains 30 speech channels, a synchronisation channel and a signalling channel. The gross line bit rate of the system is 2.048 Mb/s and it uses the A-law in the quantisation process.

The European system can be adapted for common channel signalling and if this is used then it would have 31 data channels and a single synchronisation channel. The gross line bit rate would still be 2.048 Mb/s.

Both systems use a sampling frequency of 8 kHz for each channel.

5.5 THE 30/32-CHANNEL CEPT PCM SYSTEM

The 30/32 channel PCM system uses a frame and multiframe structure. The channel allocation is as follows:

- Time slot 0: frame alignment word (FAW), frame service word (FSW).
- Time slots 1 to 15: digitised speech for channels 1 to 15.
- Time slot 16: multiframe alignment word (MFAW), multiframe service word (MFSW), signalling information.
- Time slots 17 to 31: digitised speech for channels 16 to 30.

5.5.1 The frame and multiframe structure

The frame structure

The frame consists of 32 pulse channel time slots (TS), as shown in Fig. 5.4, numbered TS 0 to TS 31. Each time slot consists of 8 bits. As shown time slots 1 to 15 are used for the digitised speech on channels 1 to 15 and time slots 17 to 31 are used for the digitised speech on channels 16 to 30.

In these time slots bit 1 is used to indicate the polarity of the sample and bits 2 to 8 indicate the amplitude of the sample. This is also shown in Fig. 5.4.

The multiframe structure

In order for signalling information (dial pulses) for all 30 channels to be transmitted, the multiframe consists of 16 frames numbered F0 to F15. This structure is shown in Fig. 5.4. Signalling for two channels is transmitted in time slot 16 in each frame.

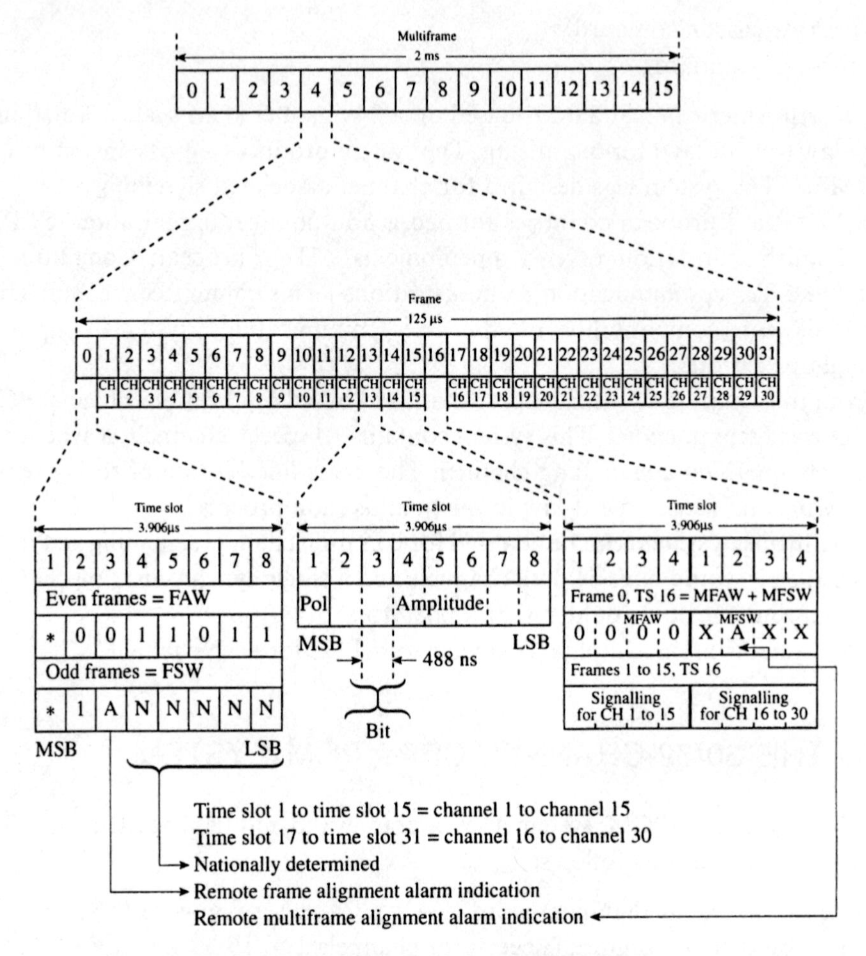

Fig. 5.4 Frame and multiframe structure for a 30/32-channel PCM system

5.5.2 Time durations

The sampling theorem states that the sampling frequency f_s must be greater than twice the highest modulating frequency.

The analogue input to each channel is the commercial speech band and this is defined as being a frequency range of 300 Hz to 3400 Hz. This means that the highest modulating frequency is 3400 Hz. The sampling frequency must be greater than 6800 Hz. The CCITT recommendation is that the 8 kHz is used as the sampling frequency.

The following calculated time durations are all shown in Fig. 5.4.

Frame duration

The frame duration is determined as follows:

$$\text{frame duration} = \frac{1}{f_s}$$

Given that f_s is 8 kHz,

$$\text{frame duration} = 125\,\mu\text{s}$$

Time slot duration

The time slot duration is determined as follows:

$$\text{time slot duration} = \frac{\text{frame duration}}{\text{time slots per frame}}$$

Given that frame duration is 125 μs and number of time slots per frame is 32,

$$\text{time slot duration} = 3.906\,\mu\text{s}$$

Bit duration

The bit duration is determined as follows:

$$\text{bit duration} = \frac{\text{time slot duration}}{\text{bits per time slot}}$$

Given that time slot is 3.906 μs and number of bits per time slot is 8,

$$\text{bit duration} = 488\,\text{ns}$$

Multiframe duration

The multiframe duration is determined as follows:

$$\text{multiframe duration} = \text{frame duration} \times \text{frames per multiframe}$$

Given that frame duration is 125 μs and number of frames per multiframe is 16,

$$\text{multiframe duration} = 2\,\text{ms}$$

Gross line bit rate

The gross line bit rate is determined as follows:

$$\text{gross line bit rate} = \frac{1}{\text{bit duration}}$$

Given that bit duration is 488 ns,

$$\text{gross line bit rate} = 2.048\,\text{Mb/s}$$

5.5.3 Time slot usage on a 30/32-channel CEPT PCM system

Table 5.1 shows the allocation of the 32 time slots in each of the 16 frames. Time slot 0 in the even frames (i.e. frames 0, 2, 4, 6, 8, 10, 12 and 14) carries the FAW, which is $X0011011_2$. Time slot 0 in the odd frames (i.e. frames 1, 3, 5, 7, 9, 11, 13 and 15) carries the FSW, which is $X1ANNNNN_2$. The structure of the time slot for these two words is shown in Fig. 5.4. The MSB (X) in both cases is a 'don't care' bit and is normally kept in the set state. Bit 2 in the FAW is always reset and bit 2 in the FSW is always set. The receiver recognises whether the received word in TS 0 of F 0 is the FSW or the FAW by detecting whether bit 2 is a logic low or a logic high.

Table 5.1 Allocation of time slots within frames

Frame	Time slots				
	0	1 to 15	16		17 to 31
0	FAW	CH 1 to 15	MFAW	MFSW	CH 16 to 30
1	FSW	CH 1 to 15	CH 1	CH 15	CH 16 to 30
2	FAW	CH 1 to 15	CH 2	CH 17	CH 16 to 30
3	FSW	CH 1 to 15	CH 3	CH 18	CH 16 to 30
4	FAW	CH 1 to 15	CH 4	CH 19	CH 16 to 30
5	FSW	CH 1 to 15	CH 5	CH 20	CH 16 to 30
6	FAW	CH 1 to 15	CH 6	CH 21	CH 16 to 30
7	FSW	CH 1 to 15	CH 7	CH 22	CH 16 to 30
8	FAW	CH 1 to 15	CH 8	CH 23	CH 16 to 30
9	FSW	CH 1 to 15	CH 9	CH 24	CH 16 to 30
10	FAW	CH 1 to 15	CH 10	CH 25	CH 16 to 30
11	FSW	CH 1 to 15	CH 11	CH 26	CH 16 to 30
12	FAW	CH 1 to 15	CH 12	CH 27	CH 16 to 30
13	FSW	CH 1 to 15	CH 13	CH 28	CH 16 to 30
14	FAW	CH 1 to 15	CH 14	CH 29	CH 16 to 30
15	FSW	CH 1 to 15	CH 15	CH 30	CH 16 to 30

In both the FSW and MFSW words the A bit is the alarm bit for the extension of remote alarm conditions. The N bits in the FSW are used for nationally determined conditions. The X bits in the MSFW are 'don't care' states and are normally placed in the set state.

Time slot 16 in frame 0 is used to carry the MFAW and MFSW words. The first 4 bits in the time slot are allocated to the MFAW and the last four bits are allocated to the MFSW. As shown in Fig. 5.4 the MFAW consists of 0000_2 and the MFSW consists of $XAXX_2$.

The purpose and usage of the FSW and the MFSW will be dealt with in Section 5.10.

Time slot 16 in frames 1 to 15 carries signalling information for the 30 speech channels. In each frame, signalling information for two channels is conveyed. This is shown in Table 5.1 as well as Fig. 5.4.

Time slots 1 to 15 carry the digitised speech information for channels 1 to 15, respectively, and time slots 17 to 31 carry the digitised speech for channels 16 to 30, respectively.

5.6 ALIASING DISTORTION

As mentioned before, the 30/32-channel PCM system must use a sampling frequency of 8 kHz, which is the ITU recommendation. The sidebands that result in the TDM signal are as shown in Fig. 5.5(a).

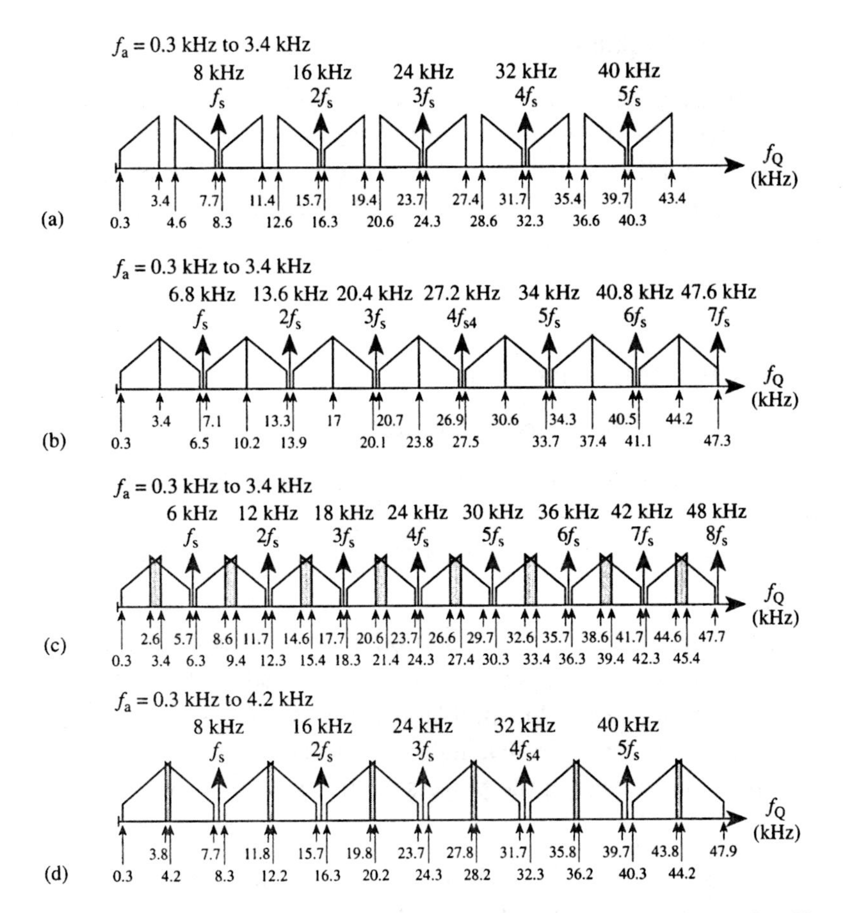

Fig. 5.5 (a) ITU recommended sampling; (b) Nyquist sampling; (c) aliasing distortion, $f_s < 2f_{a(max)}$; (d) aliasing distortion, $f_{a(max)} > 3400$ Hz

As can be seen the products of modulation are the commercial speech band, 1st LSB, 1st USB, 2nd LSB, 2nd USB, 3rd LSB, 3rd USB etc. The sampling frequencies are f_s, $2f_s$, $3f_s$ etc. Because the sampling frequency used is in fact a rectangular waveform it is rich in harmonic frequencies. The fundamental frequency is 8 kHz but the rectangular pulse has an infinite number of harmonic frequencies.

5.6.1 ITU recommended sampling

This is shown in Fig. 5.5(a), where a gap can be seen between the speech band (f_a) and the 1st LSB; likewise there is a gap between the 1st USB and the 2nd LSB etc. In this case f_a is fixed to be no higher than 3400 Hz by the input aliasing filter to the channel. Also the fundamental frequency of the sampling signal is kept fixed at 8 kHz. The resultant bands are shown in the figure; the process continues to infinity.

At the receiver the original speech can be easily extracted by means of a low-pass filter as there is a gap of 1.2 kHz between f_a and the 1st LSB.

5.6.2 Nyquist sampling

If the sampling frequency is made exactly twice the highest modulating frequency then this is referred to as Nyquist sampling. In this case there is no gap between the speech band and the 1st LSB as well as between the 1st USB and the 2nd LSB etc. This means that the original speech cannot be extracted easily by means of a simple low-pass filter. The resultants bands are shown in Fig. 5.5(b); the process continues to infinity.

5.6.3 Aliasing distortion defined

Aliasing distortion can be caused by two factors:

- When the sampling frequency is smaller than twice the highest modulating frequency.
- When the audio input frequency band extends beyond 3400 Hz.

f_s smaller than $2f_a$
This situation is shown in Fig. 5.5(c). Notice how the sidebands overlap one another. In this case the original speech cannot be extracted using a low-pass filter without interference. This problem is prevented by deriving the clock frequency and system timing from a highly stable crystal oscillator. The resultant bands are shown in the figure; the process continues to infinity.

f_a stretches beyond 3400 Hz
This situation is shown in Fig. 5.5(d). Notice how the sidebands overlap one another. In this case the original speech cannot be extracted without interference. This problem is prevented by passing the input audio through a low-pass filter at the input to the channel. This filter is referred to as an aliasing filter and limits the highest frequency that is applied to the input to the channel gate to 3400 Hz. The resultant bands are shown in the figure; the process continues to infinity.

5.7 QUANTISING AND ENCODING

The quantising and encoding process normally takes place simultaneously. However in this discussion each process will be discussed separately.

5.7.1 Quantisation

Linear quantising versus non-linear quantising
Consider the two scales shown in Fig. 5.6. Comparing sample 1 to the linear scale, the sample is increased by 0.25 V from 1.75 V to 2 V. The percentage distortion, more correctly referred to as the percentage quantisation noise, that is added to

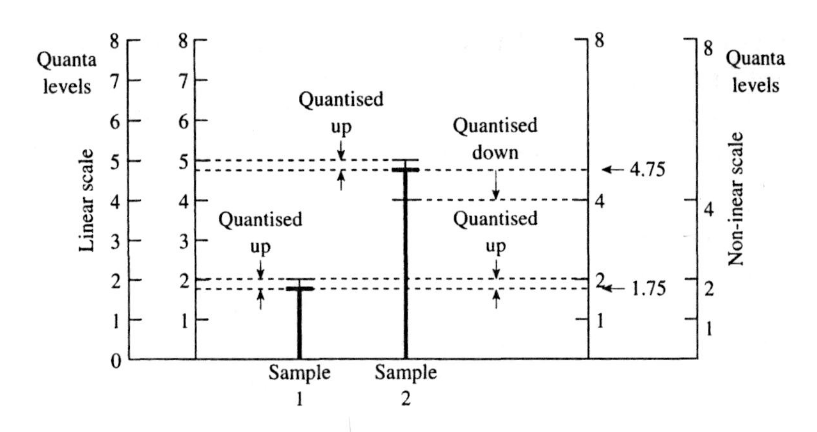

cales

ıation

ude − quantised sample amplitude| / ıde + quantised sample amplitude × 100%

merator is taken to ensure a positive value
sample amplitude is greater than the actual

sation noise added to sample 1 when com-
omparing sample 1 to the non-linear scale
f 6.7% as it is increased by 0.25 V from
at given by the linear scale.
ale means that it is increased by 0.25 V from
e quantisation noise of 2.56%. Comparing
creased in amplitude by 0.75 V from 4.75 V
age quantisation noise of 8.57%.
in the percentage quantisation noise added
pared to the linear scale, is 6.7 − 2.56%,
the percentage quantisation noise added
compared to the non-linear scale is

produces a smaller percentage difference
.87%, when compared to the percentage
scale, i.e. 4.14%.

This shows that the non-linear scale tries to add approximately the same percentage quantisation noise to both the samples. The scale tries to produce the same percentage quantisation noise irrespective of the sample amplitude. In practice this is more accurately achieved because the non-linear scale has more quanta for small amplitudes than for large amplitudes, which will reduce the percentage quantisation noise for small amplitude samples but increase the percentage quantisation noise for large amplitude samples. The linear scale, however, produces a very small distortion for large amplitude samples but a large percentage distortion for small amplitude samples.

The non-linear scale is preferred because the quantisation noise added tends to be more independent of the sample amplitude than the quantisation noise added by a linear scale.

Non-linear quantising

As mentioned earlier, the ITU has recommended two different non-linear quantising scales, the A-law and the μ-law. These quantising scales are very similar, the only difference being the mathematical equation which is used to produce the graph. These two equations are given below. The μ-law equation is

$$y = y_{max} \frac{\log_e\left(\dfrac{1+\mu|x|}{x_{max}}\right)}{\log_e(1+\mu)} \operatorname{sgn} x$$

where $\operatorname{sgn} x = \begin{cases} +1 \text{ for } x \geq 0 \\ -1 \text{ for } x < 0 \end{cases}$

To obtain a signal-to-noise ratio (SNR) >38 dB for an 8-bit time slot the standard value of μ is 255. The formula to calculate the SNR is

$$\text{SNR} = 3\left(\frac{2^n}{\log_e \mu}\right)^2$$

where n = number of bits per time slot

The equation to calculate the SNR in dB is

$$\text{SNR} = 10\log_{10}\left[3\left(\frac{2^n}{\log_e \mu}\right)^2\right] \text{dB}$$

For values of $\mu = 255$ and $n = 8$, SNR = 38.06 dB.

The A-law equation is

$$y = \begin{cases} y_{max}\left(\dfrac{A|x|/X_{max}}{1+\log_e A}\right) \operatorname{sgn} x & \text{for } 0 < \dfrac{|x|}{x_{max}} = \dfrac{1}{A} \\ y_{max}\left(\dfrac{1+\log_e(A|x|/x_{max})}{1+\log_e A}\right) \operatorname{sgn} x & \text{for } \dfrac{1}{A} < \dfrac{|x|}{x_{max}} < 1 \end{cases}$$

where $\operatorname{sgn} x = \begin{cases} +1 \text{ for } x \geq 0 \\ -1 \text{ for } x < 0 \end{cases}$

To obtain an SNR >38 dB for an 8-bit time slot the standard value of A is 87.6, which is the SNR value as a ratio. The formula to calculate the SNR in dB is

$$\text{SNR} = 20\log_{10} A \text{ dB}$$

For a value of $A = 87.6$, SNR = 38.85 dB.

The A-law and μ-law graphs for the standard values specified above are very similar and hence in the discussion given below only the A-law quantising scale will be discussed, as this is the one that the ITU recommended for use on the 30/32-channel PCM system.

Each time slot consists of 8 bits. This means that the decimal range that can be represented is from 0_{10} to 255_{10}. This can be determined as follows:

$$\text{maximum decimal number} = 2^n$$

where n = number of binary bits

Because any sample can be either negative or positive, the quantising scale is composed of a positive half and a negative half, which are mirror images of one another. As a result the scale is sometimes referred to as a folded scale.

The decimal range in each half is from 0_{10} to 127_{10}. In each time slot bit 1, which is the MSB, is used to indicate the polarity of the sample. If the sample is positive then bit 1 is set to a logic 1. If the amplitude of the sample is negative then bit 1 is reset to a logic 0. Bits 2 to 8 are used to indicate the relative amplitude of the sample in the range 0_{10} to 127_{10}. The structure of TS 1 to TS 15 and TS 17 to TS 31 is shown in Fig. 5.4.

The scale gives a smooth logarithmic curve. In practice the scale is implemented with discrete linear segments that follow the smooth curve and hence the scale is said to be quasilogarithmic. Figure 5.7 shows this scale, with straight lines between $+V_{max}$ and $+V/2$, $+V/2$ and $+V/4$, etc. through to $-V/2$ and $-V_{max}$. The scale is known as a 13-segment scale as can be seen in Fig. 5.8.

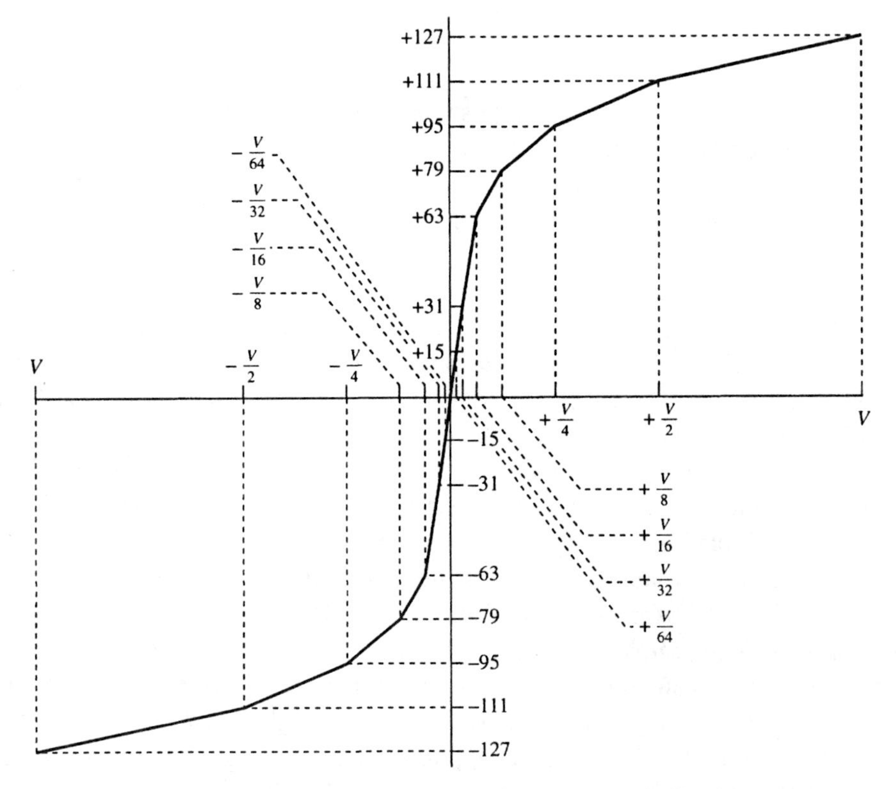

Fig. 5.7 Linear segments making up the logarithmic curve

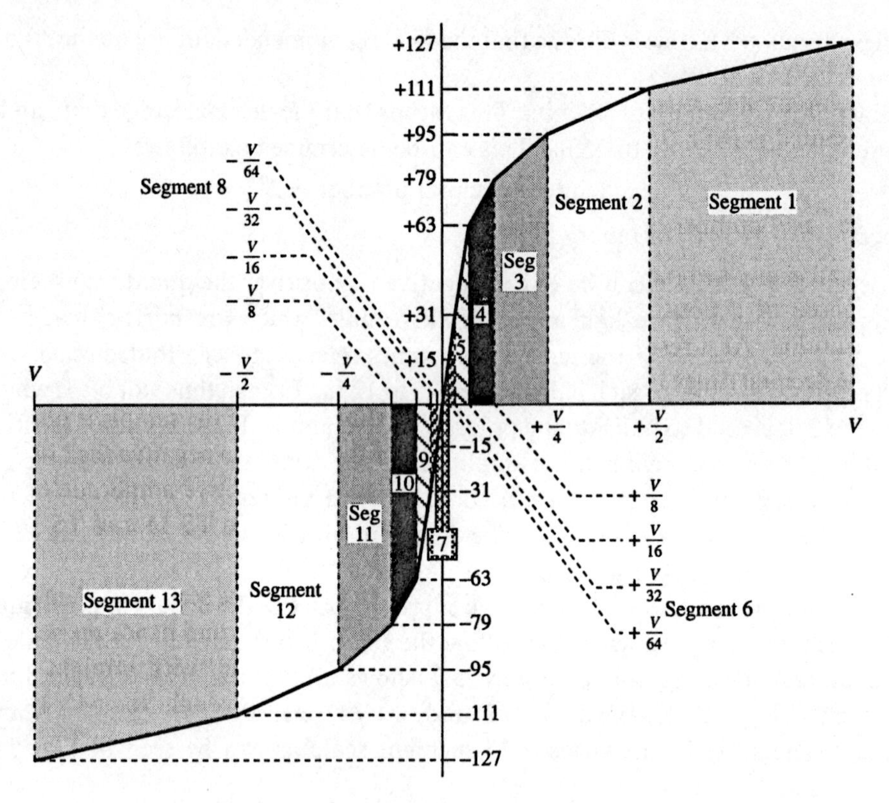

Fig. 5.8 Thirteen-segment scale

Notice that segment 7 consists of a positive section between $+V/64$ and 0, and a negative section between 0 and $-V/64$.

Each linear segment is broken up into a number of quanta. The number of quanta in each segment is given in Table 5.2. The quanta total 256, which represents the range from 0_{10} to 255_{10}.

Because the speech band has already been limited to between 300 Hz and 3400 Hz there is a loss of quality. Also the higher frequency components normally have small amplitudes. To enable these small components to be more accurately quantised, more quanta levels are allocated in the region between $+V/64$ and $-V/64$, and as shown in Table 5.2 there are 64 quanta in this range.

Figure 5.9 shows how the quantising takes place on a system. For simplicity, eight quanta have been allocated per linear section and each time slot consists of 5 bits. If a sample has an amplitude that is halfway between two quanta levels then the amplitude that is allocated is the higher quantum level. This is shown for sample 4 which has an actual amplitude of +8.5 V and is quantised to an amplitude of +9 V. The sample is then encoded to a 5-bit binary word of 11001_2.

If the sample has an amplitude that is smaller than the threshold value then it is allocated the lower amplitude. This is shown for sample 2, which has an actual amplitude of +12.3 V. The sample is quantised to an amplitude of +12 V. The sample is then encoded to a 5-bit binary word of 11100_2.

Table 5.2 Number of quanta in each segment

Segment	Voltage range	No. of quanta
1	$+V_{max}$ to $+V/2$	16
2	$+V/2$ to $+V/4$	16
3	$+V/4$ to $+V/8$	16
4	$+V/8$ to $+V/16$	16
5	$+V/16$ to $+V/32$	16
6	$+V/32$ to $+V/64$	16
7	$+V/64$ to 0	32
7	0 to $-V/64$	32
8	$-V/64$ to $-V/32$	16
9	$-V/32$ to $-V/16$	16
10	$-V/16$ to $-V/8$	16
11	$-V/8$ to $-V/4$	16
12	$-V/4$ to $-V/2$	16
13	$-V/2$ to $-V_{max}$	16

If the amplitude of the sample is higher than the threshold level then the next higher amplitude is allocated; this is shown for sample 1 which has an actual amplitude of +2.75 V. The sample is quantised to an amplitude of +3 V. The sample is then encoded to a 5-bit binary word of 10011_2.

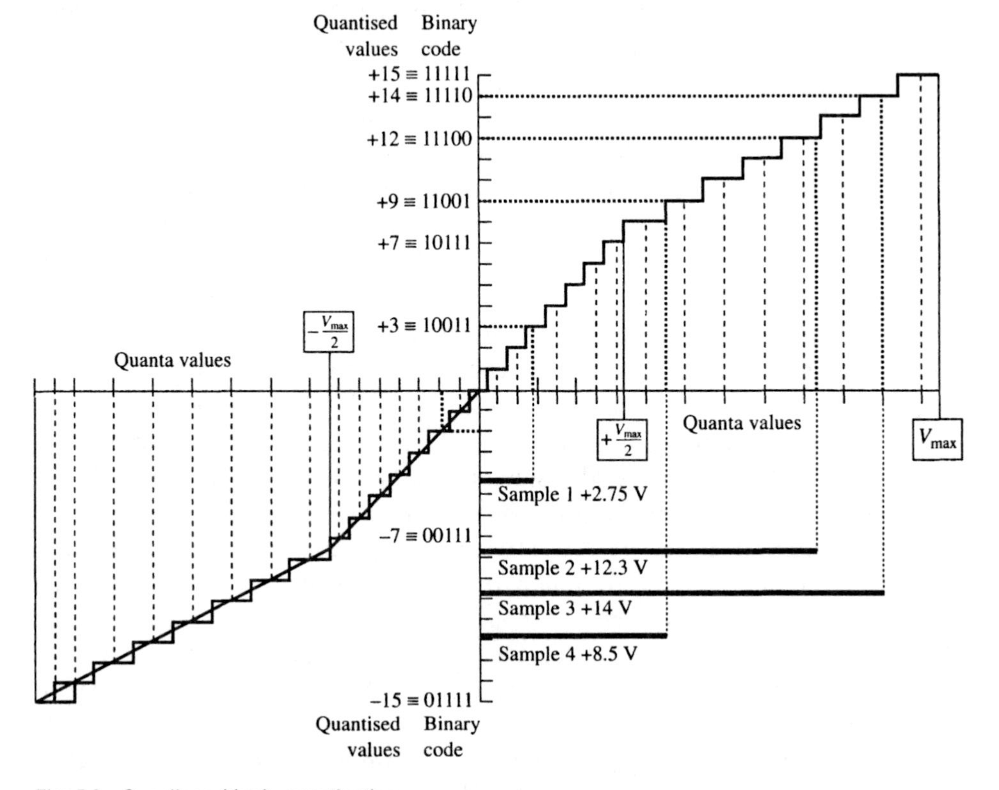

Fig. 5.9 Quasilogarithmic quantisation

Sample 3, which has an actual amplitude of +14 V, is the only sample that has the correct amplitude allocated to it as it is exactly midway between the two threshold values. It is quantised to an amplitude of +14 V and is then encoded to the 5-bit binary word of 11110_2.

In these examples it can be seen that the quantising scale does not allocate accurately the correct amplitude to a sample. This results in noise in that the sample amplitude is either slightly reduced or slightly increased. This noise is known as quantising noise. Quantising noise results due to a sample having an infinite level range being quantised against a scale having a finite level range.

Although this example only shows samples having a positive amplitude the same applies to samples having a negative amplitude. The only difference is the status of the MSB. As can be seen for the positive samples the MSB is a logic 1. For samples having a negative polarity the MSB would be coded as a logic 0.

5.7.2 Encoding

Figure 5.10 shows a block diagram of a typical encoder. Each sample is encoded into an 8-bit word in this circuit. The circuit implements the successive approximation method of analogue-to-digital conversion.

The sample is applied to the input and passes through the analogue switch. Its amplitude is then held in a store (which is normally a capacitor that is charged to the appropriate voltage) and the switch is then opened to prevent any interference. This circuit is simply a sample-and-hold circuit.

At the same time the output of the serial-input–serial-output (SISO) register is reset and the ring counter is reset by the controller circuit. The first operation performed is the determination of the sample amplitude. This is done by means of a comparator. The comparator output is connected to the controller. The controller starts the ring counter. The ring counter is used to set the output of FF 1. If the sample polarity is positive then FF 1 is left in the set state. If the polarity is negative then FF 1 is reset. The controller also connects the correct reference voltage to the weighted resistor chain.

The ring counter is clocked, which causes FF 2 to be set and the controller causes gate G2 to close, so that a reference voltage V_R is applied to the amplitude comparator, with a value $V_{ref}/2$, which is a relative amplitude of 64_{10}. Note that V_R is the total resultant voltage applied to the one input of the amplitude comparator from the output of gates G2 to G8, and V_{ref} is a fixed reference voltage that is applied via the weighted network to gates G2 to G8. The amplitude comparator compares the input voltage V_a to the reference voltage V_R. If $V_a \geq V_R$ then the controller will cause gate G2 to remain closed and cause FF 2 to remain set. If $V_a < V_R$ then the controller will cause gate G2 to open and FF 2 to be reset. Gates G3 to G8 are all open at this stage.

The controller will now cause gate G3 to close and clock the ring counter which will cause FF 3 to be set. The value of V_R becomes $V_{ref}/4$, which is a relative

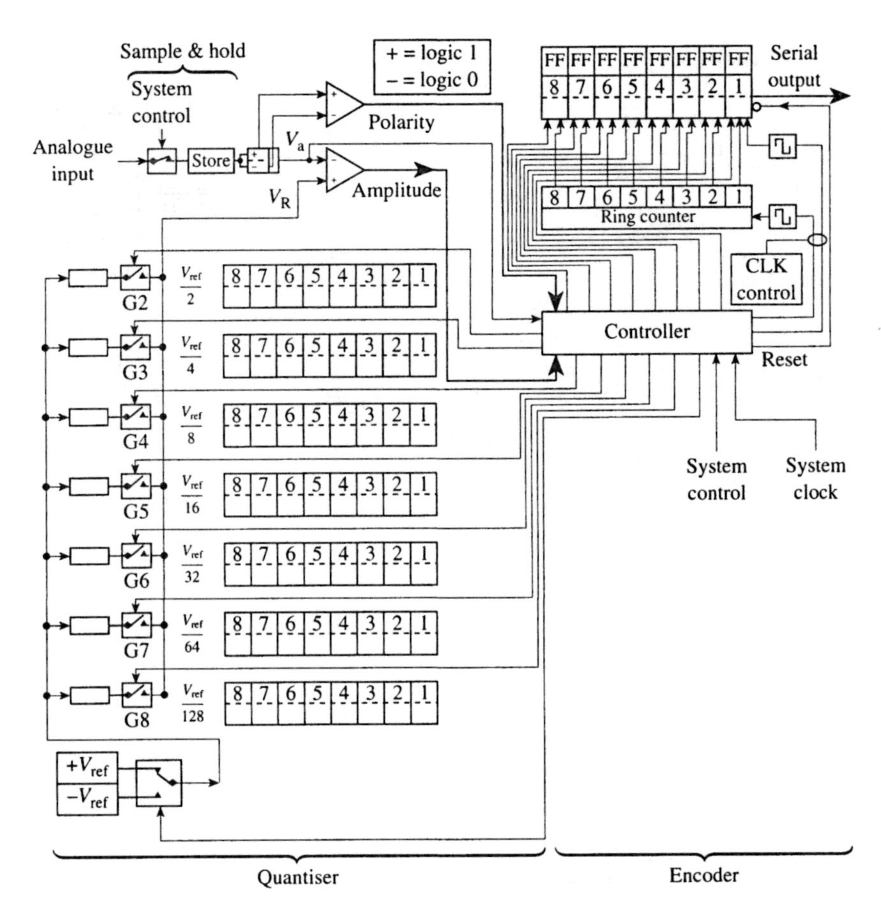

Fig. 5.10 Quantiser and encoder

amplitude of 32_{10}. If gate G2 was closed then the total reference voltage V_R would be $V_{ref}/2 + V_{ref}/4$, which has the relative amplitude of $64_{10} + 32_{10} = 96_{10}$.

The amplitude comparator compares the input voltage V_a to the reference voltage V_R. If $V_a \geq V_R$ then the controller will cause gate G3 to remain closed and cause FF 3 to remain set. If $V_a < V_R$ then the controller will cause gate G3 to open and FF 3 to be reset. Gates G4 to G8 are all open at this stage. Gate G2 is either permanently open or closed for the rest of this conversion cycle.

The cycle will continue with gates G4 to G8 with the setting and/or resetting of FF 4 to 8 in turn and comparing V_a to the new reference voltage in turn. At the end of the cycle the 8-bit binary word is stored in FF 1 to FF 8 and is output as a serial bit stream.

The process can be more easily understood by considering the following example.

Example 5.1

Using Fig. 5.11 determine the 8-bit word for a sample having a relative amplitude of -57_{10}.

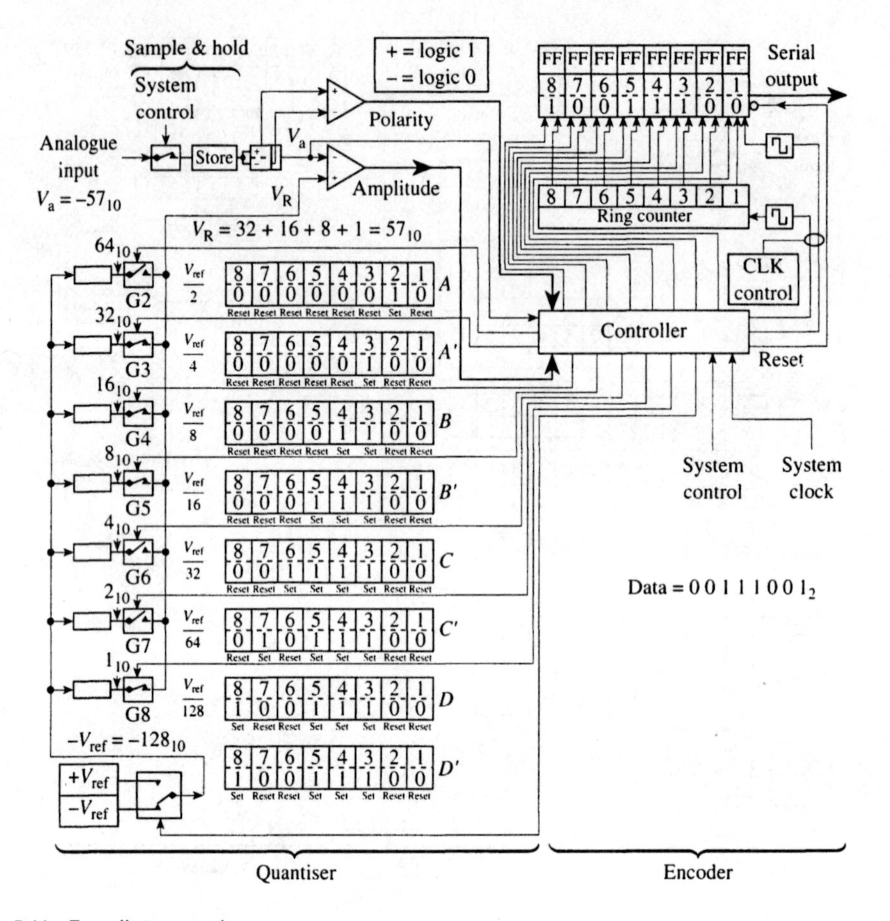

Fig. 5.11 Encoding example

Solution

Initially FF 1 to FF 8 are reset. FF 1 is then set and the polarity of the sample is checked. In this case the polarity of the sample is negative and hence FF 1 is reset. The negative reference voltage is connected to the weighted resistor chain.

FF 2 is set and gate G2 is closed. This causes V_R to be equal to the relative reference amplitude of 64_{10}. Now $57_{10} < 64_{10}$ and as a result FF 2 is reset and gate G2 is opened.

FF 3 is set and gate G3 is closed. This causes V_R to be equal to the relative reference amplitude of 32_{10}. Now $57_{10} > 32_{10}$ and as a result FF 3 remains set and gate G3 is left closed.

FF 4 is set and gate G4 is closed. This causes V_R to be equal to the relative reference amplitude of $32_{10} + 16_{10} = 48_{10}$. Now $57_{10} > 48_{10}$ and as a result FF 4 remains set and gate G4 is left closed.

FF 5 is set and gate G5 is closed. This causes V_R to be equal to the relative reference amplitude of $32_{10} + 16_{10} + 8_{10} = 56_{10}$. Now $57_{10} > 56_{10}$ and as a result FF 5 remains set and gate G5 is left closed.

FF 6 is set and gate G6 is closed. This causes V_R to be equal to the relative reference amplitude of $32_{10} + 16_{10} + 8_{10} + 4_{10} = 60_{10}$. Now $57_{10} < 60_{10}$ and as a result FF 6 is reset and gate G6 is opened. At this point V_R has a relative amplitude of 56_{10}.

FF 7 is set and gate G7 is closed. This causes V_R to be equal to the relative reference amplitude of $32_{10} + 16_{10} + 8_{10} + 2_{10} = 58_{10}$. Now $57_{10} < 58_{10}$ and as a result FF 7 is reset and gate G7 is opened. At this point V_R has a relative amplitude of 56_{10}.

FF 8 is set and gate G8 is closed. This causes V_R to be equal to the relative reference amplitude of $32_{10} + 16_{10} + 8_{10} + 1_{10} = 57_{10}$. Now $57_{10} = 57_{10}$ and as a result FF 8 remains set and gate G8 remains closed.

At this point the 8-bit word is now stored in FF 1 to FF 8 and is output as a serial data stream.

	Polarity	64	32	16	8	4	2	1
Code =	0	0	1	1	1	0	0	1_2

The following method describes the same principle using a table.

Example 5.2

Determine the 8-bit word for a sample having a relative amplitude of -57_{10} using Table 5.3.

$$57_{10} = 00111001_2$$

As can be seen in Table 5.3, a negative polarity causes bit 1 to be reset. Now the analogue input voltage is compared to the 64_{10} reference. As 57_{10} is smaller than 64_{10}, bit 2 is reset and this gate is opened. 57_{10} is now compared to 32_{10}; as it is greater than 32_{10}, bit 3 is set and this gate remains closed. 57_{10} is now compared to $32_{10} + 16_{10}$ giving 48_{10}; as 57_{10} is greater than 48_{10}, bit 4 is set and this gate remains closed. 57_{10} is now compared to $32_{10} + 16_{10} + 8_{10}$ which is 56_{10}; as 57_{10} is greater than 56_{10} bit 5 is set and this gate remains closed. 57_{10} is now compared to $32_{10} + 16_{10} + 8_{10} + 4_{10}$, which is 60_{10}; as 57_{10} is smaller than 60_{10}, bit 6 is reset

Table 5.3 Sample of relative amplitude -57_{10}

Weighting	V_R	$V_a > V_R$	Gate status	Binary	Bit number
Polarity	—	—	—	0	1
64	64	No	Open	0	2
32	32	Yes	Closed	1	3
16	48	Yes	Closed	1	4
8	56	Yes	Closed	1	5
4	60	No	Open	0	6
2	58	No	Open	0	7
1	57	Yes	Closed	1	8

and this gate is opened. 57_{10} is now compared to $32_{10} + 16_{10} + 8_{10} + 2_{10}$, which is 58_{10}; as 57_{10} is smaller than 58_{10}, bit 7 is reset and this gate is opened. 57_{10} is lastly compared to $32_{10} + 16_{10} + 8_{10} + 1_{10}$, which is 57_{10}. As $V_a = V_R$, bit 8 is set and this gate remains closed. The resultant 8-bit binary word is

Polarity	64	32	16	8	4	2	1
0	0	1	1	1	0	0	1_2

One major advantage of using this system is its ability to prevent over-modulation from taking place. Any sample having an amplitude greater than +127 will be encoded into 11111111_2 and any sample having an amplitude greater than −127 will be encoded into 01111111_2.

5.8 THE 30/32-CHANNEL CEPT PCM SYSTEM OPERATION

Refer to Fig. 5.12.

5.8.1 The transmitter

Aliasing filter

The input to each channel is the commercial speech band. This band is 300 Hz to 3400 Hz. The speech passes through the input channel filter which is a low-pass filter and called the aliasing filter. This filter ensures that no frequency greater than 3400 Hz passes to the channel switch. This is done to ensure that aliasing distortion does not take place due to the input frequency band being too large.

Channel amplifier

This amplifier is a variable gain amplifier, which enables the correct signal level to be applied to the channel switch.

Channel switch

This is an analogue switch, which is controlled by the timing circuit. Only one channel switch can be closed during a particular time slot. The switches are operated sequentially. During time slot 0 all the channel switches are open as this time slot is allocated to the frame alignment word (FAW) and frame service word (FSW). During time slot 16 all the channel switches are open as this time slot is allocated to the multiframe alignment word (MFAW) and multiframe service word (MFSW) as well as the signalling for the channels.

During time slot 1 the channel switch for channel 1 is closed, during time slot 2 the channel switch for channel 2 is closed. This process continues up time slot 15 where the channel switch for channel 15 closes.

During time slot 17 the channel switch on channel 16 is closed. During time slot 18 the channel switch of channel 17 is closed. This process continues until time slot 31 during which time the channel switch of channel 30 is closed.

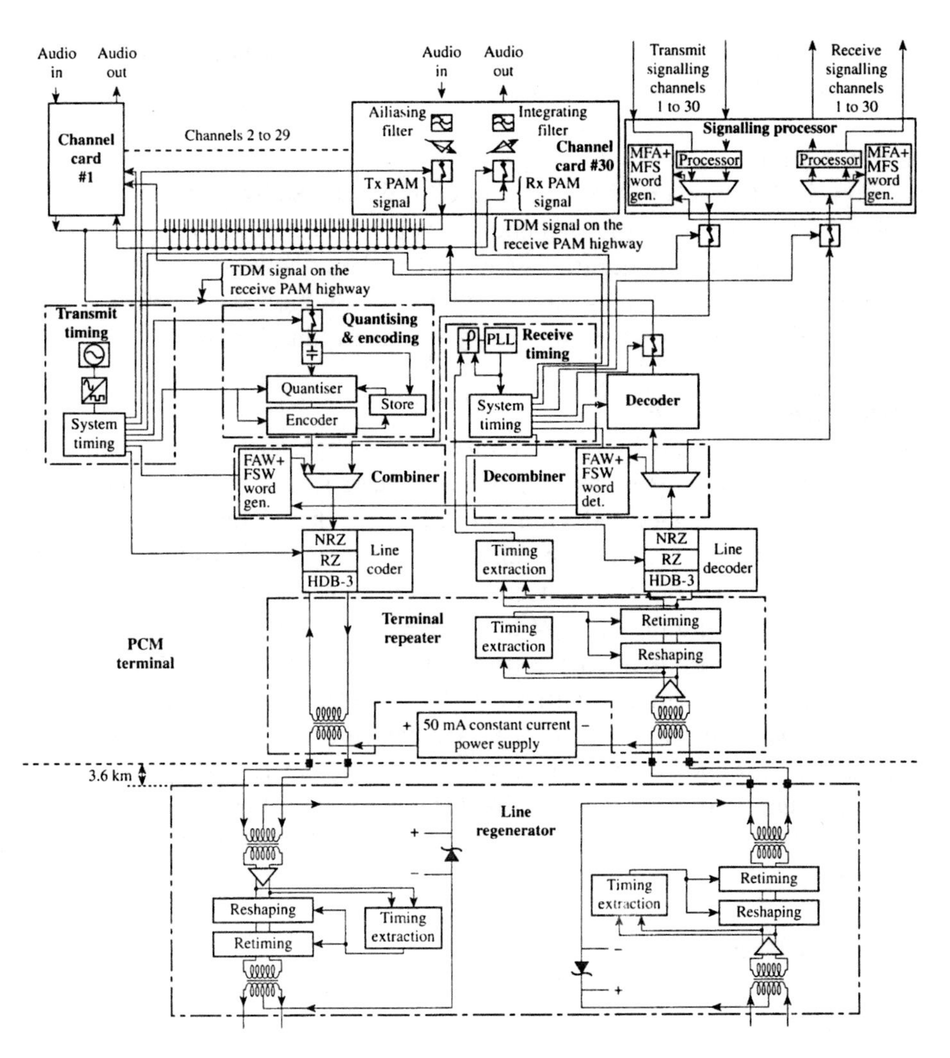

Fig. 5.12 Typical PCM system

The output signal from each individual channel switch yields a pulse amplitude modulated (PAM) signal. The outputs of the channel switches are commoned together on to the transmit PAM highway. The resultant signal on the PAM highway is a time division multiplexed (TDM) signal. This signal is still analogous in nature.

Quantising and coding

The PAM highway is connected to the input to the quantising and encoding circuit. The signal first passes through a switch, which allows the individual sample to be applied to a sample-and-hold circuit. Once the sample is stored the switch opens for the remainder of the time slot period to ensure that the next sample does not interfere with the quantising and encoding process of the present sample. The individual sample is then quantised according to the *A*-law quantising scale and encoded into an 8-bit binary word. The output of the encoder results in an NRZ-L serial data stream.

The transmit timing circuit
The main clock frequency for the transmitter is developed by a crystal controlled oscillator. The output of the oscillator is applied to a squarer circuit, which produces a square wave output. This output is then applied to divider circuits and counter circuits, including ring counters. All the necessary control signals to control the complete transmitter are developed in this circuitry.

Transmit signalling circuit
The signalling conditions from the automatic exchange, such as loop signalling, for each channel are applied to a signalling processor. The signalling condition is converted into a binary code and applied to the signalling multiplexer. The circuit containing the multiframe alignment word (MFAW) and multiframe service word (MFSW) is also connected to the signalling multiplexer. During time slot 16 of frame 0 the MFAW together with the MFSW is selected by this multiplexer and transmitted to the transmit combiner as an NRZ-L serial bit stream. During time slot 16 of frame 1 the digitised signalling conditions for channels 1 and 16 are selected by the signalling multiplexer. During time slot 16 of frame 2 the digitised signalling conditions for channel 2 and 17 are selected. This process continues until time slot 16 of frame 15 where the digitised signalling conditions for channel 15 and 30 are selected and passed to the transmit combiner.

Transmit combiner
This circuit consists of a multiplexer. The digitised speech is applied to the multiplexer on one lead. The digitised signalling is applied to the multiplexer via a switch on a separate lead. The circuit containing the frame alignment word (FAW) and frame service word (FSW) is applied to the multiplexer on another lead. During time slot 0 of each frame the output from the FAW and FSW circuit is selected by the transmit multiplexer. During time slots 1 to 15 and 16 to 31 the digitised speech input is selected. During time slot 16 the digitised signalling or MFAW and MFSW is selected. The output of the transmit multiplexer is an NRZ-L serial binary bit stream which contains the frame and multiframe structure.

Line coding
The NRZ-L serial data is now applied to the line coder circuit. The first process in the line coder is alternate digit inversion (ADI). The output of this circuit is still NRZ data. The next process is return-to-zero where the ADI data is converted into a unipolar return-to-zero (unipolar RZ) serial bit stream.

There is normally a solder strap or switch which can be changed to enable either alternate mark inversion (AMI) coding or high-density bipolar format

3 (HDB-3) coding to be implemented. The resultant AMI code or HDB-3 code is then transmitted to the transmit transformer. Both the AMI code and the HDB-3 code are bipolar RZ codes. The actual coding technique is discussed in Chapter 11.

The transmit transformer
This transformer is used for d.c. isolation between the system and the twisted copper pair line. The transformer also enables a constant current supply to be connected to the line to enable the dependent regenerators along the route to be powered from the terminal.

5.8.2 The receiver

The receive transformer
This transformer is the d.c. isolation between the twisted copper pair and the system. This transformer also enables the loop for the constant current supply, which is used to power the dependent regenerators, to be completed.

The terminal regenerator
This is used to restore the signal to its original format. The signal is first amplified by an a.c. amplifier. It is then equalised and then the timing is extracted. The extracted timing is processed to produce a clock frequency that is identical to that of the distant transmitter. The clock is used to ensure correct reshaping and retiming of the received signal. The signal is then reshaped and retimed to produce the original AMI code or HDB-3 code. This code is then connected to the line decoder.

The line decoder
The timing is extracted and this is sent to the receive timing circuit. The AMI code or HDB-3 code is decoded back to unipolar return-to-zero data and sent as a serial data stream to the return-to-zero decoder. The resultant ADI data is then decoded in the ADI decoder and the resultant NRZ-L serial data stream is sent to the receive decombiner. The actual decoding is discussed in more detail in Chapter 11.

The receive timing circuit
The extracted timing is used to control a digital phase-locked loop which produces a clock frequency that is identical to that of the distant transmitter. This clock is then divided and applied to a series of binary counter and ring counters. All the necessary timing signals for the complete receiver are generated in this circuit.

The decombiner
This consists of a demultiplexer. One output is connected to the FAW and FSW detector circuit. Initially the demultiplexer sends all the received data to this detector. The switch that connects the demultiplexed speech to the speech circuits

and the switch that connects demultiplexed digitised signalling to the signalling circuits are disabled. When the FSW and FAW have been recognised in three consecutive frames and in this order then the data in the next time slot is connected to the output going to the decoder and the speech circuits are enabled. This remains in force until time slot 16 when the data is routed to the receive signalling decombiner. The data is again routed to the decoder circuit for time slots 17 to 31.

The next time slot after time slot 31 will be time slot 0 of the next frame and as a result the data is routed to the FAW detector circuit where the FSW is expected. Once recognised, the above process will continue.

The decoder

The NRZ-L data is decoded in this circuit and the output is the TDM signal. This signal is connected to the receive PAM highway via a switch.

The channel switch

The PAM highway is commoned to the inputs of all the channel switches. Again only one switch can be closed at a time during time slots 1 to 15 and time slots 17 to 31. The appropriate switch is closed and the output of the switch yields the PAM signal for that particular channel.

The receive channel amplifier

This amplifier is an analogue amplifier with an adjustable gain. The gain is set to produce the required output level for the channel.

The integrating filter

The PAM signal is connected to the receive channel filter or integrating filter. This filter has a relatively slow response time and as a result a continuous analogue signal appears at the output.

The receive signalling circuit

The digital information in time slot 16 is passed via a switch to the receive signalling decombiner. As soon as frame alignment is achieved then this switch will be enabled. The decombiner is a demultiplexer. Initially all the data is sent to the MFAW detector and the signalling processors are disabled. On recognition of the MFAW and MFSW in three consecutive multiframes then the next time signalling data received is routed to the receive signalling processor which is now enabled. The receive signalling processor converts the binary code back to the original signalling format required by the exchange for channels 1 and 16. The next time that signalling data is received it is processed and sent to channels 2 and 17. This process continues until frame 15 at which point the received data is associated with channel 15 and 30. The next time data is received it is routed to the multiframe word detector as the multiframe word is then expected.

The system uses what is known as channel-associated signalling, as the speech path and the signalling paths are separated. This is unlike data circuits where common channel signalling is used. In common channel signalling the address is packeted together with the data.

5.9 IMPORTANCE OF FRAME AND MULTIFRAME ALIGNMENT

For a 30/32-channel PCM to operate correctly the receiver must align both the frame and multiframe correctly. Only then will all the received data be correctly decoded and sent to the correct destination.

5.9.1 Frame alignment

Figure 5.13(a) shows the frame alignment. If the time slots are correctly aligned but the frames are not then, as shown in this example, the speech on channel 3 will be connected to channel 1. This means that the wrong people speak to one another. The data on channel 15 is used for signalling information on channels 4 and 13. This results in wrong numbers.

Without frame alignment the speech channels and the signalling circuits at the receiver are disabled. Once frame alignment is achieved, the correct time slot is

Transmitted frame

FAW	CH 1	CH 2	CH 3	CH 4 to CH 13	CH 14	CH 15	MFAW MFSW	CH 17 to CH 28	CH 29	CH 30
TS 0	TS 1	TS 2	TS 3	—	TS 14	TS 15	TS 16	—	TS 30	TS 31

FAW	CH 1	CH 2	CH 3	CH 4 to CH 13	CH 14	CH 15	MFAW MFSW	CH 17 to CH 28	CH 29	CH 30
TS 0	TS 1	TS 2	TS 3	—	TS 14	TS 15	TS 16	—	TS 30	TS 31

Received frame

(a)

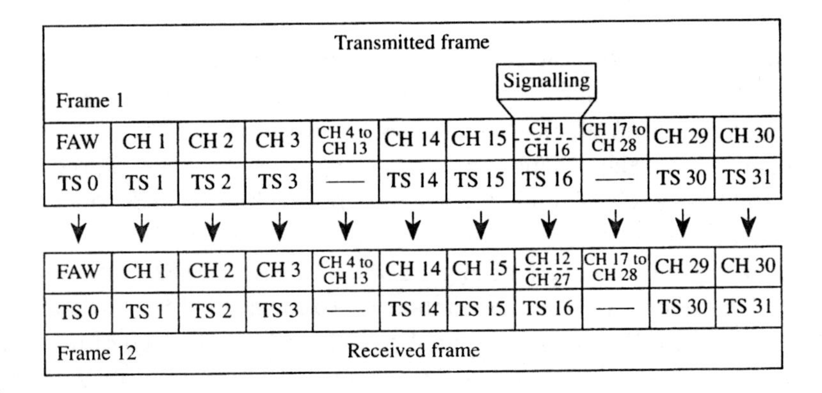

(b)

Fig. 5.13 (a) Frame alignment; (b) multiframe alignment

connected to the correct time slot. This causes the speech channels at the receiver to be enabled. However, the signalling channels are not properly aligned and the receive signalling circuits remain disabled.

If frame alignment is lost then both the speech and the signalling circuits will be affected and the speech and signalling circuits will be disabled. Once frame alignment is achieved only the speech circuits will be enabled. The signalling circuits will remain disabled.

5.9.2 Multiframe alignment

The multiframe alignment is used to align the signalling so that the correct signalling conditions are associated with the correct channel. This is shown in Fig. 5.13(b). Once multiframe alignment is achieved then the signalling circuits at the receiver are enabled.

If multiframe alignment is lost then only the signalling circuits will be disabled. In this case, established calls will be able to continue uninhibited. However, no new calls will be able to be established over the system.

5.10 ALARMS

On each terminal there are six different alarm conditions that are monitored constantly. These are:

- Loss of clock alarm (CLA).
- Bit error rate alarm (BER).
- Frame alignment alarm (FAL).
- Multiframe alignment alarm (MFAL).
- Remote frame alignment alarm (RFAL).
- Remote multiframe alignment alarm (RMFAL).

In this discussion only the FAL, RFAL, MFAL and RMFAL are going to be considered.

Referring to Fig. 5.14(a) the FAW and FSW detector has an alarm lead that is connected to the FAW and FSW generator. Bit 3 in the FSW can be set by an alarm condition placed on this lead, so as to inform the distant terminal that this terminal has an alarm. The same applies between the MFAW and MFSW detector and the MFAW and MFSW generator. In this case bit 2 of the MFSW is set by an alarm condition.

If a terminal receiver loses frame alignment the FAL and MFAL alarms on that terminal will become active and the terminal will cause bit 3 in the FSW and bit 2 in the MFSW to be set. The distant terminal receiver will be informed of these alarm conditions by checking the status of bit 3 in the FSW and bit 2 in the MFSW. Because these bits have been placed in the set state the distant terminal will cause the RFAL and RMFAL alarms to be activated. This is shown in Fig. 5.14(b).

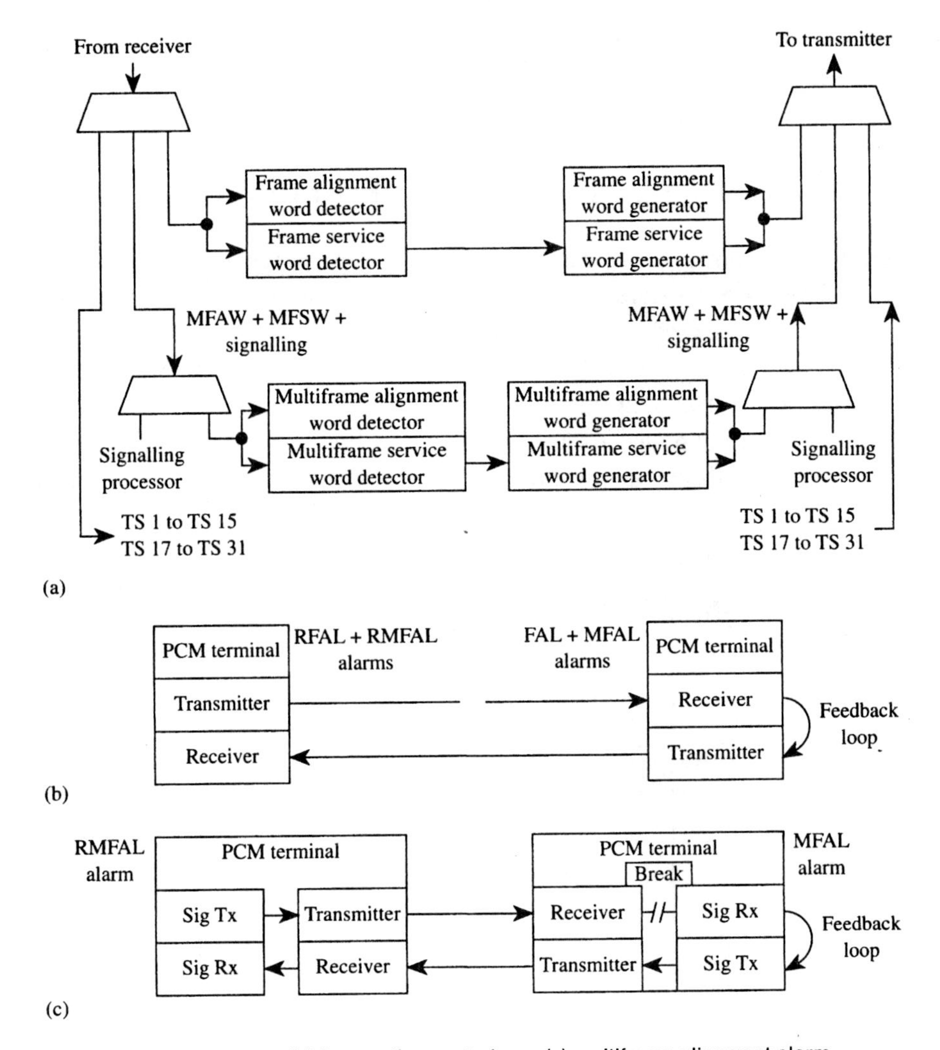

Fig. 5.14 (a) Remote alarms; (b) frame alignment alarm; (c) multiframe alignment alarm

The R indicates that a remote alarm is active. In other words the receiving terminal has lost frame alignment and the fault is either the receiver, the twisted pair linking the transmitter to the receiver or the transmitter.

If a terminal receiver loses multiframe alignment then the MFAL alarm will become active at the terminal. At the transmitter bit 3 in the FSW will remain reset but bit 2 in the MFSW will be set. At the distant terminal receiver the RMFAL alarm will become active as the receiver will detect that bit 2 in the MFSW is in the set state. This is shown in Fig. 5.14(c).

As above, the R indicates that a remote alarm is active. In other words the receiving terminal has lost multiframe alignment and the fault is either in the receive or transmit signalling circuitry in that particular direction. In this case it cannot be the twisted pair or the circuitry that carries the digitised speech and FAW at either terminal as frame alignment in that direction is still maintained.

5.11 DEPENDENT REGENERATIVE REPEATERS

Figure 5.15(a) shows a typical block diagram of a dependent regenerative repeater. These regenerators are normally housed in manholes, in sealed containers which are pressurised. If the pressure in the container is higher than outside and if a small crack in the seal should develop the pressurised air would escape and prevent water from entering into the container.

The a.c. signal is applied to an amplifier and then equalised, which reduces some of the distortion that the signal has experienced in the previous section.

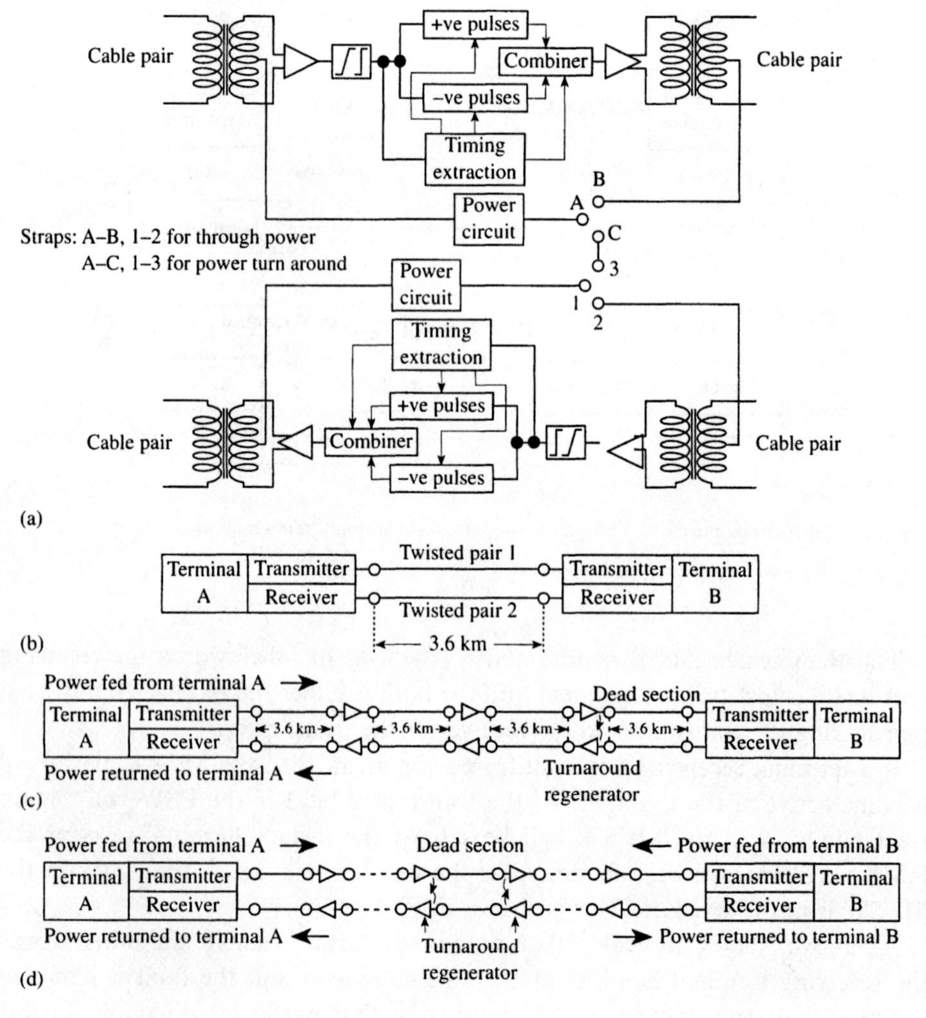

Fig. 5.15 (a) Dependent regenerative repeater; (b) no power feeding required; (c) power fed from terminal A only; (d) power fed from both terminals

The timing is extracted at this point and it is used to clock the circuits to enable reshaping and retiming of the signal. The output of the equaliser is fed to two circuits. The first one reshapes and retimes the positive pulses and the second reshapes and retimes the negative pulses. The outputs of these two circuits are combined together to form a single bipolar code. This code is the regenerated code. The bipolar code is applied to the transmission medium via a transformer.

The centre taps of the transformers are used to extract and insert the d.c. power that is fed to the regenerative repeater from the terminal. The regenerative repeater is either a through repeater or a turnaround repeater. A through repeater is a repeater where the power is fed to the regenerator and then fed to the next regenerator along the route. In this case the straps that are connected are strap A to strap B and strap 1 to strap 2.

A turnaround repeater is a repeater where the power is turned around back to the terminal that supplies the power. In this case the straps are strap A to strap C and strap 1 to strap 3.

5.12 POWER FEEDING

The repeater receives its power from one of the terminals. The terminal feeds the d.c. power via a centre-tapped transformer down the transmit pair from the terminal. The power is returned back to the terminal via the centre tap of a transformer on the receive pair. The transmit pair is the go leg and the receive pair is the return leg with respect to the d.c. power; this arrangement is shown in Fig. 5.12 as well as Fig. 5.15.

Figure 5.15(b), (c) and (d) shows three possible arrangements. When the distance between the two terminals is 3.6 km or less, no regeneration is required and no power will be supplied to the line as shown in Fig. 5.15(b). In this case the section joining the two terminals together is referred to as a dead section as it carries no d.c.

If the distance between the two terminals is greater than 3.6 km then regeneration will be required. In some cases the number of regenerators required is such that the power need be only supplied from the one terminal. This arrangement is shown in Fig. 5.15(c). In this case the last regenerator is referred to as a turnaround regenerator as the power must be turned around back to the terminal that supplies the power.

If more regenerators are required such that the power supply from one terminal is not sufficient, then power can be sent from both terminals. This arrangement is shown in Fig. 5.15(d). In this case there will be two turnaround regenerative repeaters. The section connecting these two regenerative repeaters together will be a dead section.

In the above description a distance between regenerators of 3.6 km has been quoted. This distance can only be achieved if the correct PCM Z screen cable is used. If ordinary junction cables are used then this distance drops to 1.8 km.

When the number of dependent regenerative repeaters is such that the power supplied from both terminals cannot supply the necessary power to all the required regenerative repeaters, a power feeding regenerative repeater is used. This repeater must be mounted in an equipment room where power is available. This regenerator is capable of supplying power in both directions, whereas each terminal can supply power in one direction. By using power feeding regenerative repeaters a PCM system can be made to work over extended distances.

5.13 REVIEW QUESTIONS

5.1 Describe why a non-linear quantising scale is used rather than a linear scale.

5.2 Draw the frame and multiframe structure of a CEPT 30/32-channel PCM system. Show all time durations on the diagram.

5.3 Calculate each of the following for a 38/40-channel PCM system that uses 8 bits per time slot and has a gross line bit rate of 6.556 Mb/s. The frame and multiframe structure follows the philosophy of the CEPT 30/32-channel PCM system.
(a) The bit duration.
(b) The channel time slot duration.
(c) The frame duration.
(d) The multiframe duration.

5.4 Calculate the sampling frequency of the signalling channel on the CEPT 30/32-channel PCM system.

5.5 A PCM system works between two towns A and B. The FAL alarm becomes active at town A. Describe any other alarms that would become active at town A and the alarm conditions that will become active at town B.

5.6 Describe what happens on a PCM terminal when it loses multiframe alignment. Describe the communication that take place between the two terminals and state the alarms that will become active at both terminals.

6 Noise figure and noise temperature

Effective communication opens the door to new knowledge and experiences.

6.1 INTRODUCTION

The enemy of any electronic engineer is noise. This is especially true for the transmission engineer. Noise affects both analogue and digital systems and engineers are always trying to find new methods to reduce noise in circuits and the noise introduced into the transmission media.

As a result an understanding of noise and some of the techniques used to control it are necessary. This chapter attempts to introduce the reader to the concept of noise and the importance of controlling the effect it has on transmitted and received signals.

There are two classes of noise, internal and external. Both these sources of noise affect communication systems. Internal noise affects the signal as it propagates through the system and circuits. This noise cannot be eliminated and can only be controlled by careful component selection, circuit design and circuit layout. External noise or line noise is the noise injected into the signal when the signal is propagated over the medium that connects the transmitter to the receiver. There is nothing the designer can do to control this noise. Instead the designer must take it into account this noise when designing the receiver.

6.2 INTERNAL NOISE

Internal noise can be classified under the following headings:

- Component noise.
- System noise.

6.2.1 Component noise

Component noise is caused by the following:

- Thermal agitation of electrons.
- Shot noise.
- Partition noise.
- $1/f$ noise.
- Electromagnetic and electrostatic coupling from one circuit to another.
- Contact noise.

The above types of noise are referred to as white noise. White noise is noise that has the same amplitude for all frequencies across the whole frequency spectrum.

Both thermal noise and shot noise cannot be completely eliminated. This results in set limits on possible useful amplification that can take place.

Thermal noise (Johnson noise)

Thermal noise is also referred to as Johnson noise. Thermal agitation cannot be completely eliminated but it can be reduced by cooling. At any temperature greater than 0K (−273°C) thermal agitation causes the electrons in the material to move. This movement results in thermal noise.

Any conductor/semiconductor contains a number of electrons that are free to move about constantly. This results in collisions, which cause a continuous exchange of energy. Even in a circuit where no current flows, the random motion of these free electrons produces voltage fluctuations across the terminals. These random fluctuations in voltage result in a mean square voltage $\bar{V}_0^2$. Experimental results by Nyquist showed that the available thermal noise power (P_n) is given by the empirical equation

$$P_n = KTB \text{ J/Hz}$$

where K = Boltzmann's constant
$= 1.38 \times 10^{-23}$ J/K
T = absolute temperature in kelvins
B = bandwidth

Thermal noise in a resistor

A noisy resistor can be represented by a Thévenin equivalent circuit consisting of a voltage (noise) source and a noise-free resistor in series with one another (Fig. 6.1). Under matched conditions the value of the load resistor (R_L) will be equal to the resistor (R) and be noise free. Hence the circuit consists of the voltage (noise) source in series with the noise-free resistor in series with the load, in a closed circuit.

The resultant RMS current can be determined by:

$$I_{RMS} = \frac{\sqrt{\bar{V}_0^2}}{2R}$$

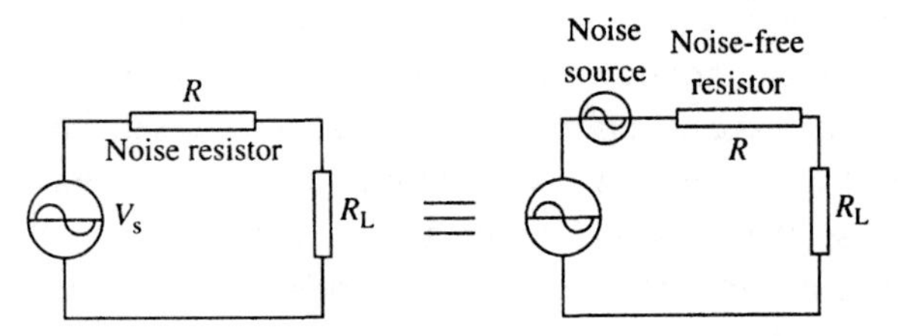

Fig. 6.1 Thévenin equivalent circuit for a noisy resistor

But

$$P_{max} = I^2R = \left(\frac{\sqrt{\bar{V}_0^2}}{2R}\right)^2 R$$

$$P_{max} = \frac{\bar{V}_0^2}{4R}$$

From Nyquist

$$P_{max} = P_n = KTB$$

$$KTB = \frac{\bar{V}_0^2}{4R}$$

$$\bar{V}_0^2 = 4KTBR$$

Example 6.1

Determine the value of the noise voltage developed in a 1 kΩ resistor which is in a circuit that must be capable of carrying a bandwidth of 10 kHz at a temperature of 17°C.

Solution

$$T = 273 + 17 = 290\text{K}$$

$$\bar{V}_0^2 = 4KTBR$$

$$\bar{V}_0^2 = 4 \times (1.38 \times 10^{-23}) \times 290 \times 10^4 \times 10^3$$

$$\bar{V}_0^2 = 1.6 \times 10^{-13}$$

$$\text{RMS noise voltage} = \sqrt{1.6 \times 10^{-13}}$$

$$= 0.4\,\mu\text{V}$$

Shot noise

The noise caused by the statistical variation of the number of free electrons contributing to the flow of current in semiconductors and valves is known as shot noise. The current flowing through a semiconductor diode is due to electrons moving across the *pn* junction, from the cathode to the anode. The number of electrons arriving at the anode at a particular instant in time is not the same as the number arriving at the next instant in time. This results in a minute alternating current. This alternating current is superimposed on the normal current through the device. Because the electrons arrive at the cathode like a shot this noise is referred to as shot noise. In transistors the shot noise is due to the variation in electrons arriving at the collector.

The fluctuating components give rise to the mean square shot noise current $\bar{i}_s^2$.

Partition noise

The input current to a transistor flows from the emitter to the base. After crossing the barrier it divides between the base terminal and the collector terminal. This base current is also subject to random fluctuations and contributes to the overall noise produced by the transistor. Because the noise is produced at the base junction the noise is referred to as partition noise.

1/*f* noise

Fluctuations in the conductivity of the semiconductor material provide a noise source which is inversely proportional to frequency. This noise is known as current or excess noise. It is usually negligible above 10 kHz and for some transistors above 1 kHz.

Electromagnetic and electrostatic induction

Electrostatic and electromagnetic coupling can be considerably reduced by proper screening of power supply circuits and the HF circuits as well as careful circuit board layout. Electrostatic coupling takes place through the capacitance effects between parallel conductors and printed circuit board tracks, whereas electromagnetic induction takes place through transformer action between parallel conductors and circuit board tracks.

Contact noise

Short breaks in the transmission in a system due to poorly adjusted relay and switch contacts cause impulse noise. This form of noise seriously affects data circuits. If the contacts are dirty then this will add to the overall attenuation of the signal in the system. Dirty contacts also produce thermal noise.

6.2.2 System noise or equipment noise

Another source of noise is intermodulation noise, which is caused by the nonlinear characteristics of components, such as diodes and transistors, as well as by circuits, such as modulators and filters. This type of noise produces frequencies in the output signal that were not present at the input. The nonlinear characteristic causes the output signal to be distorted relative to the input signal.

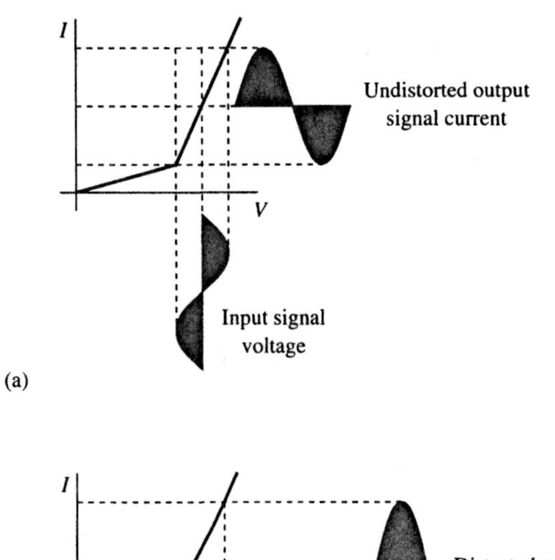

Fig. 6.2 (a) Ideal diode response; (b) typical diode response

Figure 6.2(a) shows the ideal response of a diode and the undistorted current as the result of a sinusoidal input voltage applied across the device. In practice, however, the diode characteristic is non-linear, as shown in Fig. 6.2.(b), so that the output current is distorted.

Figure 6.3(a) shows the undistorted output voltage for a transistor having an ideal response. Unfortunately the family of curves for I_B are not parallel to each other. Figure 6.3(b) shows a more realistic picture, where it can be clearly seen that the output signal voltage is distorted.

Two-tone method

A method used to determine the amount of intermodulation noise is the two-tone test. If two frequencies f_1 and f_2 are applied to the input of a non-linear device then the output signal can contain the following components:

$$f_1, f_2, f_1 + f_2, f_1 - f_2, f_1 + 2f_2, f_1 - 2f_2, f_1 + 3f_2, f_1 - 3f_2, 2f_1 + f_2, 2f_1 - f_2,$$

$$3f_1 + f_2, 3f_1 - f_2, 2f_1 + 3f_2, 2f_1 - 3f_2, 3f_1 + 2f_2, 3f_1 - 2f_2 \text{ etc.}$$

Intermodulation noise is dependent on the amplitude of the input signal and the non-linearity of the component and/or device. The exact frequency components will depend on how non-linear the device is. Devices such as diodes

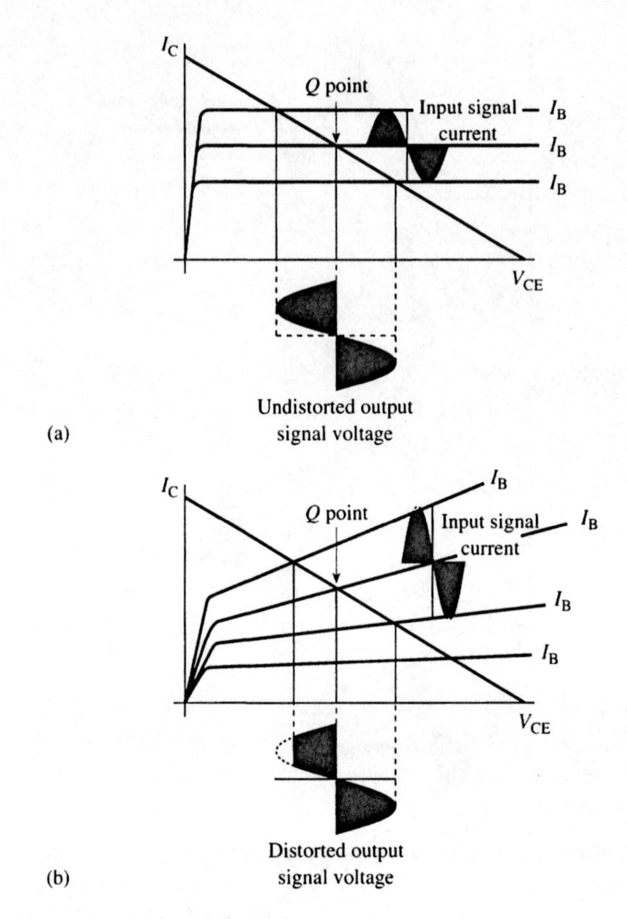

Fig. 6.3 (a) Ideal transistor response; (b) typical diode response

and transistors and circuits such as modulators exhibit non-linear or intermodulation noise.

6.3 EXTERNAL NOISE

With careful design, correct circuit board layout, proper choice of components and correct screening of RF/IF stages, the correct use of heatsinks etc., the effects of internal noise on the transmitted and received traffic can be controlled. However, once the signal is transmitted down the medium, it comes under the influences of external sources of noise that cannot be controlled.

External noise is caused by the following:

- Static interference (sand storms/dust storms).
- Radiation (sun spot activity, cosmic radiation, manmade radiation).
- Crosstalk (interference from adjacent pairs).
- Mains induction (interference from adjacent power routes).
- Voltaic interference (volcanic activity).

6.3.1 Line noise

Radiation from the medium carrying the transmitted signal can take place and this constitutes attenuation. Induction into the medium from external sources also takes place quite easily. Even properly screened cables are not totally immune to external interference. The only medium that is totally unaffected by external sources of noise is an optic fibre. However, optic fibres have their own problems.

Metallic pairs are subject to induction from adjacent circuits (crosstalk) and induction from mains routes that are parallel to the transmission route. Overhead pairs are also susceptible to static electricity due to sand, dust and thunderstorms, and induction from sources such as sunspot activity, cosmic radiation etc. Underground cables are affected by voltaic noise, which is caused by volcanic activity. Radio and microwave transmissions are also affected by atmospheric conditions such as rain, snow, hail, inversion etc., which cause fading to take place resulting in high attenuation.

All of the above sources are external sources of noise. These sources contribute to what is referred to as line noise. Line noise cannot be eliminated and hence the received signal will contain a certain amount of noise. The receiver must be able to cope with this noise in such a way that the actual information is not lost or swamped by this noise but if possible its effect is reduced. The receiver can do nothing about the internal noise arising from the equipment at the transmitter. The receiver will also contribute a certain amount of internal noise to the signal.

6.3.2 Impulse noise

Short breaks in the transmission on the transmission line, due to bad joints on open wire lines and cable pairs, seriously affect data circuits. The thermal noise produced by the transmission lines, receive aerials and aerial feeders can adversely affect analogue and digital circuits. Analogue signals will be affected worse because receive amplifiers are used. Digital signals are regenerated which means that they should not be affected to the same extent.

6.4 SYSTEM PERFORMANCE

Although digital transmission is used extensively today, many of the media used require an analogous signal. This means that the digital information must be changed into an analogue format. A typical example is radio and microwave transmission. To a large extent frequency modulation is used on radio and microwave systems. The criterion used to determine the performance of any analogue system is the signal-to-noise ratio (SNR). This is determined by means of the following equation:

$$\mathrm{SNR} = \frac{\text{wanted signal power}}{\text{unwanted noise power}} = \frac{S}{N}$$

Very often the SNR is expressed in dB:

$$\mathrm{SNR_{dB}} = 10\log_{10}\left(\frac{S}{N}\right)\ \mathrm{dB}$$

Example 6.2

Determine the signal-to-noise ratio at the output of an amplifier that produces an output signal voltage of 1.2 V if the output noise voltage is 12 mV. Express the answer in decibels.

Solution

The signal power S and noise power N are found using

$$\text{power} = \frac{(\text{voltage})^2}{\text{impedance}} = \frac{V^2}{Z}$$

therefore

$$S = \frac{V_{\mathrm{sig}}^2}{Z_{\mathrm{out}}}$$

$$N = \frac{V_{\mathrm{noise}}^2}{Z_{\mathrm{out}}}$$

$$\mathrm{SNR} = \frac{S}{N}$$

$$= \frac{V_{\mathrm{sig}}^2}{Z_0} \times \frac{Z_0}{V_{\mathrm{noise}}^2} = \frac{V_{\mathrm{sig}}^2}{V_{\mathrm{noise}}^2}$$

$$\mathrm{SNR_{dB}} = 10\log_{10}\frac{V_{\mathrm{sig}}^2}{V_{\mathrm{noise}}^2}$$

$$= 20\log_{10}\frac{V_{\mathrm{sig}}}{V_{\mathrm{noise}}}\ \mathrm{dB}$$

$$= 20\log_{10}\frac{1.2}{12\times 10^{-3}}$$

$$= 40\ \mathrm{dB}$$

6.5 NOISE FIGURE/NOISE FACTOR

A measure which is used extensively on the amplifier chain at the front end of a receiver is the noise figure or noise factor. The noise figure is a measure of the noisiness of the input stage. In other words it determines the effect that the internal noise, produced by that stage, has on the received signal. As will be shown below the more internal noise produced by a stage the worse the output signal-to-noise ratio becomes. Hence the internal noise must be controlled, especially for the input stage of the receiver where the received signal strength is already weak and a large amount of noise accompanies the signal due to the transmission medium.

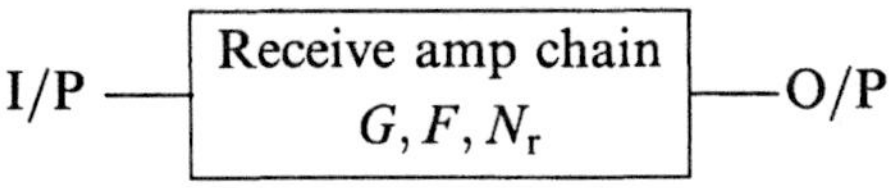

I/P — [Receive amp chain G, F, N_r] — O/P

where G = gain
F = noise figure
N_r = internally generated noise power

The noise factor is a measure of the degradation of the output signal, and is determined by

$$F = \frac{\text{SNR}_{\text{in}}}{\text{SNR}_{\text{out}}}$$

In this equation both the SNRs are given as ratios and not in dB. Now

$$\text{SNR}_{\text{in}} = \frac{S_i}{N_i} \quad \text{and} \quad \text{SNR}_{\text{out}} = \frac{S_o}{N_o}$$

$$S_o = GS_i \quad \text{and} \quad N_o = G(N_i + N_{ai})$$

$$F = \frac{\dfrac{S_i}{N_i}}{\dfrac{GS_i}{G(N_i + N_{ai})}}$$

where S_i = signal power at input
N_i = noise power at input
G = amplifier gain as a ratio
N_{ai} = amplifier noise referred to the input of the amplifier

All the internal noise generated in the amplifier is assumed to be generated by an external source at the input to the amplifier and the amplifier then becomes a noise-free device. The externally generated noise is now referred to as N_{ai}.

Therefore

$$F = \frac{S_i}{N_i} \times \frac{G(N_i + N_{ai})}{GS_i}$$

$$= \frac{N_i + N_{ai}}{N_i}$$

This is then expressed in the important equation

$$F = 1 + \frac{N_{ai}}{N_i}$$

It can be seen that if the amplifier is noise free then $N_{ai} = 0$ and the noise figure F would be equal to unity. This means that $\text{SNR}_{in} = \text{SNR}_{out}$.

An alternative derivation for an expression of N_r in terms of the noise figure is given below:

$$F = \frac{S_i}{N_i} \times \frac{G(N_i + N_{ai})}{GS_i}$$

but

$$G(N_i + N_{ai}) = N_o$$

$$F = \frac{S_i}{N_i} \times \frac{N_o}{GS_i}$$

$$= \frac{N_o}{GN_i}$$

$$N_o = GN_iF$$

We have

$$N_o = GN_i + GN_{ai}$$

but

$$GN_{ai} = N_r$$

Thus

$$N_o = GN_i + N_r$$

$$F = \frac{S_i}{N_i} \times \frac{GN_i + GN_{ai}}{GS_i}$$

$$= \frac{GN_i + N_r}{GN_i}$$

$$GFN_i = GN_i + N_r$$

$$N_r = GFN_i - GN_i$$

$$= GN_i(F - 1)$$

6.5.1 Noise in cascaded systems

Consider the following system:

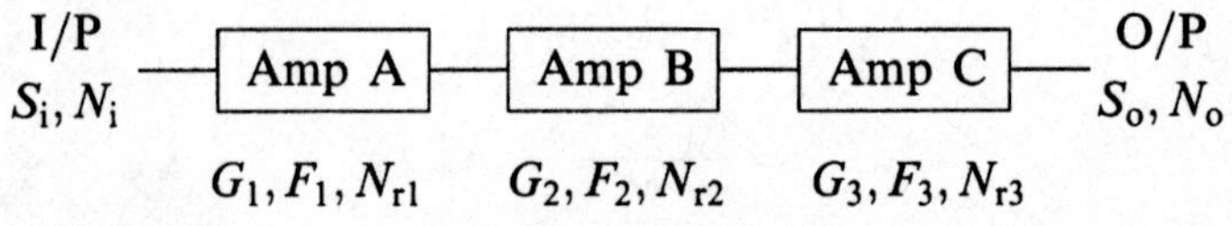

I/P S_i, N_i — Amp A — Amp B — Amp C — O/P S_o, N_o

G_1, F_1, N_{r1} G_2, F_2, N_{r2} G_3, F_3, N_{r3}

The gain is expressed as a ratio and not in dB. For each stage:

$$F = \frac{N_o}{GN_i} = \frac{GN_i + N_r}{GN_i}$$

but

$$N_r = GN_i(F-1)$$

For the cascaded system:

$$F = \frac{N_o}{GN_i} = \frac{N_o}{N_i G_1 G_2 G_3}$$

Now output noise due to N_i:

$$\text{output noise} = N_i G_A G_B G_C$$

Output noise due to amp A:

$$\begin{aligned}\text{output noise} &= N_{rA} G_B G_C \\ &= N_i G_A G_B G_C (F_A - 1)\end{aligned}$$

Output noise due to amp B:

$$\begin{aligned}\text{output noise} &= N_{rB} G_C \\ &= N_i G_B G_C (F_B - 1)\end{aligned}$$

Output noise due to amp C:

$$\begin{aligned}\text{output noise} &= N_{rC} \\ &= N_i G_C (F_C - 1)\end{aligned}$$

The total output noise is given by:

$$N_o = N_i G_A G_B G_C + N_{rA} G_B G_C + N_{rB} G_C + N_{rC}$$

$$N_o = N_i G_A G_B G_C + N_i G_A G_B G_C (F_A - 1) + N_i G_B G_C (F_B - 1) + N_i G_C (F_C - 1)$$

$$N_o = N_i G_A G_B G_C F_A + N_i G_B G_C (F_B - 1) + N_i G_C (F_C - 1)$$

$$F = \frac{N_o}{N_i G_A G_B G_C}$$

$$F = \frac{N_i G_A G_B G_C F_A}{N_i G_A G_B G_C} + \frac{N_i G_B G_C (F_B - 1)}{N_i G_A G_B G_C} + \frac{N_i G_C (F_C - 1)}{N_i G_A G_B G_C}$$

$$F = F_A + \frac{F_B - 1}{G_A} + \frac{F_C - 1}{G_A G_B}$$

Hence for a system having n cascaded stages:

$$F = F_1 + \frac{F_2 - 1}{G_1} + \frac{F_3 - 1}{G_1 G_2} + \cdots + \frac{F_n - 1}{G_1 G_2 \cdots G_{n-1}}$$

From this equation it can be seen that the first stage contributes the most to the noise figure. Any noise generated in the first stage will be subjected to the gain of all the subsequent stages, thus the first stage must have the lowest amount of internally generated noise. The first stage is normally cooled to ensure that its noise figure is as low as possible.

Normally the front end of a receiver consists of an odd number of amplifiers. In very basic systems there is only one amplifier, in more complex systems there can be up to five cascaded amplifiers. The following general rule is used when deciding the order in which these amplifiers must be connected together in the front end of a receiver:

- The first amplifier should have the lowest noise figure and the lowest gain.
- The last amplifier in the chain should have the highest gain and highest noise figure.

Example 6.3

Given the parameters for three power amplifiers as indicated below, determine the following:

1. The order in which the amplifiers must be connected.
2. The overall lowest noise figure for the cascaded system.

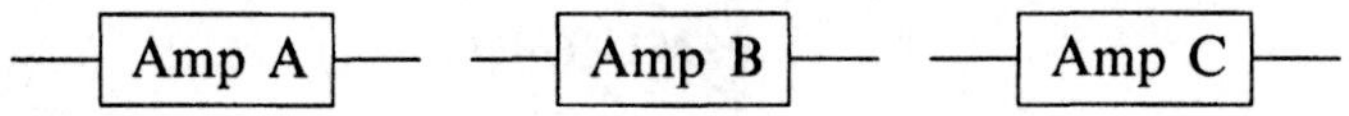

Amp A	Amp B	Amp C
$G_A = 12\,\text{dB}$	$G_B = 6\,\text{dB}$	$G_C = 20\,\text{dB}$
$F_A = 2$	$F_B = 1.7$	$F_C = 4$

Solution

Step 1. Convert all the gains in dB to ratios.

$$G_{\text{dB}} = 10 \log_{10} G_{\text{ratio}}$$

$$G_{\text{ratio}} = 10^{\frac{G_{\text{dB}}}{10}}$$

Amplifier A:

$$G_A = 10^{\frac{12\,\text{dB}}{10}} = 15.85\times$$

Amplifier B:

$$G_B = 10^{\frac{6\,\text{dB}}{10}} = 3.98\times$$

Amplifier C:

$$G_C = 10^{\frac{20\,\text{dB}}{10}} = 100\times$$

Step 2. Decide and motivate the order of connection.

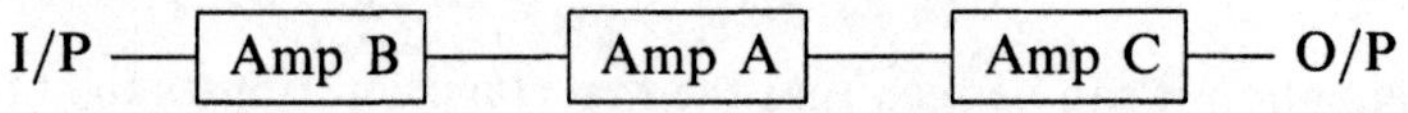

I/P — Amp B — Amp A — Amp C — O/P

Amp B	Amp A	Amp C
$G_B = 6\,\text{dB}$	$G_A = 12\,\text{dB}$	$G_C = 20\,\text{dB}$
$= 3.98\times$	$= 15.85\times$	$= 100\times$
$F_B = 1.7$	$F_A = 2$	$F_C = 4$

Amplifier B has the smallest noise figure and the smallest gain hence it is placed first. Amplifier C has the highest gain and noise figure and hence is placed last.

Step 3. Determine the overall lowest noise figure. Now

$$F_{(3)} = F_1 + \frac{F_2 - 1}{G_1} + \frac{F_3 - 1}{G_1 G_2}$$

so for the system given in step 2

$$\begin{aligned}
F_{(3)} &= F_B + \frac{F_A - 1}{G_B} + \frac{F_C - 1}{G_B G_A} \\
&= 1.7 + \frac{2 - 1}{3.981} + \frac{4 - 1}{3.981 \times 15.849} \\
&= 1.7 + \frac{1}{3.981} + \frac{3}{63.095} \\
&= 1.7 + 0.2512 + 0.0475 \\
&= 1.9987
\end{aligned}$$

Example 6.4

An amplifier has a power gain of 12 dB and has 0.118 nW of internal noise referred to the input. Determine the noise figure for the amplifier when the input signal level to the amplifier is −38 dBm and the input signal-to-noise ratio is 27 dB.

Solution

Method 1

$$\begin{aligned}
\text{Output signal level} &= \text{gain} + \text{input signal level} \\
&= 12\,\text{dB} + (-38\,\text{dBm}) = -26\,\text{dB}
\end{aligned}$$

$$\begin{aligned}
\text{Input noise level} &= \text{input signal level} - \text{SNR} \\
&= -38\,\text{dBm} - (27\,\text{dB}) = -65\,\text{dBm}
\end{aligned}$$

Input noise power:

$$\begin{aligned}
N_{\text{noise}} &= 10 \log \frac{N_{\text{noise}}\,\text{W}}{1\,\text{mW}}\,\text{dB} \\
&= 10^{\frac{N_{\text{noise}}\,\text{dBm}}{10}} \times 1 \times 10^{-3} \\
&= 10^{\frac{-65\,\text{dBm}}{10}} \times 1 \times 10^{-3} \\
&= 0.316\,\text{nW}
\end{aligned}$$

Total input noise power:

$$N_{\text{noise}} = \text{input noise power} + \text{amp noise ref input}$$
$$= 0.316\,\text{nW} + 0.118\,\text{nW}$$
$$= 0.434\,\text{nW}$$

Convert the total input noise power to a level:

$$N_{\text{noise}} = 10\log\frac{N_{\text{noise}}\,\text{W}}{1\,\text{mW}}\ \text{dBm}$$
$$= 10\log\frac{0.434\,\text{nW}}{1\,\text{mW}}\ \text{dBm}$$
$$= -63.6\,\text{dBm}$$

Total output noise level:

$$N_{\text{noise}} = \text{gain} + \text{total input noise level}$$
$$= 12\,\text{dB} + (-63.6\,\text{dBm})$$
$$= -51.6\,\text{dBm}$$

The input SNR = 27 dB. Converting to a ratio,

$$\text{SNR}_{\text{dB}} = 10\log\text{SNR}_{\text{ratio}}\ \text{dB}$$

$$\text{SNR}_{\text{ratio}} = 10^{\frac{\text{SNR}_{\text{dB}}}{10}}$$
$$= 10^{\frac{27\,\text{dB}}{10}}$$
$$= 501$$

Output SNR:

$$\text{output noise level} = \text{gain} + \text{total input noise level}$$
$$= 12\,\text{dB} + (-63.6\,\text{dBm})$$
$$= -51.6\,\text{dBm}$$

$$\text{output SNR} = \text{output signal level} - \text{output noise level}$$
$$= -26\,\text{dBm} - (-51.6\,\text{dBm})$$
$$= 25.6\,\text{dB}$$

$$\text{SNR}_{\text{ratio}} = 10^{\frac{\text{SNR}_{\text{dB}}}{10}}$$
$$= 10^{\frac{25.6\,\text{dB}}{10}}$$
$$= 363.1$$

Amplifier noise figure:

$$F = \frac{\text{input SNR}}{\text{output SNR}}$$

$$= \frac{501}{363}$$

$$= 1.38$$

Method 2

$$F = 1 + \frac{N_{\text{ai}}}{N_{\text{i}}}$$

$$= 1 + \frac{0.118\,\text{nW}}{0.316\,\text{nW}}$$

$$= 1.37$$

In practice the following equation is used when there are more than two cascaded amplifiers:

$$F_{(n)} = F_1 + \frac{F_2 - 1}{G_1}$$

In the above examples it can be seen that the higher-order terms contribute a very small amount to the final answer and as a result only the first two terms are significant.

6.6 EFFECTIVE NOISE TEMPERATURE

Consider one cascaded system having an overall noise figure of 1.87 and a second cascaded system having an overall noise figure of 2.09. Which system is better? It cannot be assumed that the system having the lower noise figure is better if they are two different systems receiving different signals. Hence some form of measurement that will enable the two systems to be compared is necessary. The measurement that is used for this is the effective noise temperature.

When a noisy resistor was examined the total noise power dissipated was given by

$$N = KTB$$

To determine the noise power in a 1 Hz bandwidth:

$$N_{\text{o}} = \frac{N}{B}\ \text{W/Hz}$$

A standard temperature of 17°C is chosen, as this temperature is suitable and convenient for human habitation. The temperature is to be expressed in kelvins, so a suitable approximation is

$$T_0 = 273 + 17 = 290\text{K}$$

Now the noise figure for an amplifier is given by

$$F = 1 + \frac{N_{ai}}{N_i}$$

$$F - 1 = \frac{N_{ai}}{N_i}$$

Hence

$$N_{ai} = N_i(F - 1)$$

Replace N_i with KT_0B:

$$N_{ai} = (F - 1)KT_0B$$

Replace N_{ai} with KT_eB:

$$KT_eB = (F - 1)KT_0B$$

$$T_e = (F - 1)T_0$$

But $T_0 = 290\text{K}$:

$$T_e = 290(F - 1)$$

For a cascaded system the overall effective noise temperature will be

$$T_{e(n)} = 290(F_{(n)} - 1)$$

The overall noise figure for a cascaded system is given by

$$F_{(n)} = F_1 + \frac{F_2 - 1}{G_1} + \frac{F_3 - 1}{G_1G_2} + \cdots + \frac{F_n - 1}{G_1G_2\cdots G_{n-1}}$$

Consequently, the overall effective noise temperature for a cascaded system is given by:

$$T_{e(n)} = T_{e1} + \frac{T_{e2}}{G_1} + \frac{T_{e3}}{G_1G_2} + \cdots + \frac{T_{en}}{G_1G_2\cdots G_{n-1}}$$

where $T_{e1}, T_{e2}, \ldots, T_{en}$ are the effective noise temperatures of the individual amplifiers all measured at the standard temperature T_0. In these equations it is assumed that the inputs and outputs of all the stages are correctly matched.

Example 6.5

Determine the effective noise temperature for each stage and the overall effective noise temperature of the cascaded system given in Example 6.3.

Solution

Amplifier A:

$$T_{eA} = T_0(F_A - 1)$$

$$= 290(2 - 1)$$

$$= 290\text{K}$$

Amplifier B:

$$T_{eB} = T_0(F_B - 1)$$
$$= 290(1.7 - 1)$$
$$= 203K$$

Amplifier C:

$$T_{eC} = T_0(F_C - 1)$$
$$= 290(4 - 1)$$
$$= 870K$$

Now the overall effective noise temperature is

$$T_{e_{(3)}} = T_{eB} + \frac{T_{eA}}{G_B} + \frac{T_{eC}}{G_B G_A}$$
$$= 203 + \frac{290}{3.98} + \frac{870}{3.98 \times 15.85}$$
$$= 289.66K$$

Check:

$$T_{e_{(3)}} = T_0(F_{(3)} - 1)$$
$$= 290(1.9987 - 1)$$
$$= 289.623K$$

Example 6.6

Two different systems have the following parameters:

System 1: Overall noise figure = 1.87
Standard operating temperature = 30°C

System 2: Overall noise figure = 1.92
Standard operating temperature = 13.5°C

Determine which system is better.

Solution

$$T_{e_{(n)}} = T_0(F_{(3)} - 1)$$

System 1:

$$T_0 = 273 + 30$$
$$= 303K$$
$$T_{e_{(n)}} = 303(1.87 - 1)$$
$$= 263.61K$$

System 2:

$$T_0 = 273 + 13.5$$
$$= 286.5\text{K}$$
$$T_{e(n)} = 286.5(1.92 - 1)$$
$$= 263.58\text{K}$$

From the above results it can be seen that there is very little difference between the two systems. Hence both systems are equally good.

In this example the designer of system 1 decided to design excellent amplifiers and the designer did not concentrate too much on the cooling and probably only used ordinary heatsinks. The designer of system 2 designed a complex cooling system to keep the amplifier chain as cool as possible but this designer did not design very good amplifiers.

6.7 VARIATION OF NOISE FIGURE WITH FREQUENCY

As frequency increases from a low frequency the noise figure starts at a high value and decays exponentially until an optimum constant value is achieved. As the frequency increases still further the noise figure increases exponentially. The response at the low frequency end is primarily due to the effect of the input noise. The response at the high frequency end is primarily due to the amplifier gain decreasing with an increase in frequency.

6.8 REVIEW QUESTIONS

6.1 Determine the value of the noise voltage that is developed in a 200 Ω resistor working at an ambient temperature of 22°C and capable of carrying the frequency range of 60 kHz to 108 kHz.

6.2 Briefly describe how internal noise can be minimised in a circuit used in a communication system.

6.3 Determine the signal-to-noise ratio on the output of an amplifier that produces an RMS output voltage of 3.2 V if the input RMS noise voltage is 2.5 mV, given that the amplifier has a voltage gain of 8 dB.

6.4 Determine the signal-to-noise ratio at the output of an amplifier that causes an RMS output current of 1.5 mA to flow through a correctly matched load, having an impedance of 1.2 kΩ, when the input RMS noise voltage is 1.85 mV. The input impedance to the amplifier is also 1.2 kΩ and the amplifier has a current gain of 17 dB.

6.5 Given the parameters for the three voltage amplifiers indicated below, determine the following:

(a) The order in which the amplifiers must be connected.
(b) The overall lowest noise figure for the cascaded system.
(c) The effective noise temperature for each amplifier.
(d) The overall effective noise temperature for the cascaded system.

Amp A	Amp B	Amp C
$G_A = 44\,\text{dB}$	$G_B = 16\,\text{dB}$	$G_C = 8\,\text{dB}$
$F_A = 2.5$	$F_B = 1.7$	$F_C = 1.2$

6.6 The received input signal level to the following system is −25 dBm. Given the parameters for three voltage amplifiers indicated below, determine the following:

(a) The noise figure for amplifier B, which must have an SNR of 38 dB.
(b) The order in which the amplifiers must be connected.
(c) The overall lowest noise figure for the cascaded system.
(d) The effective noise temperature for each amplifier.
(e) The overall effective noise temperature for the cascaded system.

Amp A	Amp B	Amp C
$G_A = 17\,\text{dB}$	$G_B = 14\,\text{dB}$	$G_C = 16\,\text{dB}$
$F_A = 2.5$	$N_{ai} = 75.15\,\text{pW}$	$F_C = 1.2$

7 Effects of noise and distortion on analogue and digital signals

Human endeavour is fraught with danger but success is very sweet.

7.1 INTRODUCTION

Having gained an understanding of the sources of noise the next important step is to understand how the noise affects analogue and digital signals that are transmitted down different media.

In this chapter the effects that noise has on a signal are discussed, as well as the effects the medium has on the signal as it propagates down the medium. In the discussion a comparison is drawn between the effects of noise on analogue and digital signals and the reasons and the advantages of digital transmission as opposed to analogue transmission.

Any signal that is propagated down a transmission medium experiences two phenomena:

- Amplitude distortion.
- Frequency distortion.

On top of these two effects line noise is added to the signal. As a result the received signal is badly distorted and contains a large amount of noise.

7.2 AMPLITUDE DISTORTION

This is the attenuation that the signal undergoes across the medium. For metallic pairs the higher the frequency the greater the attenuation.

Figure 7.1 shows the resultant waveform for a signal consisting of a fundamental frequency and the third harmonic. By observing the figure it can be seen that the third harmonic has a third of the amplitude of the fundamental frequency at the transmitter. Also when these two frequencies are added together, the beginnings of a square wave results.

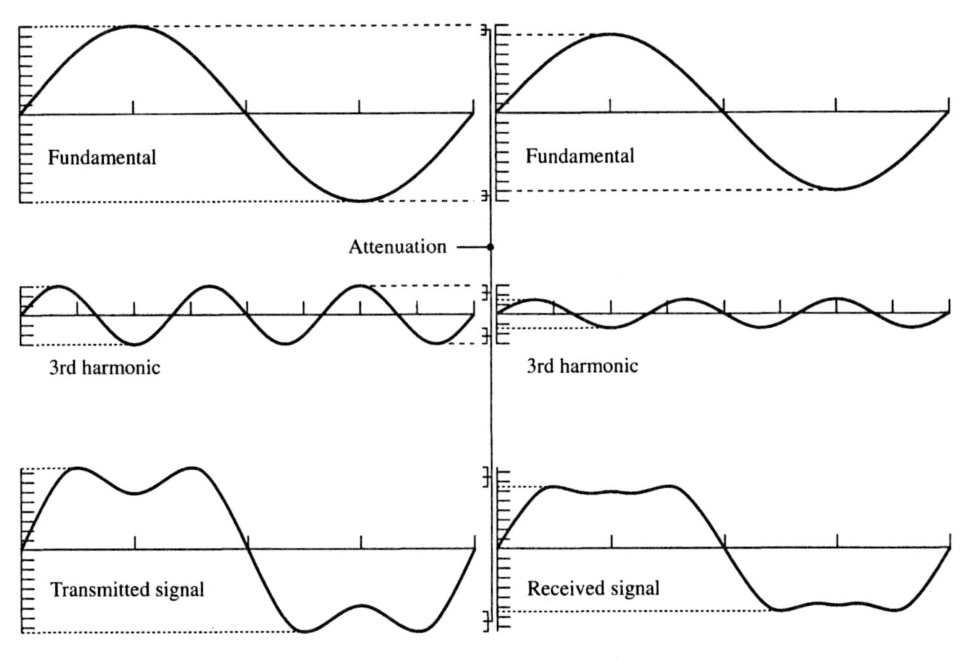

Fig. 7.1 Amplitude distortion

As can be seen, at the receiver, the higher frequency (third harmonic) experiences more attenuation than the lower frequency (fundamental). As a result the amplitude relationship between the third harmonic and the fundamental is now different as the amplitude of the third harmonic is smaller than a third of the amplitude of the fundamental. The resultant wave shape, when these two frequencies are added together, is now different to what it was at the transmitter.

7.3 FREQUENCY DISTORTION

In free space all frequencies travel at the speed of light. In any medium each frequency travels at a different velocity. This effect is referred to as frequency distortion and its effects are felt as group delay.

Consider Fig. 7.2. This figure shows a fundamental and the third harmonic. At the transmitter the two frequencies start at 0° and initially both start in a positive direction.

Now along the medium the two frequencies travel at different velocities and as a result the phase relationship changes. At the receiver it can be seen in the example that the fundamental has undergone a 360° phase change, whereas the third harmonic has undergone a phase change of 450°. The relative phase difference between the fundamental and the third harmonic is now 90°. The net effect is that this difference in velocity causes a change in phase relationship which in turn causes the shape of the resultant wave to be different to that at the transmitter.

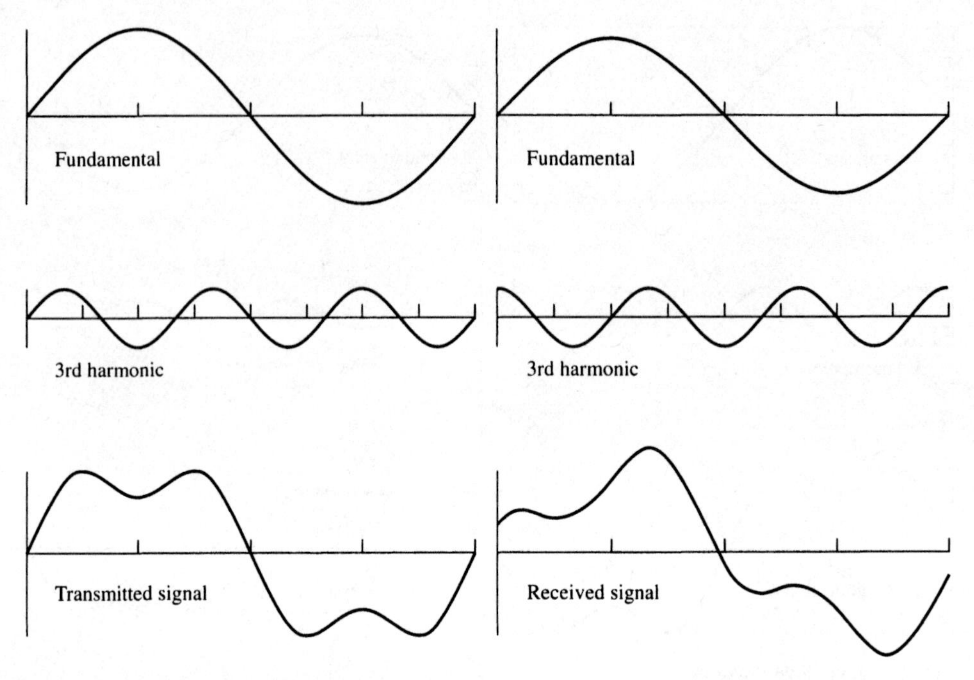

Fig. 7.2 Frequency distortion

7.4 AMPLITUDE AND FREQUENCY DISTORTION

Figure 7.3 shows the effect on the same two frequencies when both amplitude and frequency distortion take place. Notice the difference in shape of the resultant

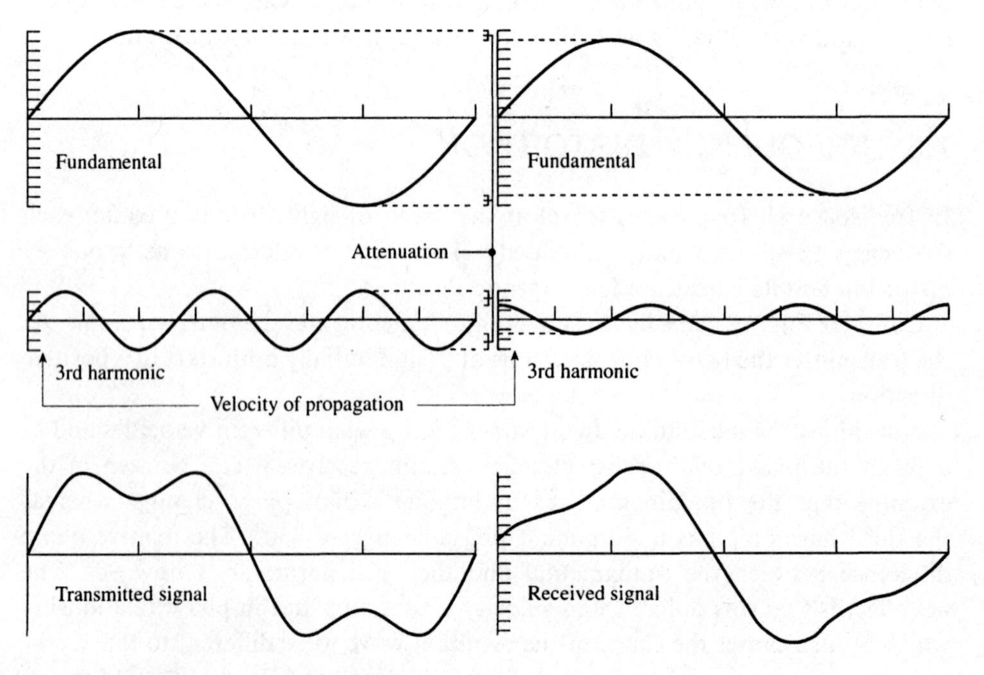

Fig. 7.3 Amplitude and frequency distortion

wave at the transmitter and receiver. The received signal looks totally different to that at the transmitter. The received signal now also contains frequency components that were not present in the original signal transmitted by the transmitter.

7.5 LIMITED BANDWIDTH

Any transmission medium can only accept a particular frequency range. A true digital signal consists of a fundamental frequency and an infinite number of harmonics. When this signal is transmitted over the medium, only those frequencies that are within the frequency range of the medium will propagate down the medium and reach the receiver. The received signal will now look different to that at the transmitter.

7.6 EFFECTS OF NOISE

Noise affects both analogue and digital signals. Digital signals, however, are more robust to the effects of noise than analogue signals.

7.6.1 Effect of noise on analogue transmission systems

As the signal is transmitted from the transmitter it is noise free. At the first repeater a certain amount of attenuation has taken place and a certain amount of noise is added. The noise is a result of external frequencies being induced into the transmission medium. As the signal passes through the repeater, the signal is amplified but so is the noise. The signal-to-noise ratio at the output of the first repeater is worse than that at the output of the transmitter.

Each section of transmission line will have a certain amount of noise induced into it. As a result the more repeaters there are, the worse the signal-to-noise ratio becomes. Eventually the signal will be totally swamped by noise. This is then the limiting factor on the total distance that the analogue system can work over.

There are two different ways of thinking about the noise and its effects on the analogue signal. Both of these philosophies are shown in Fig. 7.4. Philosophy 1 implies that the noise induced into the section can be replaced by a single noise source placed at the output of the amplifier that amplifies the signal from the previous section. The section now becomes noise free. This noise will be attenuated as it propagates down the section in the same way and by the same amount as the signal. However, each repeater amplifier will amplify the signal and the noise at the input to the amplifier, but more noise is added at the output of the amplifier, which is caused by the noise induced into that section of the medium.

As shown in Fig. 7.4 the amount of noise at the input to the next amplifier is greater than what it was at the input to the previous amplifier. Also the amplifier

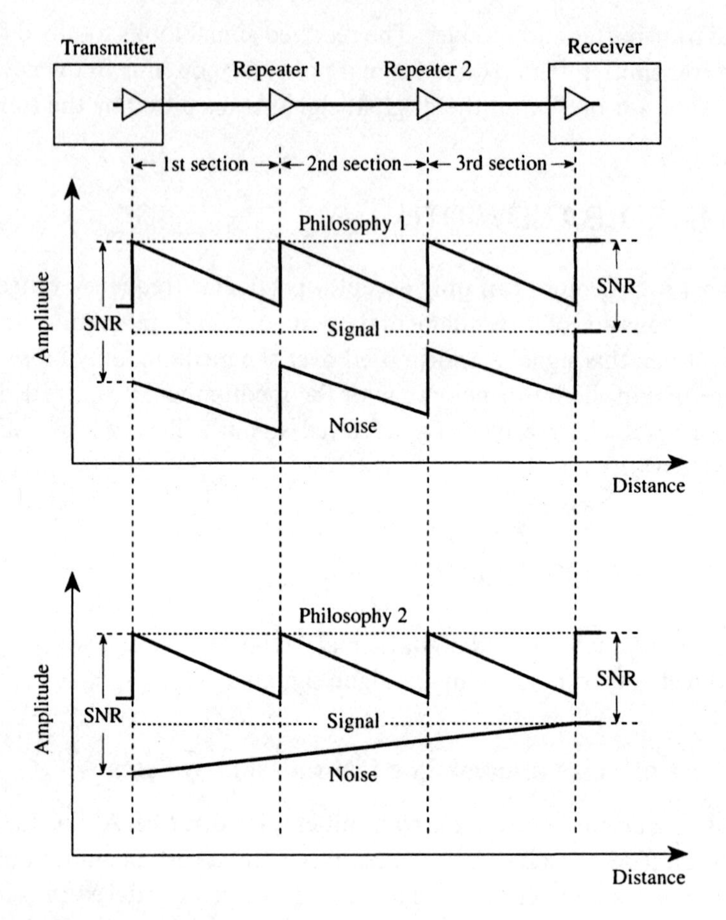

Fig. 7.4 Noise effects in an analogue system

amplifies both the signal and the noise by the same amount. The net effect is that the signal-to-noise ratio at the distant receiver is much worse that what it was at the output of the transmitter.

Philosophy 2 treats the signal in exactly the same way as it was treated in philosophy 1, as can be seen in Fig. 7.4. However, the noise can be thought of as gently increasing the further the signal moves from the transmitter. The generation of the noise can be thought of as taking place along the full length of the medium and as a result the amplitude of the noise gently increases. The increase in amplitude also reflects the effect of the amplifiers, which is now introduced along the full length of the medium instead of at discrete points. The net effect is the same in that the signal-to-noise ratio at the receiver is much worse than what it was at the output of the transmitter.

7.6.2 Effect of noise on digital transmission systems

A regenerative repeater recreates a brand new signal from the old signal but leaves behind all the problems that the old signal had. The signal undergoes attenuation

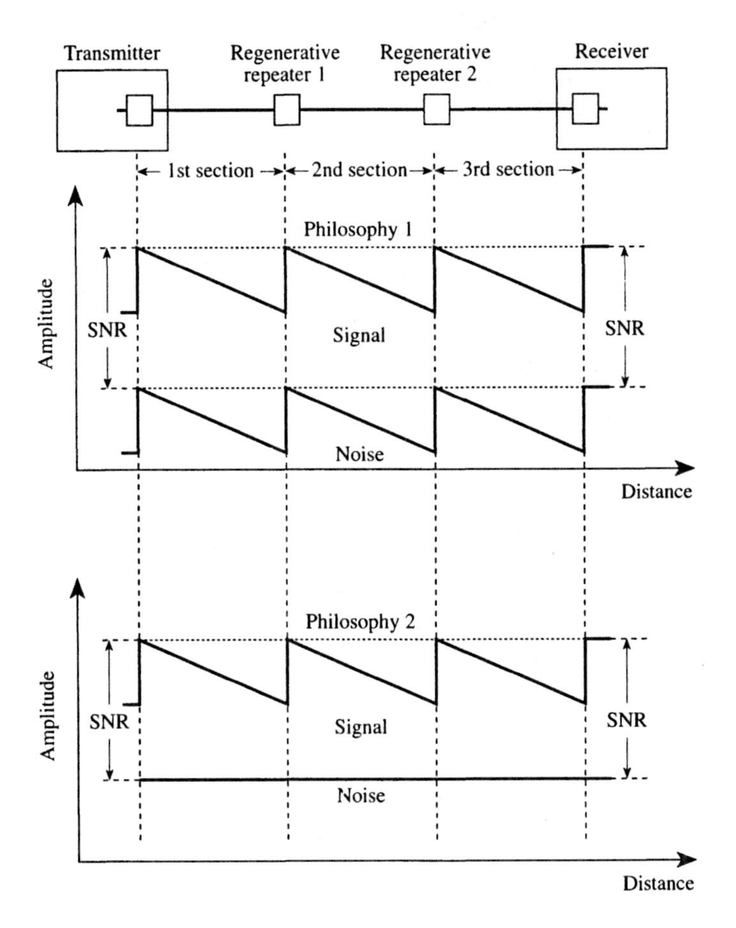

Fig. 7.5 Noise effects in a digital system

between the transmitter and the first regenerator. Along the section noise is added to the signal. At the regenerator only the signal is regenerated. This means that at the output the signal is clean and the noise in the previous section remains in that section and is not carried through the regenerator.

The result of this is that the signal-to-noise ratio at the output of the regenerator is identical to that at the output of the transmitter. Theoretically the distance over which a digital system can work is not limited by the signal-to-noise ratio as is the case in an analogue system.

This is shown in Fig. 7.5. Again the two philosophies can be applied. However, as shown, the signal-to-noise ratio at the receiver is exactly the same as what it was at the distant transmitter.

7.6.3 Transmission effects on a digital signal

Figure 7.6 shows how a digital signal is affected by a transmission medium and by noise. Notice that the attenuation affects and limited bandwidth cause the medium to act as a low-pass filter.

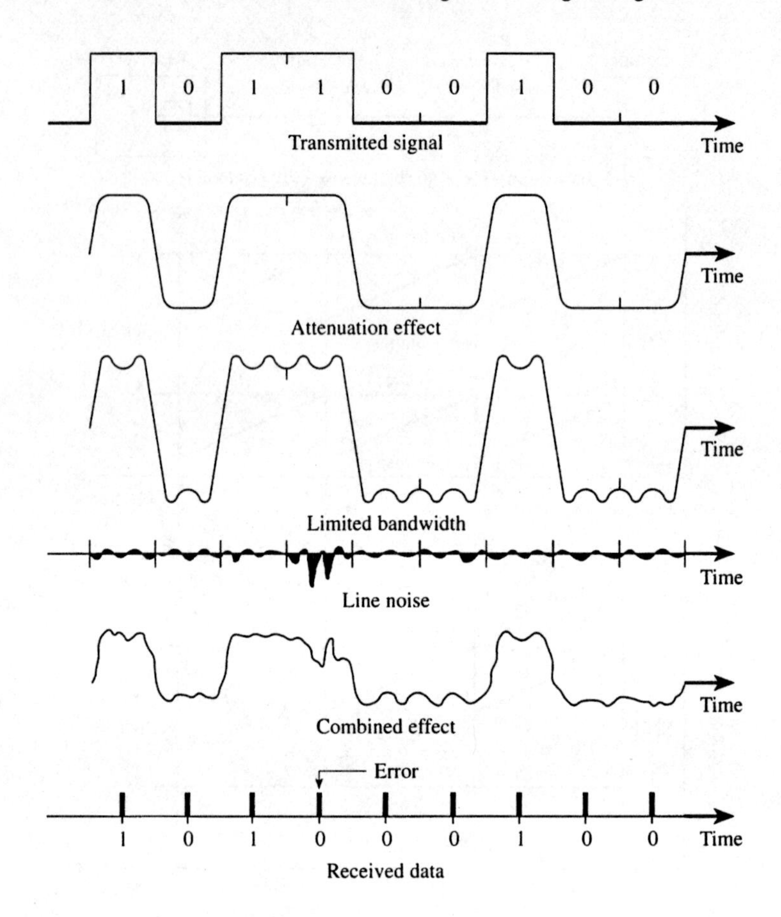

Fig. 7.6 Attenuation and distortion sources

If line noise is now added to the signal together with frequency distortion a very distorted signal is received at the receiver. As shown in Fig. 7.6, if there are high noise peaks in the correct places then the data can be decoded incorrectly creating errors in the received data stream. As shown in the example the transmitted data is 101100100_2 but because of the noise and the effects of the medium the received bit stream is 101000100_2.

7.6.4 Intersymbol interference

Another problem that confronts digital transmission is pulse spreading or pulse stretching. As a digital pulse propagates down the transmission medium it tends to be stretched out; this is due to the limited bandwidth that the signal must be conveyed through. This effect is shown in Fig. 7.7.

Pulse stretching can result in intersymbol interference. Depending on the modulation scheme used each pulse could represent a number of bits and hence is referred to as a symbol. Intersymbol interference can also result in errored data

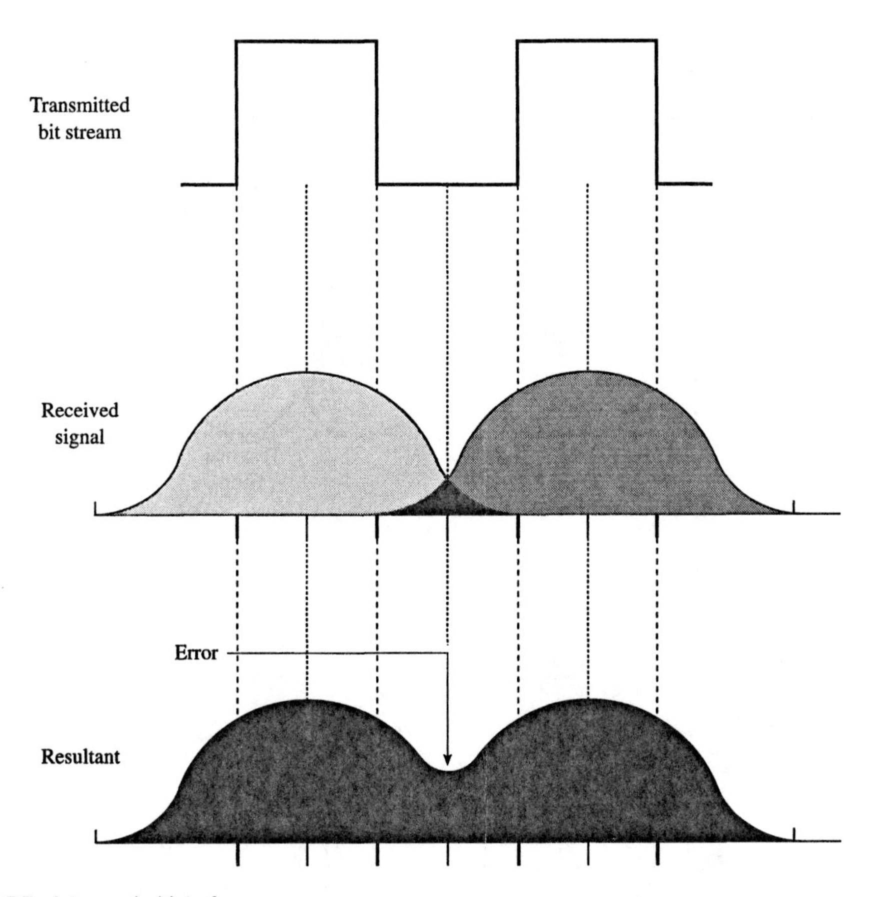

Fig. 7.7 Intersymbol interference

at the receiver because the resultant voltage level due to the stretching could cause a logic 0 to be decoded as a logic 1 instead.

7.6.5 The eye diagram

Intersymbol interference can very easily be detected in the eye diagram. Figure 7.8 shows different eye diagrams. The eye diagram is created by the two different logic states being overlaid upon one another. At the transmitter the data consists of rectangular pulses. But as the data is continually changing a square shape would result. The change in data is due to the fact that the signal used to create the data stream (i.e. speech) is continually changing.

As the signal is conveyed over a limited bandwidth channel, the elliptical shape as shown in Fig. 7.8 should result. Now if there is very little intersymbol interference then the eye diagrams would be very open; however, the eye diagram would appear to close as the intersymbol interference increased.

Most digital systems are equipped with a test point at which the eye diagram can be viewed on an oscilloscope. Intersymbol interference causes bit errors to

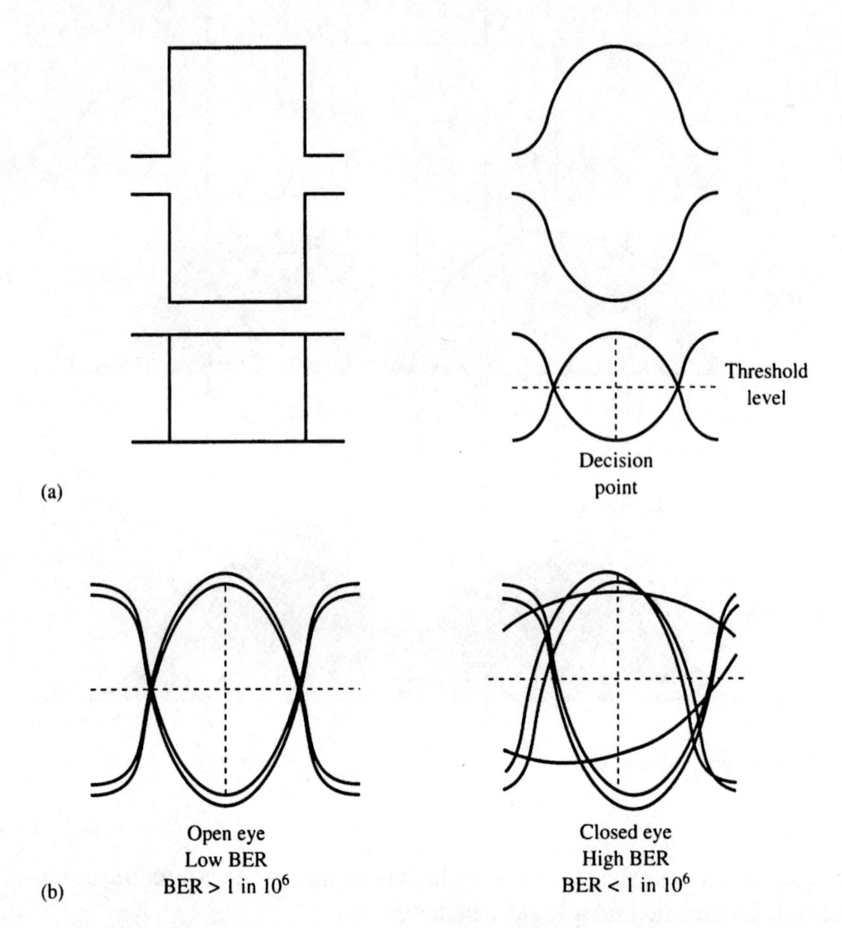

Fig. 7.8 The eye diagram: (a) construction; (b) typical diagrams

occur. A large amount of intersymbol interference causes the bit error rate (BER) to increase.

7.6 REVIEW QUESTIONS

7.1 Describe, using diagrams, what is meant by amplitude distortion.
7.2 Describe, using diagrams, what is meant by frequency distortion.
7.3 Describe, using a diagram, the effect of noise on an analogue transmission system.
7.4 Describe, using a diagram, the effect of noise on a digital transmission system.
7.5 Describe, using diagrams, what is meant by intersymbol interference.

8 Determination of bit error rates

> Noise prevents man from being able to concentrate and think of solutions to problems.

8.1 INTRODUCTION

Coding of data into a format that can efficiently be transported over a medium is the subject of this chapter. Most data streams have a certain amount of redundancy. For the efficient transmission of the actual required data it is important to remove as much redundancy as possible. In many circumstances headers, footers, error codes etc. are added to the data stream. This added data causes the redundancy of the data to increase. Thus more time is required to transport the data over the medium. This chapter introduces some of the concepts behind reducing the redundancy of the transmitted data.

In Chapter 7 the concept of noise and its effects on digital signals was introduced. This chapter extends this concept and deals with the techniques used to determine the probable bit error rate on certain types of data codes.

Any message or piece of text contains character strings, semantics and syntax:

- Characters – letters of the alphabet, numbers 0 to 9.
- Syntax – punctuation etc.
- Semantics – meaning.

Communication systems are concerned with conveying the character strings and syntax characters. A communication system does not try to understand or interpret the meaning of the text and hence it is not concerned with semantics.

All data has some redundancy. Redundancy is that data which has no meaning but accompanies the data that has meaning. An example of redundancy in the English language is the following sentence:

The boy and the dog went for a walk.

In this sentence the words conveying meaning are 'boy', 'dog', 'went', and 'walk'. The other words 'the', 'and', 'the', 'for' and 'a' are redundant in terms of the meaning of the sentence but are necessary to make the sentence grammatically correct.

8.2 ENTROPY

The entropy function is used to find the ideal amount of data that should be transmitted to enable the data to be efficiently transmitted without transmitting the redundant data.

Entropy can be defined as the average amount of information per source input. The entropy function is given as follows:

$$S(x) = -\sum_{i=1}^{n} P(x_i) \log_2 P(x_i)$$

where $P(x_i)$ = probability of the character designated as x_i.

Consider a transmission system that is to transmit the first six characters of the alphabet only. Each letter will be expressed as a digital signal. By convention each letter would be allocated 3 bits. This could be as shown in Table 8.1. Here the codes 110 and 111 are both redundant as they are not used. If there were only four characters then each character could be allocated 2 bits with no redundancy. On the other hand, if there were eight characters then each character could be allocated 3 bits with no redundancy. In this example, however, there is redundancy and $6 \times 3 = 18$ bits are required to transmit all six characters.

The entropy function is used to try to find the most economical way of transmitting the data. To use the entropy function the probability of each character must be known. Consider in the example that each character has equal probability of appearing in the text to be transmitted. This means that each character has a probability of 1/6 or 0.1667. Therefore $P(x_i) = 0.1667$.

$$\begin{aligned}
S(x) &= -\sum_{i=1}^{n} P(x_i) \log_2 P(x_i) \\
&= -\sum_{i=1}^{6} 0.1667 \log_2 0.1667 \\
&= -(6 \times 0.1667 \log_2 0.1667) \\
&= -\log_2 0.1667 \\
&= -\frac{\log_{10} 0.1667}{\log_{10} 2} \\
&= -(-2.585) \\
&= 2.585 \text{ bits/character}
\end{aligned}$$

This means that all six characters can be transmitted by a transmission system using $6 \times 2.585 = 15.51$ bits. Relative to the previous method there is a distinct saving of $18 - 15.51 = 2.49$ bits.

In practice each character in the alphabet does not have equal probability. Some characters appear in a piece of text far more often than other characters. Example 8.1 demonstrates this point.

Table 8.1 Three-bit character codes

Character	Digital code
A	000
B	001
C	010
D	011
E	100
F	101
	110
	111

Example 8.1

Determine the ideal number of bits that should be allocated to each of the following characters with the probabilities given:

x_i	$P(x_i)$
A	0.43
B	0.25
C	0.14
D	0.09
E	0.06
F	0.03

Solution

The entropy function is given as follows:

$$S(x) = -\sum_{i=1}^{n} P(x_i)\log_2 P(x_i)$$

$$= -(0.43\log_2 0.43 + 0.25\log_2 0.25 + 0.14\log_2 0.14$$

$$+ 0.09\log_2 0.09 + 0.06\log_2 0.06 + 0.03\log_2 0.03)$$

$$= 2.129 \text{ bits/character}$$

In this case the message can be most efficiently be transmitted using 2.129 bits/character.

Obviously it is not possible to transmit 0.129 of a bit. Instead only complete bits can be transmitted. In Chapter 9 coding is dealt with in more detail and the mechanisms used to enable data to be transmitted efficiently are discussed.

8.3 CAUSES OF ERRORS ON DIGITAL SIGNALS

There are a number of causes of errors on digital signals. These are:

- Noise.
- Limited bandwidth.
- Bit stream rate.

The noise that is referred to here is line noise, discussed in Chapter 7.

8.3.1 Limited bandwidth

Any transmission medium can be represented by the circuit shown in Fig. 8.1. When the inductance and capacitance effects only are considered, it can be seen that the line acts as a low-pass filter.

The input circuitry to any circuit normally has a filter. This filter limits the bandwidth of the signal that is allowed to pass on to the next stage. The input filter and the filter effect of the medium result in a band-limited channel that the digital signal passes over.

A digital signal consists of a fundamental frequency and an infinite number of harmonic frequencies. The band-limiting effect of the circuit prevents a large number of these harmonic frequencies from reaching the distant receiver.

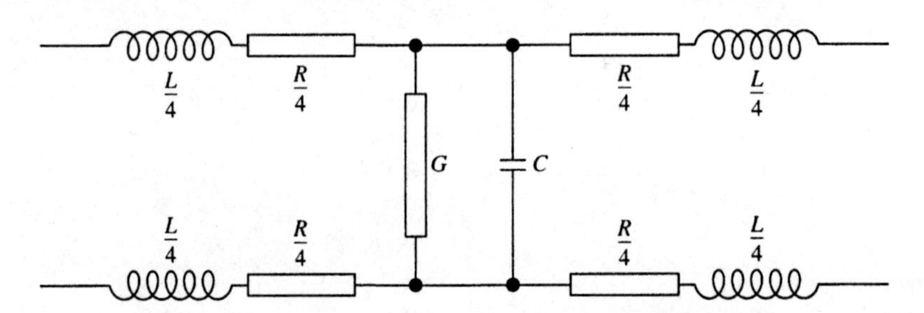

Equivalent of a 1 km section of a transmission medium

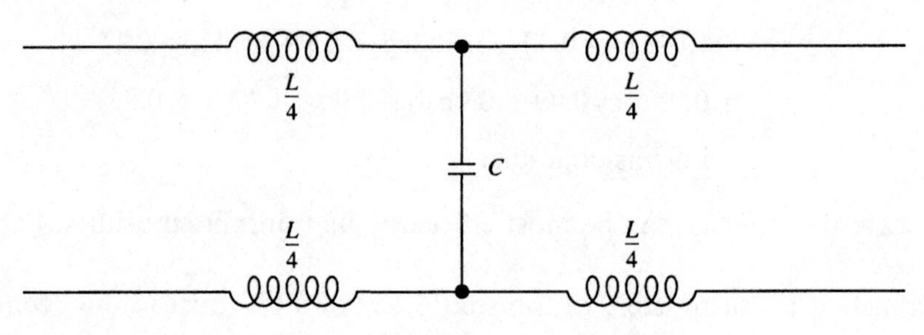

Equivalent low-pass filter

Fig. 8.1 Transmission medium

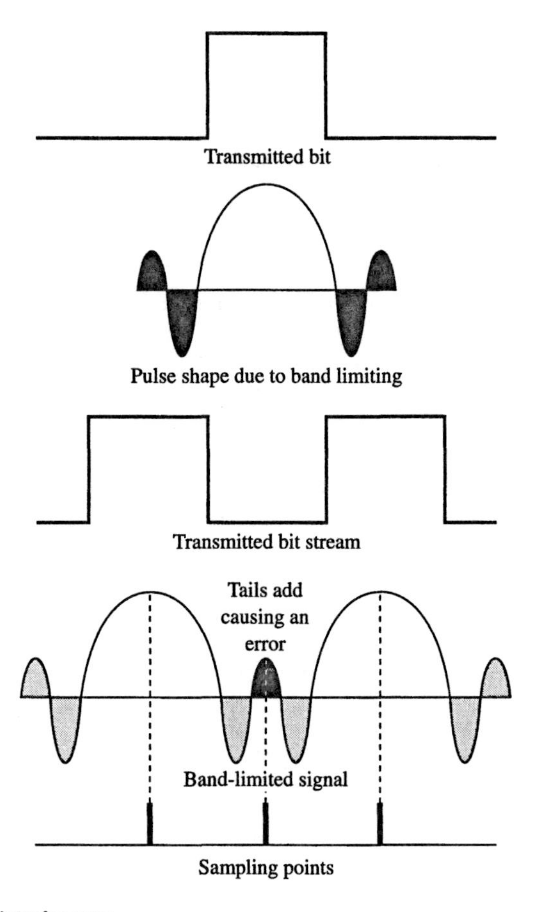

Fig. 8.2 Intersymbol interference

When the band-limiting effect of the channel is taken into account then pulse spreading results. Figure 8.2 shows how the tails of a bit affect the previous and post bits. These tails can cause errors in the bit stream.

8.4 PROBABILITY OF BIT ERROR RATE

8.4.1 Gaussian white noise

Gaussian white noise is noise that is equally spread across the whole frequency spectrum. This means that it affects all frequencies equally. However, over a transmission medium the signal suffers attenuation. Unfortunately the attenuation is not the same for all frequencies in the signal and on a metallic pair the higher the frequency the greater the attenuation.

Because of this effect the Gaussian white noise affects the higher frequencies more than the lower frequencies on a metallic pair. The Gaussian white noise is now referred to as triangular noise. Figure 8.3 shows this effect. Gaussian white

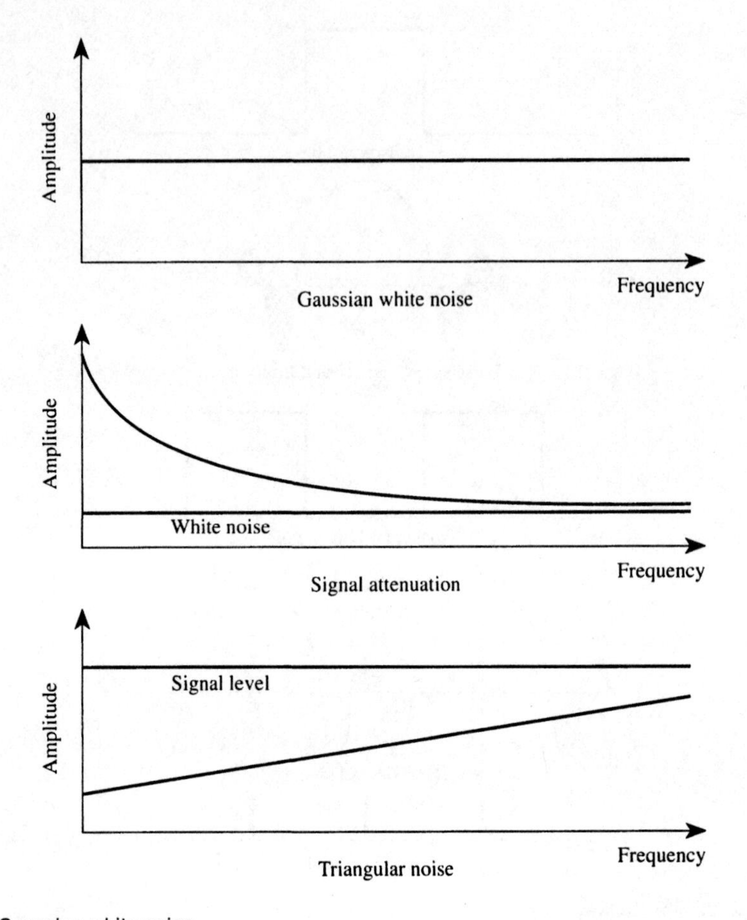

Fig. 8.3 Gaussian white noise

noise is normally measured as the amount of noise per hertz. It is then quite simple to calculate the amount of noise in the actual frequency range.

8.4.2 Probability of a bit error

Consider the following equation:

$$\frac{E_b}{N_o} = \frac{ST}{N_o}$$

where E_b = signal energy (J/bit)
N_o = Gaussian white noise (W/Hz)
S = signal power (W)
T = bit time duration (s)

Using

$$T = \frac{1}{C}$$

where C = bit rate (b/s)

$$\frac{E_b}{N_o} = \frac{S}{CN_o}$$

but

$$N = BN_o$$

where N = noise power (W)
B = Bandwidth (Hz)

Therefore

$$\frac{E_b}{N_o} = \frac{S}{C\frac{N}{B}}$$

8.4.3 Data rates and signal-to-noise ratio

The equation developed in Section 8.4.2 can be expressed as

$$\frac{E_b}{N_o} = \text{SNR}\frac{B}{C}$$

This equation can be modified to express the bit rate C in terms of the other variables:

$$C = \text{SNR}\,B\frac{N_o}{E_b}$$

In this equation the signal-to-noise ratio must be expressed as a ratio and not in dB.

Example 8.2

Given that:

$$\text{Gaussian white noise} = 1 \times 10^{-11}\ \text{W/Hz}$$
$$\text{Bandwidth} = 4\ \text{kHz}$$
$$\text{Signal energy} = 20 \times 10^{-11}\ \text{J}$$

Determine the bit rate if the signal-to-noise ratio is (a) 40 dB, (b) 20 dB and (c) 10 dB.

Solution

(a) $\text{SNR} = 40\ \text{dB} = 10\,000$

$B = 4\ \text{kHz}$

$E_b = 20 \times 10^{-11}$

$N_o = 1 \times 10^{-11}$

$$C = \text{SNR} \times B\frac{N_o}{E_b}$$

$$= 10\,000 \times 4000 \times \frac{1 \times 10^{-11}}{20 \times 10^{-11}}$$

$$= 2\,\text{Mb/s}$$

(b) $\text{SNR} = 20\,\text{dB} = 100$

$B = 4\,\text{kHz}$

$E_b = 20 \times 10^{-11}$

$N_o = 1 \times 10^{-11}$

$$C = \text{SNR} \times B\frac{N_o}{E_b}$$

$$= 100 \times 4000 \times \frac{1 \times 10^{-11}}{20 \times 10^{-11}}$$

$$= 20\,\text{kb/s}$$

(c) $\text{SNR} = 10\,\text{dB} = 10$

$B = 4\,\text{kHz}$

$E_b = 20 \times 10^{-11}$

$N_o = 1 \times 10^{-11}$

$$C = \text{SNR} \times B\frac{N_o}{E_b}$$

$$= 10 \times 4000 \times \frac{1 \times 10^{-11}}{20 \times 10^{-11}}$$

$$= 2\,\text{kb/s}$$

From this example it can be seen that the signal-to-noise ratio drastically affects the maximum bit rate that can be transmitted over the circuit.

8.4.4 Unipolar bit stream

For unipolar bit streams the following equation applies:

$$P(b) = \frac{1}{x\sqrt{2\pi}}e^{-(x^2/2)} \quad \text{for } x \geq 3$$

where

$$x = \sqrt{\frac{E_b}{N_o}}$$

$$= \sqrt{\text{SNR}\frac{B}{C}}$$

Example 8.3

A unipolar bit stream having a signal energy of 9×10^{-9} J and Gaussian white noise of 1×10^{-9} W/Hz. Determine the probable bit error rate.

Solution

$$x = \sqrt{\frac{E_b}{N_o}}$$

$$= \sqrt{\frac{9 \times 10^{-9}}{1 \times 10^{-9}}} = \sqrt{9} = 3$$

$$P(b) = \frac{1}{x\sqrt{2\pi}} e^{-(x^2/2)} \quad \text{for } x \geq 3$$

$$= \frac{1}{3\sqrt{2\pi}} e^{-(3^2/2)}$$

$$= 1.47 \times 10^{-3} \approx 1.5 \times 10^{-3}$$

This means that there is a probability that 1.5 bits out of every 1000 will be errored.

8.4.5 Bipolar bit stream

For bipolar bit streams the following equation applies:

$$P(b) = \frac{1}{x\sqrt{2\pi}} e^{-(x^2/2)} \quad \text{for } x \geq 3$$

where $x = \sqrt{\frac{2E_b}{N_o}}$

$$= \sqrt{2\,\text{SNR}\,\frac{B}{C}}$$

Example 8.4

A bipolar bit stream has a signal energy of 9×10^{-9} J and Gaussian white noise of 1×10^{-9} W/Hz. Determine the probable bit error rate.

Solution

$$x = \sqrt{\frac{2E_b}{N_o}}$$

$$= \sqrt{\frac{2 \times 9 \times 10^{-9}}{1 \times 10^{-9}}} = \sqrt{18} = 4.243$$

$$\begin{aligned} P(b) &= \frac{1}{x\sqrt{2\pi}} e^{-(x^2/2)} \quad \text{for } x \geq 3 \\ &= \frac{1}{4.243\sqrt{2\pi}} e^{-(4.243^2/2)} \\ &= 11.6 \times 10^{-6} \approx 12 \times 10^{-6} \end{aligned}$$

This means that there is a probability that 12 bits out of every 1 000 000 will be errored.

8.5 SHANNON AND HARTLEY CAPACITY THEOREM

The Shannon and Hartley capacity theorem is used to determine the maximum bit rate that can be used over a band-limited channel having a specific signal-to-noise ratio. The theorem is as follows:

$$C = B\log_2(1 + \text{SNR}) \text{ b/s}$$

where the SNR must be expressed as a ratio and not in dB.

Example 8.5

Determine the maximum bit rate that can be transmitted over a channel having a bandwidth of 4 kHz and a signal-to-noise ratio of 48 dB.

Solution

$$\begin{aligned} \text{SNR} &= 48\text{ dB} = 63\,095.73 \\ C &= B\log_2(1 + \text{SNR}) \text{ b/s} \\ &= 4000\log_2(1 + 63\,095.73) \text{ b/s} \\ &= 63.781 \text{ kb/s} \approx 64 \text{ kb/s} \end{aligned}$$

Example 8.6

Determine the maximum bit rate that can be transmitted over a channel having a bandwidth of 15 kHz with a signal-to-noise ratio at the receiver of 42 dB.

Solution

Step 1. Convert the SNR in dB to SNR as a ratio.

$$\begin{aligned} \text{SNR}_{\text{dB}} &= 10\log_{10}\text{SNR}_{\text{ratio}} \\ \text{SNR}_{\text{ratio}} &= 10^{\text{SNR}_{\text{dB}}/10} \\ &= 10^{42/10} \\ &= 15\,849 \end{aligned}$$

Step 2. Determine the maximum possible bit rate.

$$\begin{aligned} C &= B\log_2(1+\text{SNR})\ \text{b/s} \\ &= 15\,000\log_2(1+15\,849)\ \text{b/s} \\ &= 15\,000\log_2(15\,850)\ \text{b/s} \\ &= 15\,000\frac{\log_{10}(15\,850)}{\log_{10}2}\ \text{b/s} \\ &= 209.283\ \text{kb/s} \end{aligned}$$

Example 8.7

A unipolar bit stream having a bit rate of 139.264 Mb/s is transmitted down a transmission medium having a launch level of 0 dBm. At the receiver after regeneration the signal level is 0 dBm. The signal-to-noise ratio at the receiver is 30 dB.

Determine the following:

1. The minimum bandwidth required.
2. The probable bit error rate.
3. The probable bit error rate if the bandwidth is limited to 1.26 MHz.
4. The probable bit error rate if the bandwidth in (1) is used but the signal-to-noise ratio is reduced to 20 dB.

Solution

1. Determine the minimum bandwidth.

$$C = B\log_2(1+\text{SNR})$$

$$B = \frac{C}{\log_2(1+\text{SNR})}$$

$\text{SNR} = 30\ \text{dB} = 1000_{\text{ratio}}$, therefore

$$\begin{aligned} B &= \frac{139.264\times10^6}{\log_2(1+1000)} \\ &= 13.97\ \text{MHz} \approx 14\ \text{MHz} \end{aligned}$$

2. Probable bit error rate:

$$B = 14\ \text{MHz},\quad C = 139.264\ \text{Mb/s}$$

$$\begin{aligned} x &= \sqrt{\frac{E_b}{N_o}} = \sqrt{\text{SNR}\frac{B}{C}} \\ &= \sqrt{1000\frac{14\times10^6}{139.264\times10^6}} \\ &= 10.026 \end{aligned}$$

$$P(b) = \frac{1}{x\sqrt{2\pi}} e^{-(x^2/2)}$$

$$= \frac{1}{10.026\sqrt{2\pi}} e^{-(10.026^2/2)}$$

$$= 5.92 \times 10^{-24} \approx 6 \times 10^{-24}$$

3. Probable bit error rate:

$$B = 1.26\,\text{MHz}, \quad \text{SNR} = 30\,\text{dB} = 1000$$

$$x = \sqrt{\frac{E_b}{N_o}} = \sqrt{\text{SNR}\frac{B}{C}}$$

$$= \sqrt{1000\frac{1.26 \times 10^6}{139.264 \times 10^6}}$$

$$= 3.008$$

$$P(b) = \frac{1}{x\sqrt{2\pi}} e^{-(x^2/2)}$$

$$= 1.44 \times 10^{-3}$$

4. Probable bit error rate:

$$B = 14\,\text{MHz}, \quad \text{SNR} = 20\,\text{dB} = 100$$

$$x = \sqrt{\frac{E_b}{N_o}} = \sqrt{\text{SNR}\frac{B}{C}}$$

$$= \sqrt{100\frac{14 \times 10^6}{139.264 \times 10^6}}$$

$$= 3.17$$

$$P(b) = \frac{1}{x\sqrt{2\pi}} e^{-(x^2/2)}$$

$$= \frac{1}{3.17\sqrt{2\pi}} e^{-(3.17^2/2)}$$

$$= 0.827 \times 10^{-3} \approx 0.83 \times 10^{-3}$$

Example 8.8

A bipolar bit stream having a bit rate of 139.264 Mb/s is transmitted down a transmission medium having a launch level of 0 dBm. At the receiver after regeneration the signal level is 0 dBm. The signal-to-noise ratio at the receiver is 30 dB.

Determine the following:

1. The minimum bandwidth required.
2. The probable bit error rate.
3. The probable bit error rate if the bandwidth is limited to 1.26 MHz.
4. The probable bit error rate if the bandwidth in (1) is used but the signal-to-noise ratio is reduced to 20 dB.

Solution

1. Determine the minimum bandwidth.

$$C = B\log_2(1 + \text{SNR})$$

$$B = \frac{C}{\log_2(1 + \text{SNR})}$$

$\text{SNR} = 30\,\text{dB} = 1000_{\text{ratio}}$, therefore

$$B = \frac{139.264 \times 10^6}{\log_2(1 + 1000)}$$

$$= 13.97\,\text{MHz} \approx 14\,\text{MHz}$$

2. Probable bit error rate:

$$B = 14\,\text{MHz}, \quad C = 139.264\,\text{Mb/s}$$

$$x = \sqrt{\frac{2E_\text{b}}{N_\text{o}}} = \sqrt{2\text{SNR}\frac{B}{C}}$$

$$= \sqrt{2 \times 1000\frac{14 \times 10^6 B}{139.264 \times 10^6}}$$

$$= 14.18$$

$$P(b) = \frac{1}{x\sqrt{2\pi}}\text{e}^{-(x^2/2)}$$

$$= \frac{1}{14.18\sqrt{2\pi}}\text{e}^{-(14.18^2/2)}$$

$$= 612.23 \times 10^{-48} \approx 612 \times 10^{-48}$$

3. Probable bit error rate:

$$B = 1.26\,\text{MHz}, \quad \text{SNR} = 30\,\text{dB} = 1000$$

$$x = \sqrt{\frac{2E_\text{b}}{N_\text{o}}} = \sqrt{2\text{SNR}\frac{B}{C}}$$

$$= \sqrt{2 \times 1000\frac{1.26 \times 10^6}{139.264 \times 10^6}}$$

$$= 4.254$$

$$P(b) = \frac{1}{x\sqrt{2\pi}} e^{-(x^2/2)}$$

$$= \frac{1}{4.254\sqrt{2\pi}} e^{-(4.254^2/2)}$$

$$= 11.03 \times 10^{-6} \approx 11 \times 10^{-6}$$

4. Probable bit error rate:

$$B = 14\,\text{MHz}, \quad \text{SNR} = 20\,\text{dB} = 100$$

$$x = \sqrt{\frac{2E_b}{N_o}} = \sqrt{2\text{SNR}\frac{B}{C}}$$

$$= \sqrt{2 \times 100 \frac{14 \times 10^6 B}{139.264 \times 10^6}}$$

$$= 4.484$$

$$P(b) = \frac{1}{x\sqrt{2\pi}} e^{-(x^2/2)}$$

$$= \frac{1}{4.484\sqrt{2\pi}} e^{-(4.484^2/2)}$$

$$= 3.83 \times 10^{-6} \approx 4 \times 10^{-6}$$

8.6 COMPARISON OF UNIPOLAR AND BIPOLAR BIT STREAMS

Table 8.2 shows the comparison between results from Examples 8.7 and 8.8. As can be seen, the bipolar bit stream performs far better than the unipolar bit stream. By comparing different values for the SNR it can be seen that, for both codes, the worse the SNR becomes the worse the probable bit error rate becomes. Hence in practice the received SNR does play a part in the quality of the received signal and cannot be ignored.

Table 8.2 Comparison of results from Examples 8.7 and 8.8

Bit stream	SNR (dB)	*B* (MHz)	*P*(*b*)
Unipolar	30	14	6×10^{-24}
	30	1.26	1.4×10^{-3}
	20	14	0.83×10^{-3}
Bipolar	30	14	612×10^{-48}
	30	1.26	11×10^{-6}
	20	14	4×10^{-6}

ate the effects of limiting the bandwidth too
ıe SNR the probable bit error rate increases

m, what is meant by Gaussian white noise
o as triangular noise.
bits that should be allocated to each of
nber of times the character appears in a
brackets after the character.

A (25), B (32), C (12), D (4), E (64), F (18)

8.3 Given that:

$$\text{Gaussian white noise} = 1 \times 10^{-11}\ \text{W/Hz}$$

$$\text{Bandwidth} = 4\,\text{kHz}$$

$$\text{Signal energy} = 20 \times 10^{-11}\ \text{J}$$

determine the bit rate if the signal-to-noise ratio is (a) 35 dB, (b) 15 dB and (c) 8 dB.

8.4 A unipolar bit stream has a signal energy of 4.5×10^{-9} J/bit and Gaussian white noise of 3.5×10^{-10} W/Hz. Determine the probable bit error rate.

8.5 A bipolar bit stream has a signal energy of 4.5×10^{-9} J/bit and Gaussian white noise of 3.5×10^{-10} W/Hz. Determine the probable bit error rate.

8.6 Determine, using the Shannon and Hartley capacity theorem, the maximum bit rate that can be transmitted over a channel having a bandwidth of 4 kHz and a signal-to-noise ratio of 38 dB.

9 Source coding techniques

> Mankind must be able to make sense of received information for it to be meaningful.

9.1 INTRODUCTION

Coding of data into a format that can efficiently be transported over a medium is the subject of this chapter. Most data streams have a certain amount of redundancy. For data to be efficiently transported all the redundancy should be removed.

The second important criterion is the accuracy of the received data. It is no good receiving a compressed data stream if, during transmission, the data becomes errored, resulting in a high error rate at the receiver. Some of the error detection and error correction schemes that are employed in modern communication systems are discussed in this chapter.

Data can be transferred asynchronously or synchronously. In both these methods error detection is very necessary. There are many techniques that are used to enable error detection to take place. Some methods enable error correction to take place as well.

All error detection schemes have some disadvantages and all error detection schemes fail to detect all types of errors. However, some methods outperform others with respect to redundancy and efficiency.

9.2 ASYNCHRONOUS AND SYNCHRONOUS TRANSMISSION

Asynchronous transmission means that the transmitted data stream is not continuous. In other words the data stream is only present when there is data to transmit. Even then the data stream need not be continuous. Each data block can be transmitted separately with a break before the next data block is transmitted. The break between the data blocks need not be constant and can be of any duration. The data block can consist of a page of text, a paragraph of text, a single word or even a single character as in the case of the transmission of ASCII characters over an asynchronous, serial port.

9.3 CODES USED ON COMPUTERS

Over the years many different types of coding techniques have been used to encode alphanumeric characters. Currently there are two accepted codes that are specifically used on computers. These two types are ASCII and EBCDIC.

9.3.1 American Standards Institute Code for Information Interchange (ASCII)

The American Standards Institute Code for Information Interchange or ASCII was developed specifically to enable standardisation in the computer industry.

Table 9.1 The ASCII code

	Hex code					0	1	2	3	4	5	6	7
	Bits				5	0	1	0	1	0	1	0	1
Hex code					6	0	0	1	1	0	0	1	1
	4	3	2	1	7	0	0	0	0	1	1	1	1
0	0	0	0	0		NUL	DLE	SP	0	@	P	\	p
1	0	0	0	1		SOH	DC1	!	1	A	Q	a	q
2	0	0	1	0		STX	DC2	"	2	B	R	b	r
3	0	0	1	1		ETX	DC3	#	3	C	S	c	s
4	0	1	0	0		EOT	DC4	$	4	D	T	d	t
5	0	1	0	1		ENQ	NAK	%	5	E	U	e	u
6	0	1	1	0		ACK	SYN	&	6	F	V	f	v
7	0	1	1	1		BEL	ETB	'	7	G	W	g	w
8	1	0	0	0		BS	CAN	(	8	H	X	h	x
9	1	0	0	1		HT	EM	)	9	I	Y	i	y
A	1	0	1	0		LF	SUB	*	:	J	Z	j	z
B	1	0	1	1		VT	ESC	+	;	K	[	k	{
C	1	1	0	0		FF	FS	,	<	L	\	l	|
D	1	1	0	1		CR	GS	-	=	M	]	m	}
E	1	1	1	0		SO	RS	.	>	N	^	n	~
F	1	1	1	1		SI	US	/	?	O	_	o	DEL

NUL = null, or all zeros; SOH = start of header; STX = start of text; ETX = end of text; EOT = end of transmission; ENQ = enquiry; ACK = acknowledge; BEL = bell or alarm; BS = backspace; HT = horizontal tabulation; LF = line feed; VT = vertical tabulation; FF = form feed; CR = carriage return; SO = shift out; SI = shift in;
DLE = data link escape; DC1 = device control 1; DC2 = device control 2; DC3 = device control 3; DC4 = device control 4; NAK = negative acknowledge; SYN = synchronous idle; ETB = end of transmission block; CAN = cancel; SUB = substitute; EM = end of medium; DEL = delete; ESC = escape; FS = file separator; GS = group separator; RS = record separator; US = unit separator; SP = space.

The ASCII code is widely used by many computer manufacturers worldwide. It is a 7-bit code with the possibility of a parity bit being added to form a byte. Table 9.1 indicates the coding for the various alphanumeric characters as well as control characters. The code can support up to 128 different characters and all of these 128 states are used in the coding.

The table is read from bit 7 to bit 1, where bit 7 is the MSB and bit 1 is the LSB. When the code is transmitted from a PC over an asynchronous port such as an RS232C then the LSB is transmitted first and the MSB last. A parity bit is normally added and sent out between the MSB and the stop bits. The 8-bit data stream is packaged between a single start bit and 1, 1.5 or 2 stop bits. The start bit is normally a logic 0 and the stop bits are normally a logic 1.

When the data is transmitted over a synchronous port, the data is not packaged between start and stop bits but is packaged in a frame between the start of text (STX) and end of text (ETX) commands. With this method the LSB is still transmitted first and the MSB last. Again a parity bit can be included after the MSB to make up an 8-bit binary word for each character.

Table 9.2 The EBCDIC code

BITS				b3	0	0	0	0	0	0	0	0	1	1	1	1	1	1	1	1
				b2	0	0	0	0	1	1	1	1	0	0	0	0	1	1	1	1
				b1	0	0	1	1	0	0	1	1	0	0	1	1	0	0	1	1
b7	b6	b5	b4	b0	0	1	0	1	0	1	0	1	0	1	0	1	0	1	0	1
0	0	0	0		NUL	SOH	STX	ETX		HT		DEL				VT	FF	CR	SO	SI
0	0	0	1		DLE	DC1	DC2	DC3	RES	NL	BS		CAN				IFS	IGS	IRS	IUS
0	0	1	0						BYP	LF	EOB							ENQ	ACK	BEL
0	0	1	1				SYN			DC4		EOT						NAK		SUB
0	1	0	0		SP										¢	.	<	(	+	
0	1	0	1		&										!	$	*	)	;	`
0	1	1	0		-	/										,	%	_	>	?
0	1	1	1												:	#	@	/	=	"
1	0	0	0			a	b	c	d	e	f	g	h	i		{				
1	0	0	1			j	k	l	m	n	o	p	q	r		}				
1	0	1	0				s	t	u	v	w	x	y	z				[		
1	0	1	1																	
1	1	0	0			A	B	C	D	E	F	G	H	I						
1	1	0	1			J	K	L	M	N	O	P	Q	R				]		
1	1	1	0				S	T	U	V	W	X	Y	Z						
1	1	1	1		0	1	2	3	4	5	6	7	8	9						

RES, NL not specifically allocated.

NUL = null, or all zeros; SOH = start of header; STX = start of text; ETX = end of text; HT = horizontal tabulation; DEL = delete; VT = vertical tabulation; FF = form feed; CR = carriage return; SO = shift out; SI = shift in.

DLE = data link escape; DC1 = device control 1; DC2 = device control 2; DC3 = device control 3; BS = backspace; CAN = cancel; IFS = file separator; IGS = group separator; IRS = record separator; IUS = unit separator.

BYP = bypass (escape); LF = line feed; EOB = end of block; ENQ = enquiry; ACK = acknowledge; BEL = bell.

SYN = synchronous idle; DC4 = device control 4; EOT = end of transmission; NAK = negative acknowledge; SUB = substitute; SP = space.

9.3.2 Extended binary coded decimal code (EBCDIC)

The extended binary coded decimal code (EBCBIC) was developed by IBM for standardisation on all their computers. This code is extensively used on many large computers today. The EBCDIC code is an 8-bit code that enables encoding of 256 different alphanumeric characters, control characters etc. Currently not all the codes are specifically allocated, as indicated in Table 9.2.

The code has one major disadvantage that there is no parity bit. In order for an error to be detected the same data needs to be transferred twice and the two received data streams must be compared.

9.4 HUFFMANN CODING

The problem with the ASCII and EBCDIC codes is that they still have redundancy. When trying to store or transmit data efficiently it is necessary to try to remove as much redundancy as possible. Huffmann coding is a technique that is used to achieve this. The technique is more commonly employed when storing data, known as 'zipping' the data up. Obviously the data must be 'unzipped' when it is retrieved.

Huffmann coding uses the entropy function, a tree diagram, a state diagram and a table. The following examples demonstrate the Huffmann coding technique.

Example 9.1

The following six letters, having the probabilities shown, must be digitally transmitted.

x_i	$P(x_i)$
A, B	0.3
C	0.13
D, E, F	0.09

Determine the following:

1. The minimum number of bits per character.
2. The Huffmann codes.
3. The compression rate.
4. The efficiency.

Solution

1. Minimum number of bits per character.

$$S(x) = -\sum_{i=1}^{n} P(x_i)\log_2 P(x_i)$$

$$S(x) = -(2 \times 0.3\log_2 0.3 + 0.13\log_2 0.13 + 3 \times 0.09\log_2 0.09)$$
$$= 2.363 \text{ bit/character}$$

2. The Huffmann codes are found from Fig. 9.1. A summary is given in Table 9.3. The codes given average out at 2.4 bits per character.

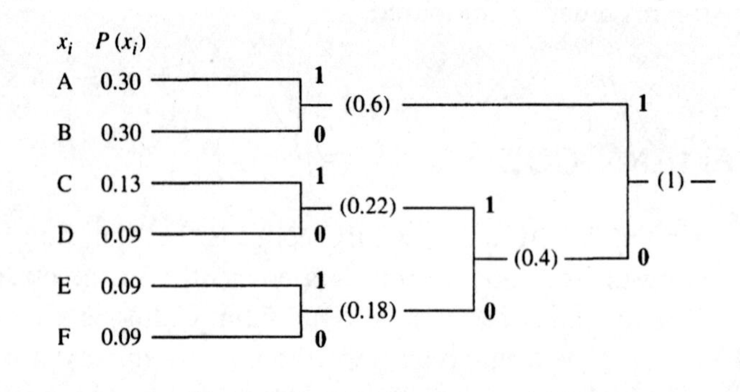

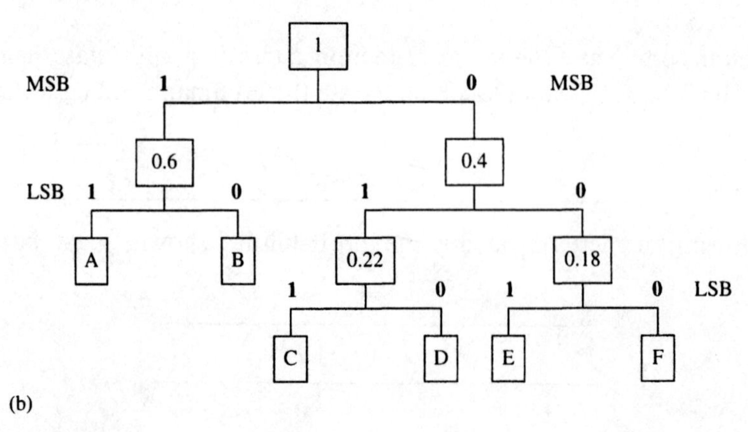

Fig. 9.1 (a) Tree diagram; (b) state diagram

Table 9.3 The Huffmann codes

x_i	$P(x_i)$	Code	n_i	$n_iP(x_i)$
A	0.3	11	2	0.6
B	0.3	10	2	0.6
C	0.13	011	3	0.39
D	0.09	010	3	0.27
E	0.09	001	3	0.27
F	0.09	000	3	0.27
		Sum $n_iP(x_i)$		2.4

The state diagram is read from the top to the bottom. In other words the MSB is at the top of the diagram and the LSB is at the bottom of the diagram for each individual character.

3. Compression rate: conventionally for six characters, three bits would be allocated to each character.

$$\text{Compression rate} = \frac{\text{conventional no bits/char}}{\text{Huffmann average}}$$

$$= \frac{3}{2.4}$$

$$= 1.25$$

4. Efficiency:

$$\text{Efficiency } (\eta) = \frac{S(x)}{\sum n_i P(x_i)} \times 100\%$$

$$= \frac{2.363}{2.4} \times 100\%$$

$$= 98.46\%$$

Figure 9.2 shows a flow chart to decode the Huffmann codes received for Example 9.1. As can be seen, if the first bit in the stream is a logic 1 then the character will be either A or B. To determine which character the status of the next bit must be determined. If the next bit is a logic 1 then the character is A, but if the next bit is a logic 0 then the character is B.

If the first bit is a logic 0 then the next 2 bits must be examined to determine whether the character is C, D, E, or F. If the next bit is a logic 1 then the character is either C or D. If the last bit is a logic 1 then the character is C. However, if the last bit is a logic 0 then the character is D.

If the first bit is a logic 0 and the next bit is a logic 0 then the character is either E or F. If the last bit is a logic 1 then the character is E and if the last bit is a logic 0 then the character is F.

Example 9.2

In a piece of text only the following letters are used. The number of times each letter appears is given in brackets after the letter.

A (183), B (183), C (183), D (108), E (108), F (67), G (67), H (67), I (34)

Determine the following:

1. The minimum number of bits per character.
2. The Huffmann codes.
3. The compression rate.
4. The efficiency.

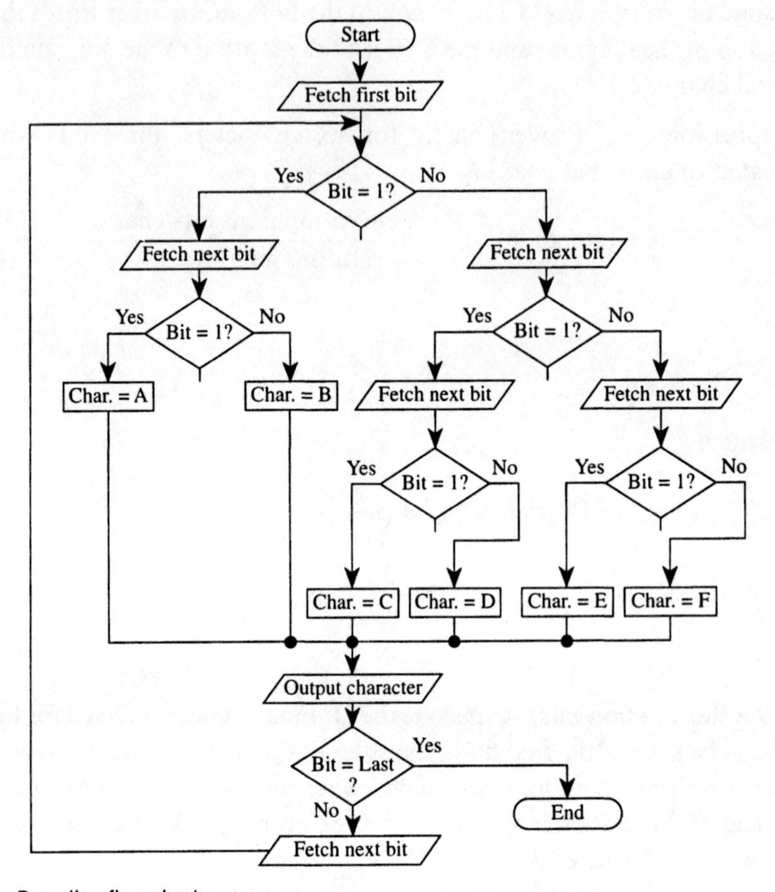

Fig. 9.2 Decoding flow chart

Solution

Before doing anything determine the total number of characters and then determine the probability for each character.

$$\text{Total No.} = 3 \times 183 + 2 \times 108 + 3 \times 67 + 34$$
$$= 1000$$

Probability for characters:

$$P(x_i) = \frac{\text{No. of characters in text}}{\text{Total No. of characters in text}}$$

Probability for characters A, B and C:

$$P(x_i) = \frac{183}{1000}$$
$$= 0.183$$

Probability for characters D and E:

$$P(x_i) = \frac{108}{1000} = 0.108$$

Probability for characters F, G and H:

$$P(x_i) = \frac{67}{1000} = 0.067$$

Probability for character I:

$$P(x_i) = \frac{34}{1000} = 0.034$$

1.
$$S(x) = -\sum_{i=1}^{n} P(x_i)\log_2 P(x_i)$$

$$\begin{aligned} S(x) &= -(3 \times 0.183 \log_2 0.183 + 2 \times 0.108 \log_2 0.108 \\ &\quad + 3 \times 0.067 \log_2 0.067 + 0.034 \log_2 0.034) \\ &= 2.9884 \text{ bits/character} \end{aligned}$$

2. The Huffmann codes are determined using Fig. 9.3. They are shown in Table 9.4.
3. Compression rate: conventionally for nine characters, four bits would be allocated to each character.

$$\text{Compression rate} = \frac{\text{conventional No. bits/char}}{\text{Huffmann average}} = \frac{4}{3.052} = 1.311$$

4. Efficiency:

$$\text{Efficiency } (\eta) = \frac{S(x)}{\sum n_i P(x_i)} \times 100\% = \frac{2.9884}{3.052} \times 100\% = 97.92\%$$

9.4.1 Usage of Huffmann coding

Huffmann coding is only of use as source code as a computer program is required to compress the characters. The program will be written in to follow the steps as indicated in the flow chart given in Fig. 9.2. As a result it is only carried out in the front-end processor.

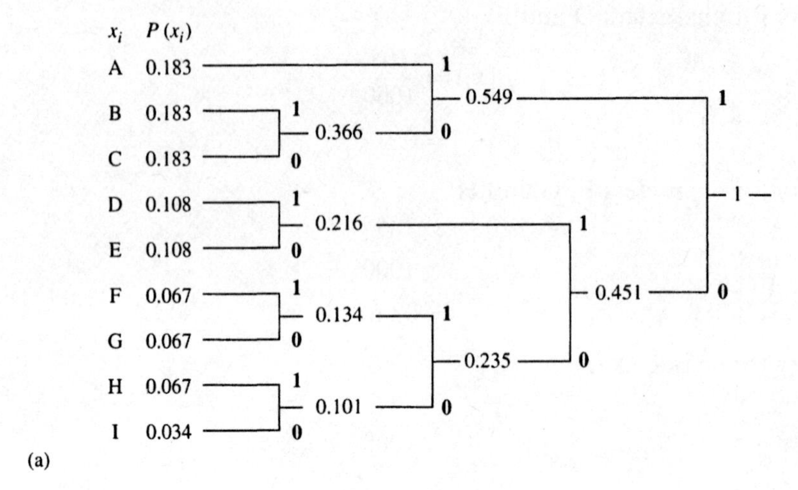

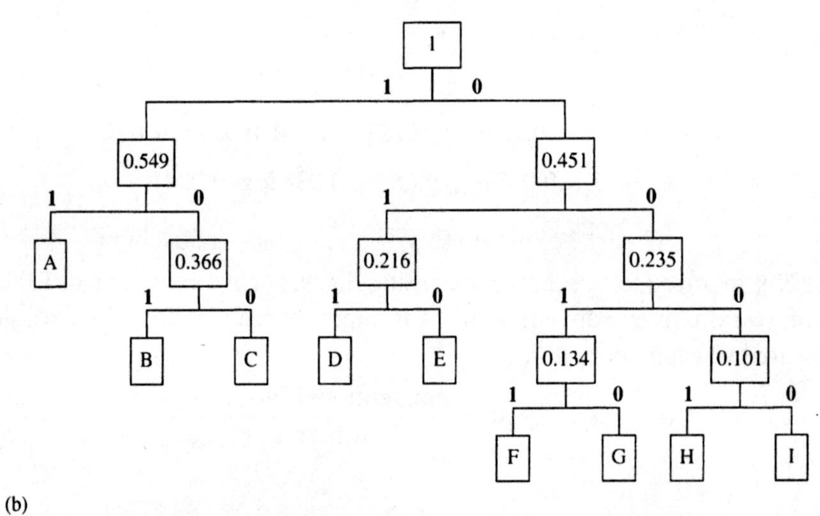

Fig. 9.3 (a) Tree diagram; (b) state diagram

Table 9.4 Huffman coding

x_i	$P(x_i)$	Code	n_i	$n_iP(x_i)$
A	0.183	11	2	0.366
B	0.183	101	3	0.549
C	0.183	100	3	0.549
D	0.108	011	3	0.324
E	0.108	010	3	0.324
F	0.067	0011	4	0.268
G	0.067	0010	4	0.268
H	0.067	0001	4	0.268
I	0.034	0000	4	0.136
		Sum $n_iP(x_i)$		3.052

9.4.2 Advantages and disadvantages of Huffmann coding

The advantage of this coding method is that less memory is required to store the compressed data. This technique is carried out when files are zipped up for storage on floppy disks.

The disadvantage of this method on live changing data is that the data to be transmitted must first be stored and then compressed and then transmitted. At the receiver the data must be stored, decoded and then passed on to the rest of the circuitry. The storage of the data adds a propagation delay to the process and as a result this coding method is not used on digitised speech.

9.5 HAMMING CODING

Another coding method is Hamming coding. This is a data-based coding technique. The number of Hamming bits required is directly proportional to the size of the data block. The Hamming bits are used for error detection and correction at the distant receiver. The method described below is for the detection and correction of single-bit errors.

The number of Hamming bits required is determined as follows.

$$2^r \geq k$$

where $r = 1, 2, 3 \ldots$

$k =$ number of message bits

The number of Hamming bits (H) required is

$$H = r + 1$$

The total number of bits that are transmitted in the block is

$$\text{block length } (n) = H + k$$

This coding technique is also referred to as an (n, k) block check code. The Hamming bit positions must be specified and the transmitter and receiver must know their positions. By convention the Hamming bit positions are as shown in Table 9.5. Note that the bit positions are numbered from 1 and not from 0.

The Hamming bits are determined by means of the following. For all the data bits, those bits that yield a logic 1, their bit position is converted into the binary

Table 9.5 Hamming bit positions

Bit number																
17[a]	16	15	14	13	12	11	10	9	8	7	6	5	4	3	2	1[b]
D	*H*	*D*	*D*	*D*	*D*	*D*	*D*	*D*	*H*	*D*	*D*	*D*	*H*	*D*	*H*	*H*

[a] MSB.
[b] LSB.
D = data bit.
H = Hamming bit.

equivalent. All the binary codes are then modulo 2 added to produce the Hamming codes. This process is shown in Example 9.3.

The distant receiver determines the Hamming codes and then compares this with the received Hamming codes. If the modulo 2 sum of the received Hamming codes and the receiver-generated Hamming codes is all logic 0 then the data stream is unerrored. If there is a difference then the actual binary code indicates which bit is errored.

The Hamming distance is the minimum number of bit positions in which two valid code words differ. Consider the following 7-bit code words using even parity:

Word 1	0000000	0
Word 2	0000001	1
Word 3	0000010	1
Word 4	0000011	0

Comparing words 1 and 4, there is a 2-bit change and these words both yield the same parity status. Comparing words 2 and 3, there is a 2-bit change and these words both yield the same parity status. Hence the Hamming distance is 2. This means that the code will only detect and correct single-bit errors.

For the code to detect x errors, the code must have a Hamming distance of $x + 1$. In the above example $x = 1$ resulting in a Hamming distance of 2. It enables the code to detect and correct x errors then a Hamming distance of $2x + 1$ is required.

The value k/n is known as the code rate or code efficiency and the value $(1 - k/n)$ is known as the redundancy.

Example 9.3

The Hamming bits are to be inserted into bit positions 1, 2, 4, 8 etc. Determine each of the following:

1. The number of Hamming bits required for the data block given below.
2. The (n, k) block check code.
3. The Hamming code for the data block given below.
4. The code efficiency.
5. The code redundancy.

$$\text{Binary word} = 01010000001_2$$

Solution

1. Message bits $(k) = 11$.

$$2^r \geq k$$

$$\text{Let } r = 1: \quad 2^1 = 2 < 11$$
$$\text{Let } r = 2: \quad 2^2 = 4 < 11$$
$$\text{Let } r = 3: \quad 2^3 = 8 < 11$$
$$\text{Let } r = 4: \quad 2^4 = 16 > 11$$

Therefore

$$H = r + 1$$
$$= 4 + 1$$
$$= 5$$

2. Total number of bits in the block:

$$n = k + H$$
$$= 11 + 5$$
$$= 16$$

Therefore a (16,11) block check code is required.

3.

MSB															LSB
16	15	14	13	12	11	10	9	8	7	6	5	4	3	2	1
H	0	1	0	1	0	0	0	*H*	0	0	0	*H*	1	*H*	*H*

$$3 = 00011$$
$$12 = 01100$$
$$14 = \underline{01110}$$
Modulo 2 sum $= 00001$

H								*H*				*H*		*H*	*H*
16	15	14	13	12	11	10	9	8	7	6	5	4	3	2	1
0	0	1	0	1	0	0	0	0	0	0	0	0	1	0	1_2

4. Code efficiency:

$$\text{Efficiency} = \frac{k}{n} = \frac{11}{16} = 0.6875$$

This equates to an efficiency of 68.75%.

5. Code redundancy:

$$\text{Redundancy} = 1 - \frac{k}{n} = 1 - \frac{11}{16} = 1 - 0.6875 = 0.3125$$

Example 9.4

The Hamming bits appear in bit positions 1, 2, 4, 8 and 16. The following (16,11) block check code is received at a receiver. Determine whether the data is error free or not.

$$0010110011100111_2$$

Solution

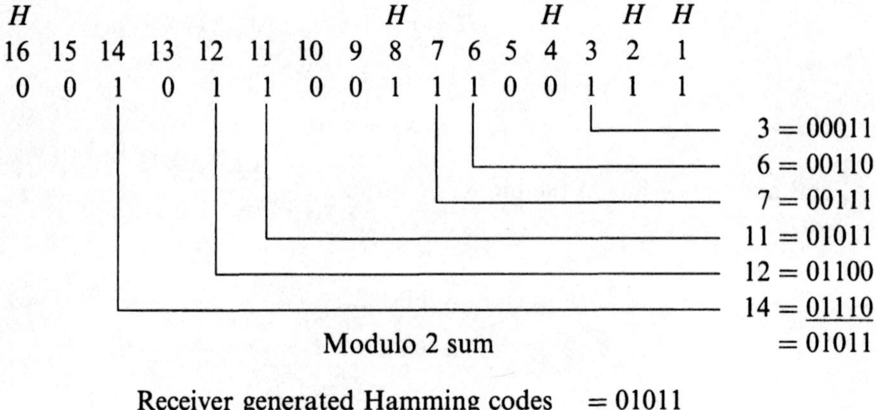

Receiver generated Hamming codes	= 01011
Received Hamming codes	= 01011
Modulo 2 sum	= 00000

The received bit stream is error free.

Example 9.5

The Hamming bits appear in bit positions 1, 2, 4, 8 and 16. The following (16,11) block check code is received at a receiver. Determine whether the data is error free or not.

$$0010111010101100_2$$

Solution

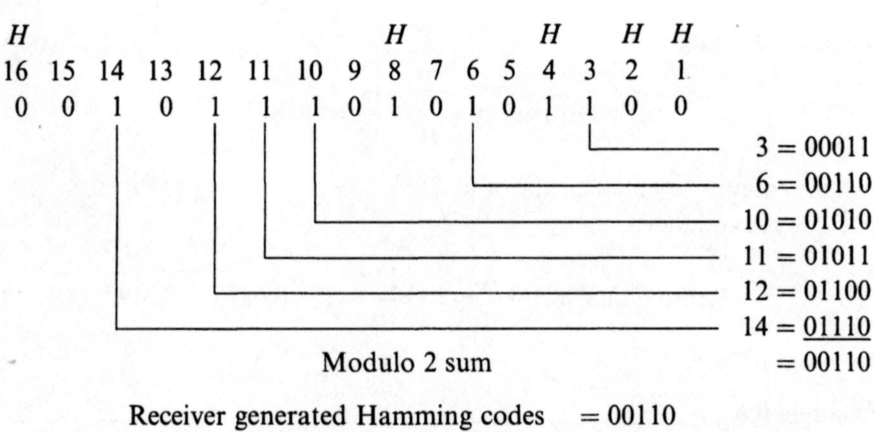

Receiver generated Hamming codes	= 00110
Received Hamming codes	= 01100
Modulo 2 sum	= 01010
The error code	= 01010_2

This equates to bit 10, hence bit 10 is errored.

The corrected code is:

H								*H*				*H*		*H*	*H*
16	15	14	13	12	11	10	9	8	7	6	5	4	3	2	1
0	0	1	0	1	1	0	0	1	0	1	0	1	1	0	0

The correct message data stream is:

$$01011000101_2$$

9.5.1 Usage of Hamming coding

Hamming coding is only of use as source code as a computer program is required to compress the characters. As a result it is only carried out in the front-end processor.

9.5.2 Advantages and disadvantages of Hamming coding

The advantage of this coding method is that it is extremely effective on networks where the data streams are prone to single-bit errors.

The disadvantage of this method is that for a single-bit detection and correction code, if multiple bits are errored then the errors are detected but the resultant could cause another bit that is correct to be changed, causing the data to be further errored. Consider the following example.

Example 9.6

An (11,7) block check code is received. The Hamming bits are in bit positions 1, 2, 4 and 8. Determine whether the data is errored and which bit or bits are errored.

Transmitted data block = 10010110000_2
Received data block = 1001010**01**00
H *H HH*

Solution

Receiver-generated Hamming bits:

Bit 3 = 0011
Bit 6 = 0110
Bit 11 = 1011
Modulo 2 add = 1110
Received Hamming code = 1000
Modulo 2 add = 0110

This indicates that bit 6 is errored, which is wrong as bits 3 and 5 were errored.

9.6 REVIEW QUESTIONS

9.1 State the full name for the following mnemonics:
(a) ASCII
(b) EBCDIC

9.2 The following six letters, having the probabilities shown, must be digitally transmitted.

x_i	$P(x_i)$
A, B	0.305
C	0.15
D, E, F	0.08

Determine the following:
(a) The minimum number of bits per character.
(b) The Huffman codes.
(c) The compression rate.
(d) The efficiency.

9.3 In a piece of text only the following letters are used. The number of times each letter appears is given in brackets after the letter.

A (205), B (183), C (178), D (178), E (84), F (84), G (53), H (53), I (32)

Determine the following:
(a) The minimum number of bits per character.
(b) The Huffman codes.
(c) The compression rate.
(d) The efficiency.

9.4 The Hamming bits are to be inserted into bit positions 1, 2, 4, 8 etc. Determine each of the following:
(a) The number of Hamming bits required for the data block given below.
(b) The (n, k) block check code.
(c) The Hamming code for the data block given below.
(d) The code efficiency.
(e) The code redundancy.

$$\text{Binary word} = 100110001000_2$$

9.5 The Hamming bits appear in bit positions 1, 2, 4, 8 and 16. The following (16,11) block check code is received at a receiver. Determine whether the data is error free or not.

$$0010110011110111_2$$

10 Bit error detection and correction

Mankind has relied on coding systems to make complex problems understandable.

10.1 INTRODUCTION

Coding systems within a transmission system are the subject of this chapter. Basic systems such as parity bits for error detection and block check codes for error detection and single-bit error correction are discussed. The advantages and disadvantages of these systems are covered.

More complex systems such as cyclic redundancy checks are discussed in some detail from a mathematical point of view as well as from a practical perspective.

The chapter ends with an in-depth discussion on trellis coding, how it works and why it is used on large-scale systems today.

10.2 ERROR DETECTION USING PARITY BITS

10.2.1 Parity determination

Parity bits are used on asynchronous data streams to determine whether the received data has been errored or not. The transmitter determines the status of the parity bit and inserts it into the data stream, normally at the back of the stream. The receiver determines what the parity should be using the data bits only and then compares this to the status of the received parity bit. If the two states are the same then the receiver assumes that the received data is error free. If, however, the two states are different then the receiver assumes that the data was errored.

Consider an ASCII character that is transmitted via an RS232C port. The ASCII character consists of 7 bits. These 7 bits are loaded into a UART (universal asynchronous receiver transmitter). The UART adds a single start bit and 1.5 to two stop bits. The UART also determines the parity bit of the 7-bit data code.

Even parity is where there are an even number of logic 1s in the binary word. The word 1100110_2 has four logic 1 states and hence the parity is even. If the system was using even parity then the transmitter would insert a parity bit having a logic 0 state into the data stream.

The word 1100100_2 has only three logic 1 states and hence it has odd parity. If the system was using even parity then the transmitter would insert a parity bit having a logic 1 state into the bit stream. This would cause the 8-bit word to have even parity.

If the word 1100110_2, which has even parity, was transmitted over a system using odd parity then the transmitter would insert a parity bit having a logic 1 state. Likewise for the word 1100100_2 transmitted over a system using odd parity then the transmitter would insert a parity bit having a logic 0 state.

Example 10.1

Consider the ASCII code for the character 'h'.

Character	Hex code	Binary code
h	68	1101000

Assume even parity is used and two stop bits are used. The start bit is always a logic 0, the stop bits are always logic 1 and the parity bit for H is 1. The transmitted data stream for the h is as follows:

S	LSB						MSB	P	Stp	Stp
0	0	0	0	1	0	1	1	1	1	1

where S = stop
LSB = least significant bit
P = parity
MSB = most significant bit
Stp = stop

As can be seen above, the LSB is transmitted first and the MSB last when the data is transferred asynchronously.

Figure 10.1(a) shows one method used to determine the parity of the ASCII character. The switch shown must be placed in the appropriate position according to the parity to be transmitted. For even parity XOR gates are used, whereas for odd parity the last gate is an XNOR gate. The output of the last gate together with the input data is then clocked into the output buffer.

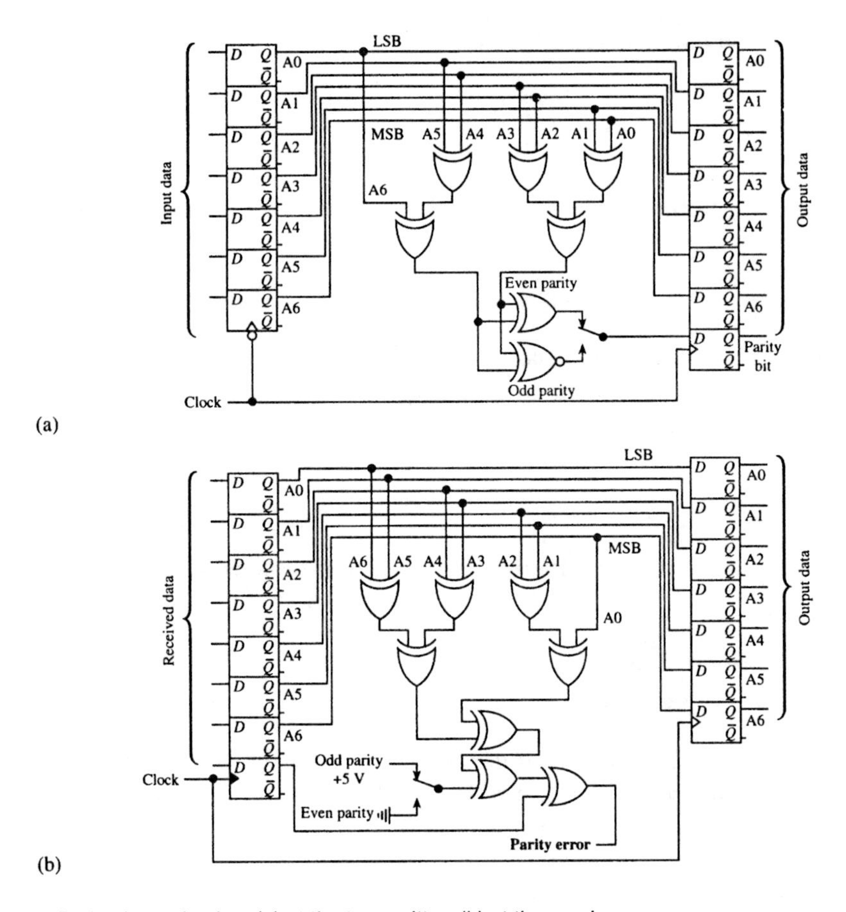

Fig. 10.1 Parity determination: (a) at the transmitter; (b) at the receiver

10.2.2 Error detection using the parity bit

Figure 10.1(b) shows how the parity bit is used to determine whether the received data is errored or not. The switch is in the position for odd parity to be detected. This shows that the data applied to the last gate is exclusive ORed with a logic 1. With the switch placed in the position for the detection of even parity, the data applied to the last gate is exclusive ORed with a logic 0.

Example 10.2

Consider the following asynchronous bit stream that arrives at a receiver:

$$01010011111_2$$

Stripping off the start bit and two stop bits first:

$$10100111_2$$

Identifying the parity bit and the data bits:

$$\text{Data bits} = 1010011_2$$

$$\text{Parity bit} = 1_2$$

The receiver determines the parity of the data and compares it to the received parity bit:

$$\text{Receiver determined parity} = 0_2$$

This is not the same as that received and hence the data has been errored. The receiver does not know whether the parity bit was errored or whether, as in this example, the data was errored.

Example 10.3

Consider the following transmitted character and received character:

		LSB						MSB	P
Transmitted character	=	1	0	0	0	0	1	1	1_2
Received character	=	1	0	0	0	1	0	1	1_2

In this case two bits have been errored, but when the receiver determines the parity and compares this to the received parity bit, no difference is found and hence the receiver assumes that the data is correct. This is clearly not the case.

Example 10.4

The following asynchronous data stream is received at an RS232C port. The data has a single start bit and two stop bits. A parity bit is incorporated with each character.

$$096A6C6D833F426_H$$

Determine the serial binary code that has transmitted and determine the ASCII characters.

Solution

Determine the binary code:

0	9	6	A	6	C	6	D	8	3	F	4	2	6_H
0000	1001	0110	1010	0110	1100	0110	1101	1000	0011	1111	0100	0010	0110_2

Group the bits in groups of 11_{10}, i.e. one start bit, one parity bit, two stop bits and seven data bits:

$$00001001011 \quad 01010011011 \quad 00011011011 \quad 00000111111 \quad 01000010011\ 0_2$$

List each character separately:

	MSB										LSB
Word 1	0	0	0	0	1	0	0	1	0	1	1
Word 2	0	1	0	1	0	0	1	1	0	1	1
Word 3	0	0	0	1	1	0	1	1	0	1	1
Word 4	0	0	0	0	0	1	1	1	1	1	1
Word 5	0	1	0	0	0	0	1	0	0	1	1

Strip off the start and stop bits and separate the parity bit:

Data	Parity
0001001	0
1010011	0
0011011	0
0000111	1
1000010	0

Even parity is used.

Remember that the LSB of each character is sent out first, so reverse the order:

$$1001000 = 48_{16} = \text{H}$$
$$1100101 = 65_{16} = \text{e}$$
$$1101100 = 6\text{C}_{16} = \text{l}$$
$$1110000 = 70_{16} = \text{p}$$
$$0100001 = 21_{16} = !$$

10.2.3 Advantages and disadvantages of using parity bits

When used for error detection parity bits will:

- Detect errored data where an odd number of bits have been errored.
- Not detect errored data where an even number of data bits have been errored.

10.2.4 Disadvantages of using asynchronous transmission to send each character individually

As can be seen in the above examples, to transmit 7 bits, 10.5 to 11 bits must be transmitted. The efficiency is determined by the following equation:

$$\eta = \frac{\text{number of message bits}}{\text{total number of transmitted bits}} \times 100\%$$

The redundancy is determined by

$$\text{Redundancy} = 100\% - \eta$$

This means that the ASCII character that is transmitted asynchronously has an efficiency of 66.67% when 1.5 stop bits are used, giving a redundancy of

33.33%. If two stop bits are used then the efficiency drops to 63.64% which yields a redundancy of 36.36%.

A second disadvantage is that if the parity bit is errored during the transmission but the data is not errored then the receiver will detect that the data has been errored.

10.3 BLOCK CHECK CODES

Parity bits can be used in another way for error detection and single-bit error correction.

Consider the block of data shown in Table 10.1, consisting of EBCDIC codes, which is to be transmitted asynchronously

The data will be sent out in the following frame format:

SOF	Word 1 ... Word 8	VRC	HRC	EOF

where SOF = start of frame
EOF = end of frame

The receiver will determine the VRC and HRC on the data and then compare it to the received VRC and HRC. If there is a difference between the received VRC/HRC and the receiver-determined VRC/HRC then the receiver might be able to correct the error.

Table 10.1 EBCDIC data

	LSB							MSB	VRC
Word 1 = T	0	0	1	1	0	1	1	1	1
Word 2 = r	1	0	0	1	1	0	0	1	0
Word 3 = y	1	0	0	0	0	1	0	1	1
Word 4 = SP	0	0	0	0	0	0	1	0	1
Word 5 = t	0	0	1	1	0	1	0	1	0
Word 6 = h	1	0	0	0	0	0	0	1	0
Word 7 = i	1	0	0	1	0	0	0	1	1
Word 8 = s	0	0	1	0	0	1	0	1	1
LRC or HRC	0	0	1	0	1	0	0	1	

VRC = vertical redundancy check.
HRC or LRC = horizontal redundancy check or longitudinal redundancy check.

Example 10.5

The transmitted data is received from a distant transmitter as shown in Table 10.2. In this example the receiver is able to correct the error as the comparison between the VRCs and HRCs yields a logic 1 in only one column and one row. The errored bit is shown emboldened in the table.

Table 10.2 Received data, Example 10.5

	LSB Bit 1	Bit 2	Bit 3	Bit 4	Bit 5	Bit 6	Bit 7	MSB Bit 8	Received VRC	Receiver VRC	XOR
Word 1	0	0	1	1	0	1	1	1	1	1	0
Word 2	1	0	1	1	1	0	0	1	0	1	**1**
Word 3	1	0	0	0	0	1	0	1	1	1	0
Word 4	0	0	0	0	0	0	1	0	1	1	0
Word 5	0	0	1	1	0	1	0	1	0	0	0
Word 6	1	0	0	0	0	0	0	1	0	0	0
Word 7	1	0	0	1	0	0	0	1	1	1	0
Word 8	0	0	1	0	0	1	0	1	1	1	0
	0	0	1	0	1	0	0	1	Received HRC		
	0	0	0	0	1	0	0	1	Receiver HRC		
	0	0	**1**	0	0	0	0	0	XOR		

Example 10.6

The transmitted data is received from a distant transmitter as shown in Table 10.3. In this example there are two separate errors that are in two different columns and two different rows. As a result the receiver can identify that there are two errors but it cannot uniquely identify which data bits are errored.

Looking at the table, is bit 1 or bit 5 of word 3 errored and is bit 1 or bit 5 of word 7 errored? The system is only able to determine that either bit 1 or bit 5 of word 3 was errored and either bit 1 or bit 5 of word 7 was errored.

Table 10.3 Received data, Example 10.6

	LSB Bit 1	Bit 2	Bit 3	Bit 4	Bit 5	Bit 6	Bit 7	MSB Bit 8	Received VRC	Receiver VRC	XOR
Word 1	0	0	1	1	0	1	1	1	1	1	0
Word 2	1	0	0	1	1	0	0	1	0	0	0
Word 3	1	0	0	0	1	1	0	1	1	0	**1**
Word 4	0	0	0	0	0	0	1	0	1	1	0
Word 5	0	0	1	1	0	1	0	1	0	0	0
Word 6	1	0	0	0	0	0	0	1	0	0	0
Word 7	0	0	0	1	0	0	0	1	1	0	**1**
Word 8	0	0	1	0	0	1	0	1	1	1	0
	0	0	1	0	1	0	0	1	Received HRC		
	1	0	0	0	0	0	0	1	Receiver HRC		
	1	0	0	0	**1**	0	0	0	XOR		

This shows that the system can correct single-bit errors but can only detect multiple-bit errors and identify the errored words.

Example 10.7

In Table 10.4 the received data has two errors in the same row. As a result the error is reflected in two columns but because two bits are errored there is no indication in which row. As a result the receiver is unable to correct these two errors. Again the system is able to identify the errored rows and is unable to correct the errored bits.

The same would result if two bits were errored in the same column. In this case the errors would be indicated in two different rows (words).

Table 10.4 Received data, Example 10.7

	LSB Bit 1	Bit 2	Bit 3	Bit 4	Bit 5	Bit 6	Bit 7	MSB Bit 8	Received VRC	Receiver VRC	XOR
Word 1	0	0	1	1	0	1	1	1	1	1	0
Word 2	1	0	0	1	1	0	0	1	0	0	0
Word 3	1	0	0	0	0	1	0	1	1	1	0
Word 4	0	0	0	0	0	0	1	0	1	1	0
Word 5	0	0	1	0	1	1	0	1	0	0	0
Word 6	1	0	0	0	0	0	0	1	0	0	0
Word 7	1	0	0	1	0	0	0	1	1	1	0
Word 8	0	0	1	0	0	1	0	1	1	1	0
	0	0	1	0	1	0	0	1	Received HRC		
	0	0	0	1	0	0	0	1	Receiver HRC		
	0	0	0	**1**	**1**	0	0	0	XOR		

10.3.1 Advantages of block check codes

The advantages of block check codes are:

- The detection of all single- and multiple-bit errors.
- The correction of all single-bit errors.

10.3.2 Disadvantages of block check codes

The disadvantages of block check codes are:

- The inability to correct multiple errors that occur in the same row or column.

- The inability to correct multiple errors that occur in different columns and different rows.
- The high redundancy for small matrices.

Consider the above examples. In each case an 8-word by 8-bit matrix is used. This means that the actual data consists of the following:

Start of text	= 1 × 8-bit word	= 8 bits
Data	= 8 × 8-bit word	= 64 bits
VRC	= 1 × 8-bit word	= 8 bits
HRC	= 1 × 8-bit word	= 8 bits
End of text	= 1 × 8-bit word	= 8 bits
	Total	= 96 bits

This gives an efficiency of 66.67% and a redundancy of 33.33%.

10.4 FRAME CHECK SEQUENCE OR CYCLIC REDUNDANCY CHECK

When dealing with live data on a transmission system, which takes place dynamically, there is no time to store the data and then retransmit it. In this case it is more important to be able to determine whether the data has been errored or not. Not only must single-bit errors be detected but error bursts as well.

10.4.1 Error bursts

An error burst begins and ends with a bit that is errored; the bits between the erroneous bits may or may not be errored.

The length of the error burst is determined by the following:

- The number of bits including the two errored bits.
- The number of bits between the last erroneous bit and the first erroneous bit in the next error burst must be greater than or equal to the number of bits between the two successive erroneous bits.

Example 10.8

Transmitted message = 01 011101 110110 01 0110 1100 10
Received message = 01 111100 110110 01 1111 1100 10
=

6-bit error burst (011101 / 111100)
Min. 6 error-free bits (110110)
4-bit error burst (0110 / 1111)
Min. 4 error free bits (1100)

10.4.2 Advantages of cyclic redundancy checks

A generator polynomial of R bits will detect:

- All single-bit errors.
- Most double-bit errors.
- Most odd numbers of bit errors.
- All error bursts $<R$.
- Most error bursts $>R$.

The error code that is incorporated into the data stream is called either a frame check sequence (FCS) or cyclic redundancy check (CRC). The placement of the FCS or CRC is incorporated into the frame at the end of the frame, i.e. immediately after the data.

10.4.3 Common cyclic redundancy code polynomials

The generator polynomial is chosen in such a way as to enable the error sequences indicated above to be detected. The generator polynomial is a prime binary number which means that it is only divisible by itself and by 1_2 without leaving a remainder.

There are three common CRC codes used; these are CRC-16, CRC-CCITT and CRC-32. The binary representation of the CRC-16 and CRC-CCITT codes is given below.

Weighting	2^{16}	2^{15}	2^{14}	2^{13}	2^{12}	2^{11}	2^{10}	2^{9}	2^{8}	2^{7}	2^{6}	2^{5}	2^{4}	2^{3}	2^{2}	2^{1}	2^{0}
CRC-16	1	1	0	0	0	0	0	0	0	0	0	0	0	0	1	0	1_2
CRC-CCITT	1	0	0	0	1	0	0	0	0	0	0	1	0	0	0	0	1_2

The polynomial representation of the CRC-16, CRC-CCITT and CRC-32 codes is given below.

$$\text{CRC-16} = X^{16} + X^{15} + X^{2} + 1$$

$$\text{CRC-CCITT} = X^{16} + X^{12} + X^{5} + 1$$

$$\text{CRC-32} = X^{32} + X^{26} + X^{23} + X^{16} + X^{12} + X^{11} + X^{10} + X^{8} + X^{7} + X^{5} + X^{4} + X^{2} + X + 1$$

10.4.4 Binary prime number

To explain this more fully a generator polynomial $G(x)$ of 11001_2 will be used. This code will develop a CRC-4 code and this will be shown later. Is the binary code 11001_2 only divisible by 1_2 and by 11001_2 or not? In other words is 11001_2 a binary prime number or not? The proof that this number is indeed a binary prime number is shown in Fig. 10.2. The addition that is done at each successive point in the solution is modulo 2 addition. This is simply the XOR of the two bits:

0 XOR 0 = 0, 0 XOR 1 = 1, 1 XOR 0 = 1, 1 XOR 1 = 0

```
    1 1 0 0 1 r 0
 1 |1 1 0 0 1
   |1 1 0 0 1
   |---------
   |• • • • •
```

```
      1 1 0 0 r 1
1 0 |1 1 0 0 1
    |1 0
    |---------
    |  • 1 0
    |    1 0
    |  -------
    |    • 0 0
    |      0 0
    |    -------
    |      • 0 1
```

```
      1 0 0 0 r 1 1
1 1 |1 1 0 0 1
    |1 1
    |---------
    |• 0 0
    |  0 0
    |  -------
    |  • 0 0
    |    0 0
    |  -------
    |    • 0 1
    |      0 0
    |  -------
    |      • 1
```

```
          1 1 0 r 0 1
1 0 0 |1 1 0 0 1
      |1 0 0
      |---------
      |• 1 0 0
      |  1 0 0
      |  -------
      |  • 0 0 1
      |    0 0 0
      |  -------
      |    • 0 1
```

```
            1 1 1 r 1 1
1 0 1 |1 1 0 0 1
      |1 0 1
      |---------
      |• 1 1 0
      |  1 0 1
      |  -------
      |  • 1 1 1
      |    1 0 1
      |  -------
      |    • 1 1
```

```
            1 0 0 r 0 1
1 1 0 |1 1 0 0 1
      |1 1 0
      |---------
      |• 0 0 0
      |  0 0 0
      |  -------
      |  • 0 0 1
      |    0 0 0
      |    -------
      |    • 0 1
```

```
            1 0 1 r 1 0
1 1 1 |1 1 0 0 1
      |1 1 1
      |---------
      |• 0 1 0
      |  0 0 0
      |  -------
      |  • 1 0 1
      |    1 1 1
      |  -------
      |    • 1 0
```

```
                1 1 r 0 0 1
1 0 0 0 |1 1 0 0 1
        |1 0 0 0
        |---------
        |• 1 0 0 1
        |  1 0 0 0
        |  -------
        |  • 0 0 1
```

```
                1 1 r 0 1 0
1 0 0 1 |1 1 0 0 1
        |1 0 0 1
        |---------
        |• 1 0 1 1
        |  1 0 0 1
        |  -------
        |  • 0 1 0
```

```
                1 1 r 1 1 1
1 0 1 0 |1 1 0 0 1
        |1 0 1 0
        |---------
        |• 1 1 0 1
        |  1 0 1 0
        |  -------
        |  • 1 1 1
```

```
                1 1 r 1 0 0
1 0 1 1 |1 1 0 0 1
        |1 0 1 1
        |---------
        |• 1 1 1 1
        |  1 0 1 1
        |  -------
        |  • 1 0 0
```

```
                1 0 r 0 0 1
1 1 0 0 |1 1 0 0 1
        |1 1 0 0
        |---------
        |• 0 0 0 1
        |  0 0 0 0
        |  -------
        |  • 0 0 1
```

Fig. 10.2 Proof that 11001 is a prime number (continues)

```
              1 0 r 0 1 1
1 1 1 0 1 | 1 1 0 0 1
            1 1 0 1
          • 0 0 1 1
            0 0 0 0
            • 0 1 1
```

```
             1 0 r 1 0 1
1 1 1 0 | 1 1 0 0 1
          1 1 1 0
        • 0 1 0 1
          0 0 0 0
          • 1 0 1
```

```
            1 0 r 1 1 1
1 1 1 1 | 1 1 0 0 1
          1 1 1 1
        • 0 1 1 1
          0 0 0 0
          • 1 1 1
```

Fig. 10.2 (Continued)

It is obvious that when the following 5-bit codes are divided into 11001_2, they will yield a remainder, 10000_2, 10001_2, 10010_2, 10011_2, 10100_2, 10101_2, 10101_2, 10110_2, 10111_2, 11000_2, 11010_2, 11011_2, 111002, 11101_2, 11110_2, 11111_2. The only 5-bit code that will yield a remainder of 0 is 11001_2.

In the above example it is clear that the code 11001_2 is a binary prime number. Now the decimal equivalent of 11001_2 is 25_{10} which is not a decimal prime number.

10.5 THE CRC PROCESS

At the transmitter the CRC is determined on the whole frame on a bit-by-bit basis. The input binary bit stream is applied to a serial shift register and the division process takes place serially.

At the receiver the received bit stream is again divided on a bit-by-bit basis through a serial shift register. At the end of the frame the resultant CRC should be all zeros. If this is not the case then the data has been errored. The CRC or FCS circuit does not try to correct the errored data.

The circuits and description given below relate to a data stream of only 8 bits and a CRC code of 4 bits. In practice the data streams are much longer.

10.5.1 The CRC generation process at the transmitter

Using the binary notation

Assume the generator polynomial $G(x)$ and the input data are as given below:

$$G(x) = 11001_2$$

$$\text{Data} = 11100110$$

Add four zeros to the back of the data stream:

$$\text{Data} = 11100110 \quad 0000_2$$

The calculation is shown in Fig. 10.3.

$$\text{Transmitted data stream} = 111001100110_2$$

Using the polynomial notation

$$\text{Data stream} = 11100110_2$$

Add four zeros to the back of the data:

$$\text{Data stream} = 11100110\quad 0000_2$$

Weighting	=	x^{11}	x^{10}	x^9	x^8	x^7	x^6	x^5	x^4	x^3	x^2	x^1	x^0
Data	=	1	1	1	0	0	1	1	0	0	0	0	0
$G(x)$	=								1	1	0	0	1

Hence

$$\text{Data} = x^{11} + x^{10} + x^9 + x^6 + x^5$$

$$G(x) = x^4 + x^3 + 1$$

The calculation is shown in Fig. 10.4.

$$\text{Remainder} = x^2 + x$$

This equates to 0110_2

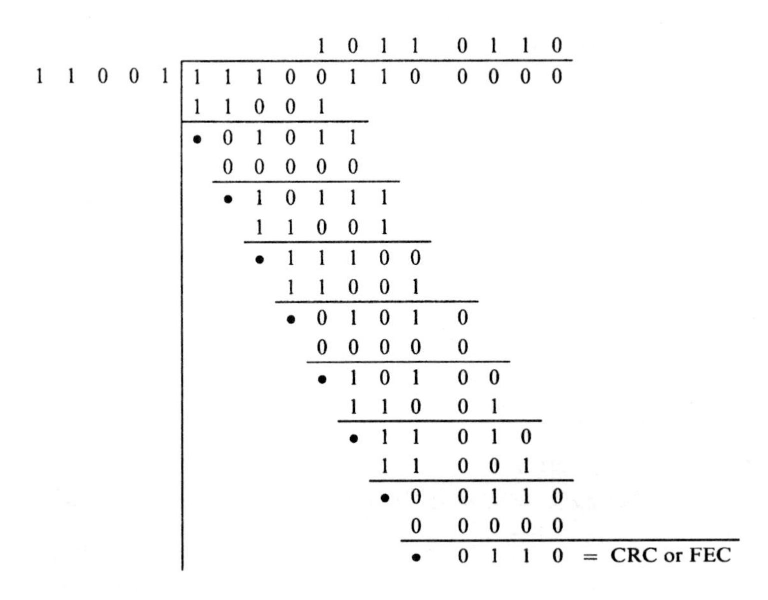

Fig. 10.3 CRC generation using binary notation

CRC circuit

The CRC or FEC code can be developed using software or hardware. A typical CRC circuit is shown in Fig. 10.5(a). The circuit works as follows:

$$
\begin{array}{r|l}
 & x^7 + x^5 + x^4 + x^2 + x \\
x^4 + x^3 + 1 & x^{11} + x^{10} + x^9 + x^6 + x^5 \\
 & x^{11} + x^{10} + x^7 \\
\hline
 & x^9 + x^7 + x^6 + x^5 \\
 & x^9 + x^8 + x^5 \\
\hline
 & x^8 + x^7 + x^6 \\
 & x^8 + x^7 + x^4 \\
\hline
 & x^6 + x^4 \\
 & x^6 + x^5 + x^2 \\
\hline
 & x^5 + x^4 + x^2 \\
 & x^5 + x^4 + x \\
\hline
 & x^2 + x
\end{array}
$$

Fig. 10.4 CRC generation using polynomial notation

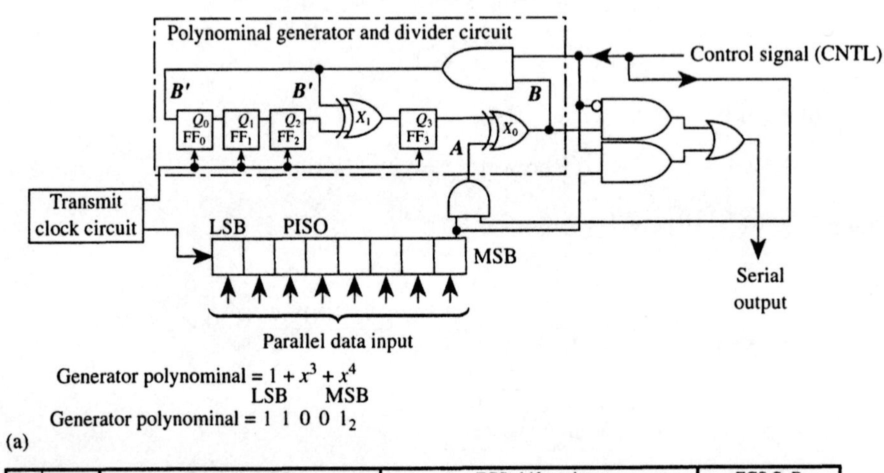

Tx Clk	CNTL	Transmit PISO								FCS shift register							FCS S–R			
		LSB							MSB	A XOR Q_3	Q_0	Q_1	Q_2	B' XOR Q_2	Q_3	B	Q_0	Q_1	Q_2	Q_3
0	1	0	1	1	0	0	1	1	1	1	0	0	0	1	0	1	0	0	0	0
1	1	0	0	1	1	0	0	1	1	0	1	0	0	0	1	0	1	0	0	1
2	1	0	0	0	1	1	0	0	1	1	0	1	0	1	0	1	0	1	0	0
3	1	0	0	0	0	1	1	0	0	1	1	0	1	0	1	1	1	0	1	1
4	1	0	0	0	0	0	1	1	0	0	1	1	0	0	0	0	1	1	0	0
5	1		0	0	0	0	0	1	1	1	0	1	1	0	0	1	0	1	1	0
6	1			0	0	0	0	0	1	1	1	0	1	0	0	1	1	0	1	0
7	1				0	0	0	0	0	0	1	1	0	0	0	0	1	1	0	0
										0 XOR Q_3				0 XOR Q_2						
8	0					0	0	0	0	0	0	1	1	1	0	0	0	1	1	0
9	0						0	0	0	1	0	0	1	1	1	1	0	0	1	1
10	0							0	0	1	1	0	0	0	1	1	1	0	0	1
11	0								0	0	1	1	0	0	0	0	1	1	0	0

Time →

Transmitted data stream

MSB LSB

← 1 1 1 0 0 1 1 1 0 0 1 1 0

Data CRC or FCS

(b)

Fig. 10.5 (a) CRC-4 generator circuit; (b) circuit operation at the transmitter

- Initially all the flipflops are reset.
- The data is loaded into the parallel in serial out (PISO) register.
- A logic 1 is placed on to the control lead (CNTL).
- The MSB in the PISO register is applied to the first input of the XOR gate X_0. The Q_3 output of FF_3 is applied to the second input of the XOR gate X_0.
- The output of the XOR gate X_0 is fed to the first input of XOR gate X_1 via the AND gate enabled by the logic 1 on the CNTL lead. The Q2 output of FF_2 is fed to the second input of XOR gate X_1.
- The data at the input to each flipflop is clocked through to the Q outputs.
- Each data bit is passed through to the serial output via the AND & OR combinational logic circuit.
- The complete cycle continues as above for all the data bits in the frame. In this example an 8-bit data stream has been used.
- After the last data bit has passed through the data level on the control lead (CNTL) is made low.
- The A input to XOR gate X_0 is now a permanent logic 0. The input to FF_0 as well as the first input to the XOR gate X_1 is also a logic 0 as the logic condition on B′ is low.
- The resultant status of each of the four flipflops is now clocked to the output serial lead via the AND & OR combinational circuitry.

This process is shown in Fig. 10.5(b).

10.5.2 The CRC checking process at the receiver

Using the binary notation

$$\text{Received data stream} = 11100110 \quad 0110_2 \text{ (no errors)}$$

The generator polynomial that is used at the receiver must be the same as that used at the transmitter:

$$G(x) = 11001_2$$

The calculation is shown in Fig. 10.6.

$$\text{Received data stream} = 11100110 \quad 0110_2 \text{ (no errors)}$$

Assume the received data is errored (errored bit is emboldened):

$$\text{Received data stream} = \mathbf{10100101} \quad 0110_2$$

The calculation is shown in Fig. 10.7. The remainder is not 0000_2; this informs the receiver that the received data is errored (remember that the errored data is in a frame).

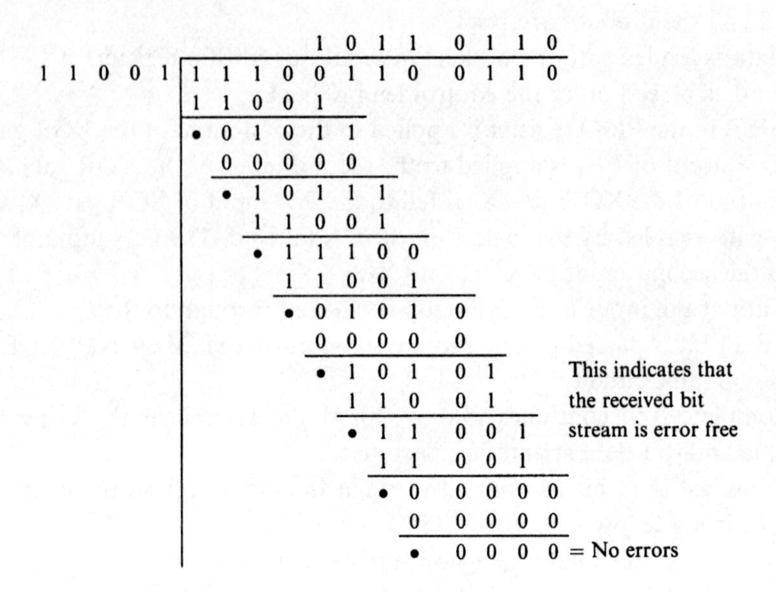

Fig. 10.6 CRC checking using binary notation

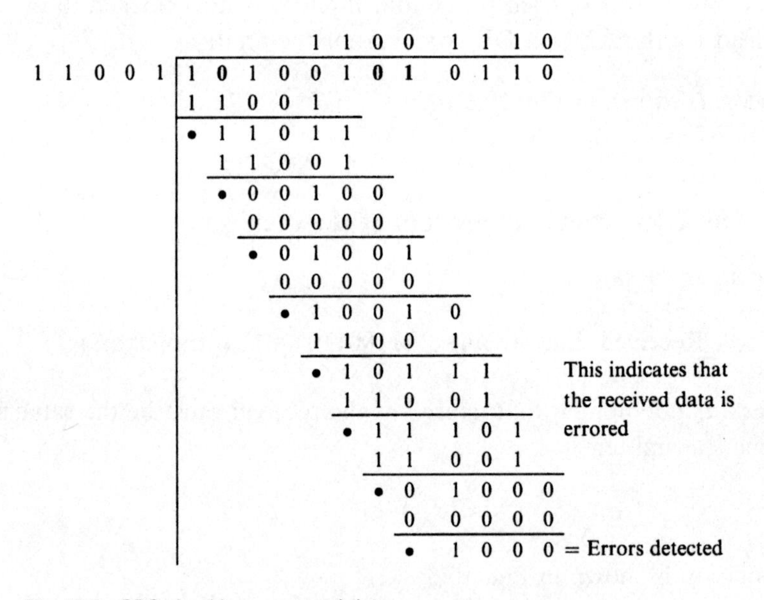

Fig. 10.7 CRC checking – errored data

Using polynomial notation

$$G(x) = 11001_2$$

$$\text{Received data} = 111001100110_2 \text{ (no errors)}$$

$$G(x) = x^4 + x^3 + 1$$

$$\text{Data} = x^{11} + x^{10} + x^9 + x^6 + x^5 + x^2 + x$$

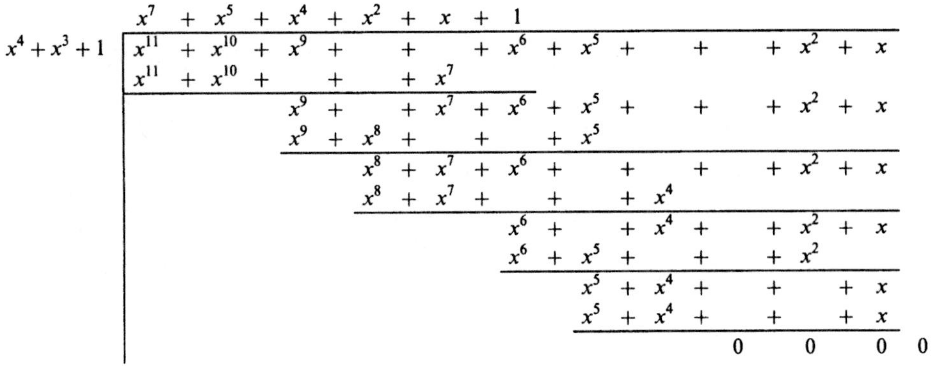

$$
\begin{array}{r|l}
 & x^7 + x^5 + x^4 + x^2 + x + 1 \\
\hline
x^4+x^3+1 & x^{11} + x^{10} + x^9 + x^6 + x^5 + x^2 + x \\
 & x^{11} + x^{10} + x^7 \\
\cline{2-2}
 & x^9 + x^7 + x^6 + x^5 + x^2 + x \\
 & x^9 + x^8 + x^5 \\
\cline{2-2}
 & x^8 + x^7 + x^6 + x^2 + x \\
 & x^8 + x^7 + x^4 \\
\cline{2-2}
 & x^6 + x^4 + x^2 + x \\
 & x^6 + x^5 + x^2 \\
\cline{2-2}
 & x^5 + x^4 + x \\
 & x^5 + x^4 + x \\
\cline{2-2}
 & 0 \quad 0 \quad 0 \quad 0
\end{array}
$$

Fig. 10.8 CRC checking using polynomial notation

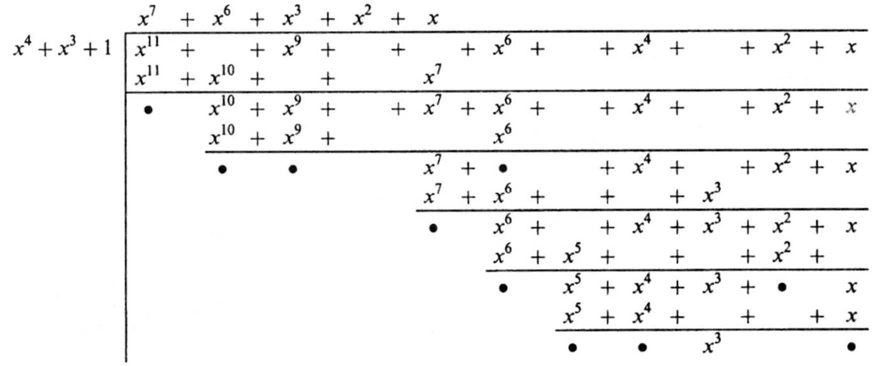

$$
\begin{array}{r|l}
 & x^7 + x^6 + x^3 + x^2 + x \\
\hline
x^4+x^3+1 & x^{11} + x^9 + x^6 + x^4 + x^2 + x \\
 & x^{11} + x^{10} + x^7 \\
\cline{2-2}
 & \bullet \quad x^{10} + x^9 + x^7 + x^6 + x^4 + x^2 + x \\
 & x^{10} + x^9 + x^6 \\
\cline{2-2}
 & \bullet \quad \bullet \quad x^7 + \bullet + x^4 + x^2 + x \\
 & x^7 + x^6 + x^3 \\
\cline{2-2}
 & \bullet \quad x^6 + x^4 + x^3 + x^2 + x \\
 & x^6 + x^5 + x^2 \\
\cline{2-2}
 & \bullet \quad x^5 + x^4 + x^3 + \bullet \quad x \\
 & x^5 + x^4 + x \\
\cline{2-2}
 & \bullet \quad \bullet \quad x^3 \quad \bullet
\end{array}
$$

Fig. 10.9 CRC checking – errored data

The calculation is shown in Fig. 10.8. The remainder is zero and hence the received data is error free.

Now assume that the data stream is errored (errored bit shown in bold):

$$\text{Received data stream} = 10100\mathbf{1}01 \quad 0110_2$$

The calculation is shown in Fig. 10.9. In this case a non-zero remainder results, 0111; hence this indicates that the received data is errored.

CRC circuit

A typical CRC checking circuit is shown in Fig. 10.10(a). The circuit works as follows:

- Initially all the flipflops are reset.
- The serial data stream is clocked into the serial input parallel output (SIPO) register.
- A logic 1 is placed on to the control lead.
- Each received bit is applied to the first input of the XOR gate X_0 on the D_I lead. The Q_3 output of FF_3 is applied to the second input of the XOR gate X_0 as well as the first input of the XOR gate X_1.
- The Q_2 output of FF_2 is applied to the second input of the XOR gate X_1.

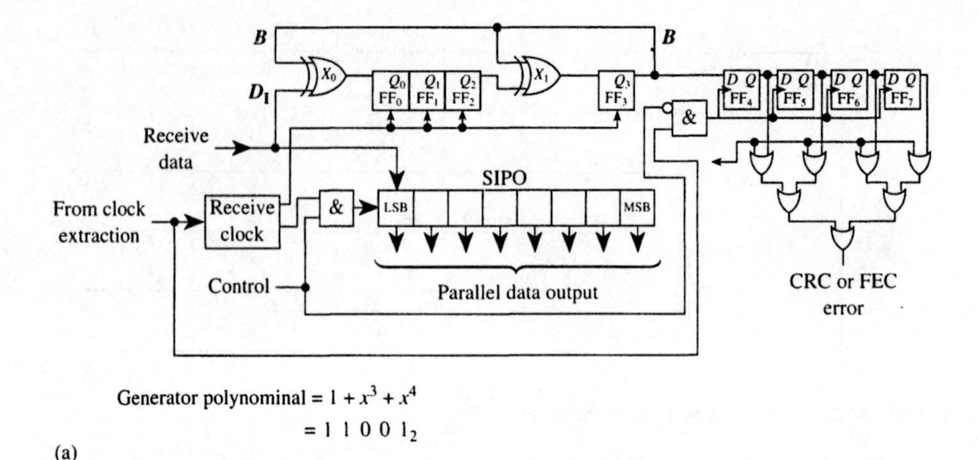

Rx Clk	Rx data	Receive SIPO S–R								FCS shift register							FCS S–R			
		LSB							MSB	B XOR D_I	Q_0	Q_1	Q_2	B XOR Q_2	Q_3	B	Q_0	Q_1	Q_2	Q_3
0	1	0	0	0	0	0	0	0	0	1	0	0	0	0	0	0	0	0	0	0
1	1	1	0	0	0	0	0	0	0	1	1	0	0	0	0	0	1	0	0	0
2	1	1	1	0	0	0	0	0	0	1	1	1	0	0	0	0	1	1	0	0
3	0	1	1	1	0	0	0	0	0	0	1	1	1	1	0	0	1	1	1	0
4	0	0	1	1	1	0	0	0	0	1	0	1	1	0	1	1	0	1	1	1
5	1	0	0	1	1	1	0	0	0	1	1	0	1	1	0	0	1	0	1	0
6	1	1	0	0	1	1	1	0	0	0	1	1	0	1	1	1	1	1	0	1
7	0	1	1	0	0	1	1	1	0	1	0	1	1	0	1	1	0	1	1	1
8	0	0	1	1	0	0	1	1	1	0	1	0	1	1	0	0	1	0	1	0
9	1									0	0	1	0	1	1	1	0	1	0	1
10	1									0	0	0	1	0	1	1	0	0	1	1
11	0									0	0	0	0	0	0	0	0	0	0	0
12											0	0	0	0	0	0	0	0	0	0
																	CRC			

Time ↓

(b)

Fig. 10.10 (a) CRC-4 checking circuit; (b) circuit operation at the receiver

- The data at the input to each flipflop is clocked through to the Q outputs.
- The input data bits are clocked into the SIPO register serially and then clocked out in parallel. The data is also applied to the CRC checking circuit serially.
- When the last data bit has been applied, the logic level on the control is made low.
- The four CRC bits are now applied to the input of the XOR gate X_0 serially.
- The resultant output states of the FF_0 to FF_3 are clocked out into FF_4 to FF_7.
- The Q outputs are ORed together through a combinational logic circuit consisting of OR gates.
- If the final output from this combinational circuit is a logic 0 at the end of the sequence then the received data stream is error free.
- If, however, the final output of the combinational circuit yields a logic 1 the received data and/or the CRC is errored.

The procedure discussed above is shown in Fig. 10.10(b) and applies to an 8-bit data stream.

10.6 CONVOLUTIONAL ENCODING

Convolutional encoding is used in systems for the same reasons as differential coding, as discussed in Chapter 4. The difference is that with convolutional decoding errored data is more likely to be corrected than with differential decoding. This method is sometimes referred to as forward error correction as the encoding method actually improves the error performance of the digital system.

The encoding can be represented in the following ways:

- Connection pictorial.
- Connection vectors or polynomials.
- State and tree diagrams.
- Trellis diagram.

10.6.1 Connection pictorial

Figure 10.11(a) shows the circuit for a typical convolutional encoder. Two generator polynomials are used. The first one is g_1, which is the polynomial for the top branch:

$$g_1 = 111_2 = 1 + x + x^2$$

The second polynomial, g_2, is for the bottom branch:

$$g_2 = 101_2 = 1 + x^2$$

Now consider a message vector (m) as follows:

$$m = 101_2 = 1 + x^2$$

In this encoder there are three registers; to clear the registers after data has been input, two logic 0s must be added to the end of the data stream. This means that for the above message vector the data becomes

$$m = 10100_2$$

Now considering that modulo 2 addition is employed, the following process takes place to produce an output vector U, which consists of u_1 and u_2. When the message vector is shifted into the encoder, the truth table is as shown in Table 10.5. Figure 10.11(b) shows how this truth table is developed in the encoder circuit.

The circuit operation can be verified as shown below. This shows the multiplication of the two words. Remember the following:

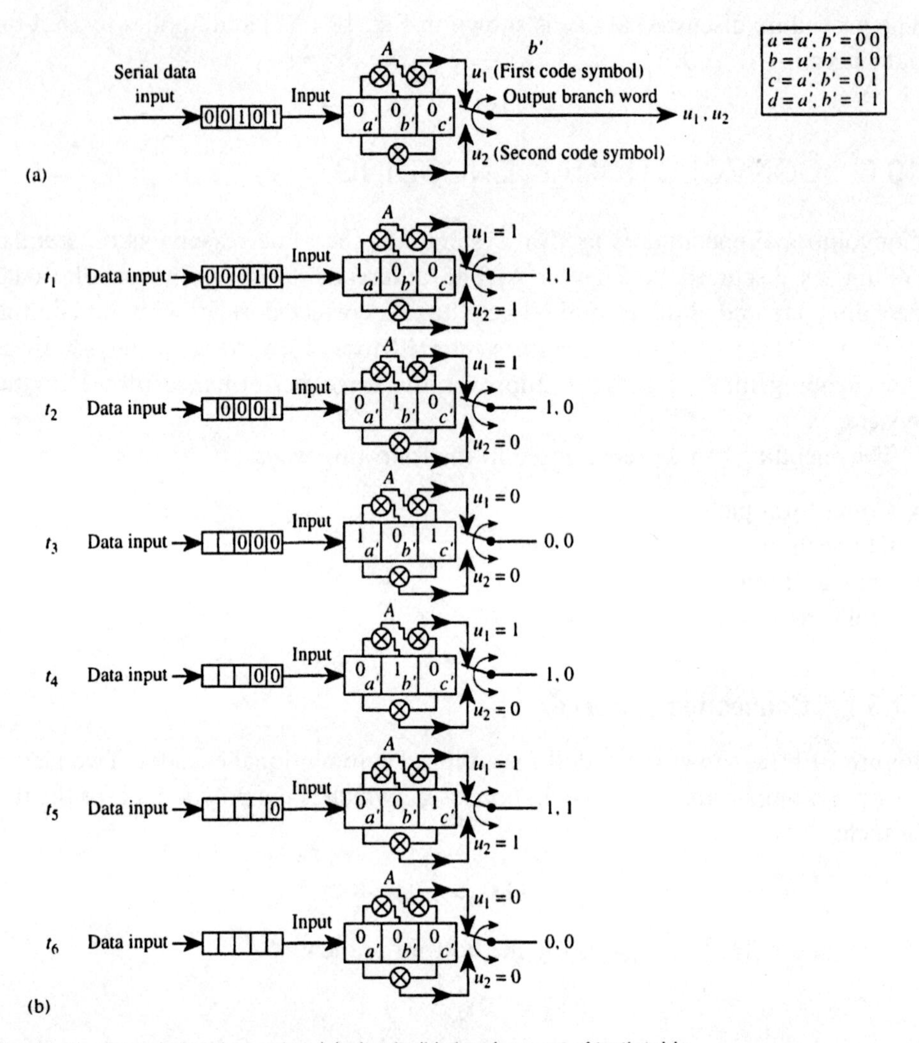

Fig. 10.11 Convolutional encoder: (a) circuit; (b) development of truth table

Table 10.5 Truth table

Serial input						a'	b'	c'	a'	b'	A	c'	u_1	a'	c'	u_2	U	
0	0	0	1	0	1	0	0	0	0	0	0	0	0	0	0	0	0	0
	0	0	0	1	0	1	0	0	1	0	1	0	1	1	0	1	1	1
		0	0	0	1	0	1	0	0	1	1	0	1	0	0	0	1	0
			0	0	0	1	0	1	1	0	1	1	0	1	1	0	0	0
				0	0	0	1	0	0	1	1	0	1	0	0	0	1	0
					0	0	0	1	0	0	0	1	1	0	1	1	1	1
						0	0	0	0	0	0	0	0	0	0	0	0	0

$$0 \times 0 = 0, \quad 0 \times 1 = 1, \quad 1 \times 0 = 0, \quad 1 \times 1 = 1$$

$$\begin{array}{rr}
g_1(x) \cdot m(x) = & 1\,1\,1_2 \\
 & 1\,0\,1\,0\,0_2 \\
\hline
 & 1\,1\,1\,0\,0 \\
 & 0\,0\,0\,0\,0\,0 \\
 & 1\,1\,1\,0\,0\,0\,0 \\
\hline
u_1(x) = & 1\,1\,0\,1\,1\,0\,0
\end{array} \quad \text{(modulo 2 addition)}$$

$$\begin{array}{rr}
g_2(x) \cdot m(x) = & 1\,0\,1_2 \\
 & 1\,0\,1\,0\,0_2 \\
\hline
 & 1\,0\,1\,0\,0 \\
 & 0\,0\,0\,0\,0\,0 \\
 & 1\,0\,1\,0\,0\,0\,0 \\
\hline
u_2(x) = & 1\,0\,0\,0\,1\,0\,0
\end{array} \quad \text{(modulo 2 addition)}$$

Now $u_1(x)$ and $u_2(x)$ yield

$$\begin{array}{rccccccc}
u_1(x) = & 1 & 1 & 0 & 1 & 1 & 0 & 0 \\
u_2(x) = & 1 & 0 & 0 & 0 & 1 & 0 & 0 \\
\hline
U(x) = & 1,1 & 1,0 & 0,0 & 1,0 & 1,1 & 0,0 & 0,0
\end{array}$$

10.6.2 Connection vectors or polynomials

Another method of describing the convolutional operation is by means of polynomials. As shown in Section 10.6.1, the following polynomials result:

$$g_1(x) = 111_2 = 1 + x + x^2$$

$$g_2(x) = 101_2 = 1 + \quad + x^2$$

$$m(x) = 10100_2 = x^2 + x^4$$

Now

$$m(x) \cdot g_1(x) = (x^2 + x^4) \cdot (1 + x + x^2)$$

This yields

$$\begin{array}{r}
x^2 + x^1 + 1 \\
x^4 + x^2 \\
\hline
x^4 + x^3 + x^2 \\
x^6 + x^5 + x^4 \qquad\qquad \\
\hline
x^6 + x^5 + \quad + x^3 + x^2 \\
\hline
\end{array}$$

$$u_1(x) = x^6 + x^5 + x^3 + x^2 \quad \text{(modulo 2 addition)}$$

Therefore

$$u_1(x) = x^6 + x^5 + 0 \cdot x^4 + x^3 + x^2 + 0 \cdot x^1 + 0$$

$$m(x) \cdot g_2(x) = (x^2 + x^4) \cdot (1 + x^2)$$

This yields

$$\begin{array}{r} x^2 + 1 \\ x^4 + x^2 \\ \hline x^4 + \quad + x^2 \\ x^6 + \quad + x^4 \qquad\quad \\ \hline x^6 + \quad + x^4 + \quad + x^2 \\ \hline \end{array}$$

$$u_2(x) = x^6 + x^4 + x^2 \quad \text{(modulo 2 addition)}$$

Therefore

$$u_2(x) = x^6 + 0 \cdot x^5 + 0 \cdot x^4 + 0 \cdot x^3 + x^2 + 0 \cdot x^1 + 0$$

Now $u_1(x)$ and $u_2(x)$ yield

$$\begin{array}{rccccccc} u_1(x) = & x^6 + & x^5 + 0 \cdot x^4 + & x^3 + x^2 + 0 \cdot x^1 + & 0 \\ u_2(x) = & x^6 + 0 \cdot x^5 + 0 \cdot x^4 + 0 \cdot x^3 + x^2 + 0 \cdot x^1 + & 0 \\ \hline U(x) = 1,1 & 1,0 & 0,0 & 1,0 & 1,1 & 0,0 & 0,0 \end{array}$$

Although this method clearly describes the operation of the encoder, it does not display the operation with respect to time.

10.6.3 State and tree diagrams

Figure 10.12(a) shows a convolutional encoder and Fig. 10.12(b) shows the resultant state diagram.

Example 10.9

Determine the output code words produced when the data 10100_2 is applied to the system, using the state diagram given in Fig. 10.12(b).

Solution

$$\text{Bit} = 1234567$$

$$\text{Input data} = 1010000_2$$

Two logic 0 states are added to the back of the data to flush the registers:

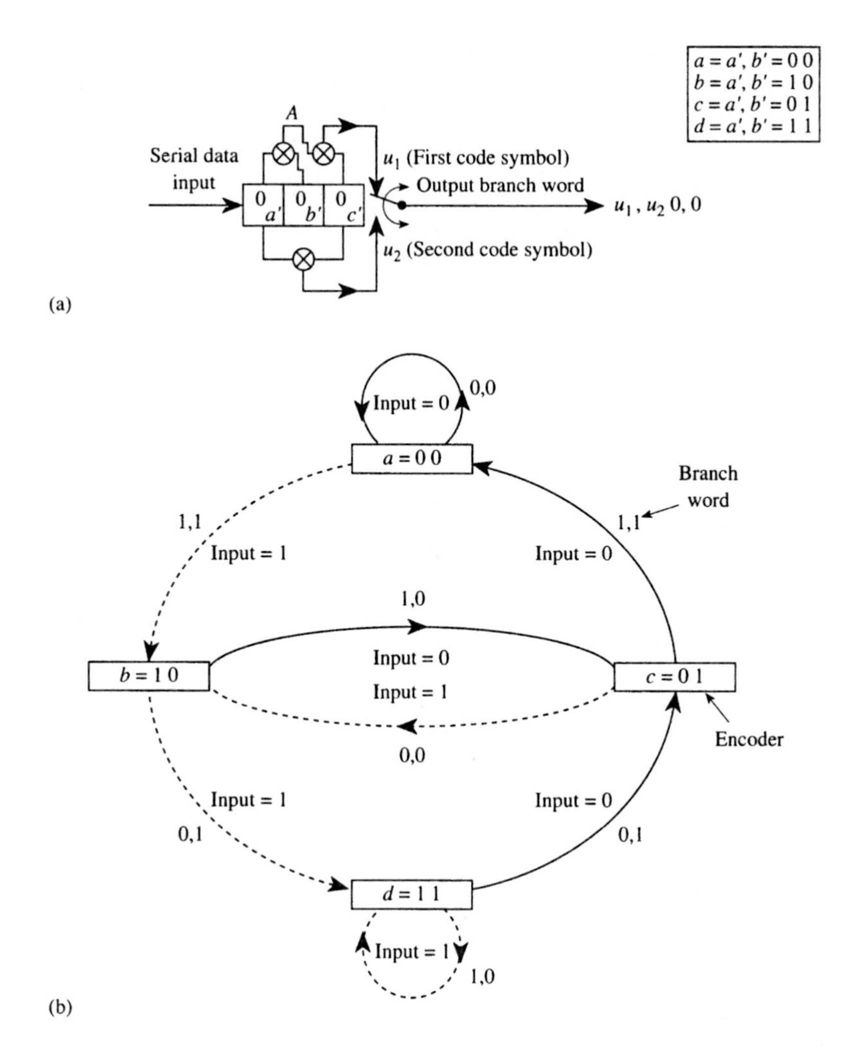

Fig. 10.12 Convolutional encoder: (a) circuit; (b) state diagram

Bit 1 = 1:	Code word = 1,1:	Branch = *a* to *b*
Bit 2 = 0:	Code word = 1,0:	Branch = *b* to *c*
Bit 3 = 1:	Code word = 0,0:	Branch = *c* to *b*
Bit 4 = 0:	Code word = 1,0:	Branch = *b* to *c*
Bit 5 = 0:	Code word = 1,1:	Branch = *c* to *a*
Bit 6 = 0:	Code word = 0,0:	Branch = *a* to *a*
Bit 7 = 0:	Code word = 0,0:	Branch = *a* to *a*

Therefore:

$$\text{Input data} = 1\ 0\ 1\ 0\ 0\ 0\ 0_2$$

$$\text{Output code words} = 11\ 10\ 00\ 10\ 11\ 00\ 00$$

From this state diagram a tree diagram can be drawn as shown in Fig. 10.13. Note that after time t_4 the tree repeats itself.

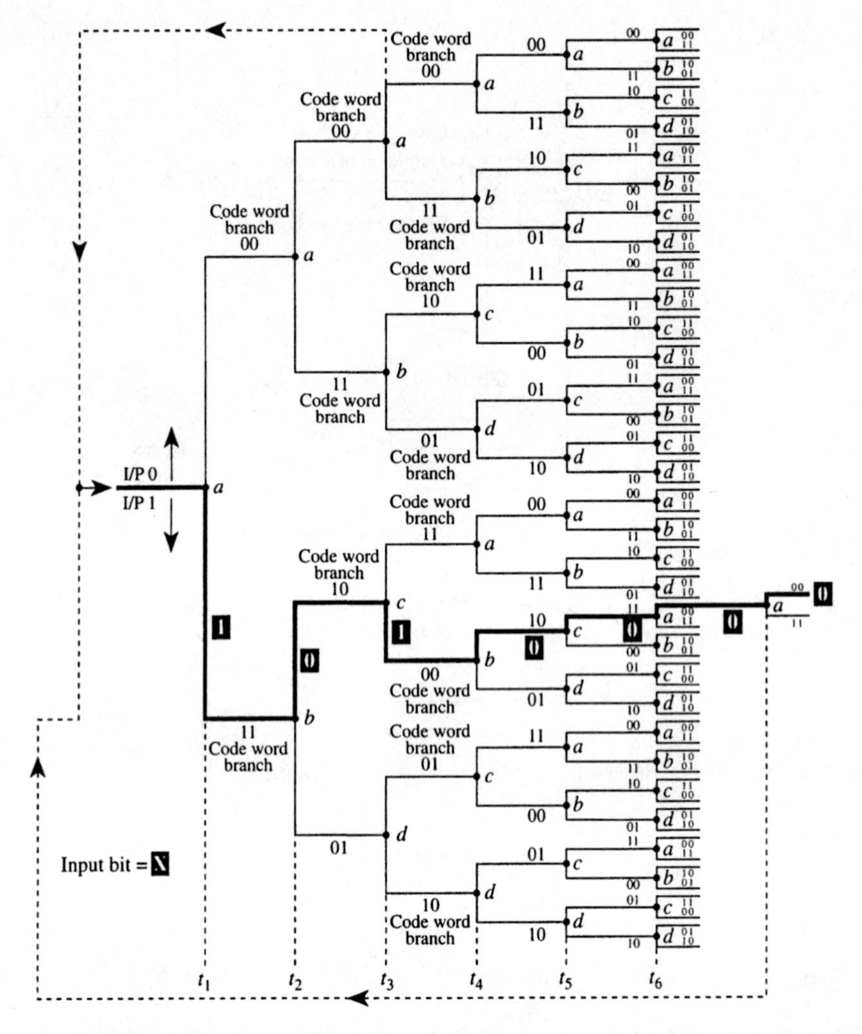

Fig. 10.13 Tree diagram

With respect to the tree diagram note that when a logic 1 is input into the system, the bottom branch is taken; when a logic 0 is input, the upper branch is taken. The tree repeats itself and hence can be used for larger systems than the example used.

When the message vector 10100_2 is entered then the first 1 causes the lower branch to be taken giving a branch code word of 1,1. The 0 then causes the upper branch to be taken resulting in a branch code word of 1,0 to be output. The next 1 causes the lower branch to be taken resulting in a branch code word of 00. The next 0 causes the upper branch to be taken producing the branch code word of 10. The next 0 causes the upper branch to be taken producing the branch code word of 11.

To flush the registers a further two logic 0 states must be applied. The first of these causes the upper branch to be taken, producing a branch code word of 00. The second causes the upper branch to be taken, producing a branch code word of 00.

Figure 10.13 shows the process when the input data 1010000_2 is applied. By following the path through the tree diagram the branch code words are as follows:

MSB: Input 1 = 11 → Branch a to b
Input 0 = 10 → Branch b to c
Input 1 = 00 → Branch c to b
Input 0 = 10 → Branch b to c
Input 0 = 11 → Branch c to a
Input 0 = 00 → Branch a to a
LSB: Input 0 = 00 → Branch a to a

Therefore the data stream causes the output branch code words to be produced as shown below:

$$\text{Input data} = 1\ 0\ 1\ 0\ 0\ 0\ 0_2$$

$$\text{Code words} = 11\ 10\ 00\ 10\ 11\ 00\ 00$$

The same result was obtained from the methods given in Sections 10.6.1 and 10.6.2. However, this method introduces the element of time into the operation.

10.6.4 The trellis diagram

The trellis diagram is better than the state diagram and tree diagram as it combines both of these diagrams into one. Figure 10.14 shows the trellis diagram. The branching for an input logic 1 bit is shown as a dotted line and the branching for an input logic 0 bit is shown as a solid line.

Example 10.10

The input data to a convolutional encoder is 10100_2. Determine the output code for this encoder.

Solution

To clear the encoder, two zeros must be added to the data. In the example an extra logic 0 has been added. The encoding process is shown in Fig. 10.15.

$$\text{Input data} = 1010000_2$$

$$\text{Bit number} = 1234567$$

- Bit 1 is a logic 1 and causes the output code to be 11. The path is from a to b.
- Bit 2 is a logic 0 and causes the output code to be 10. The path is from b to c.
- Bit 3 is a logic 1 and causes the output code to be 00. The path is from c to b.

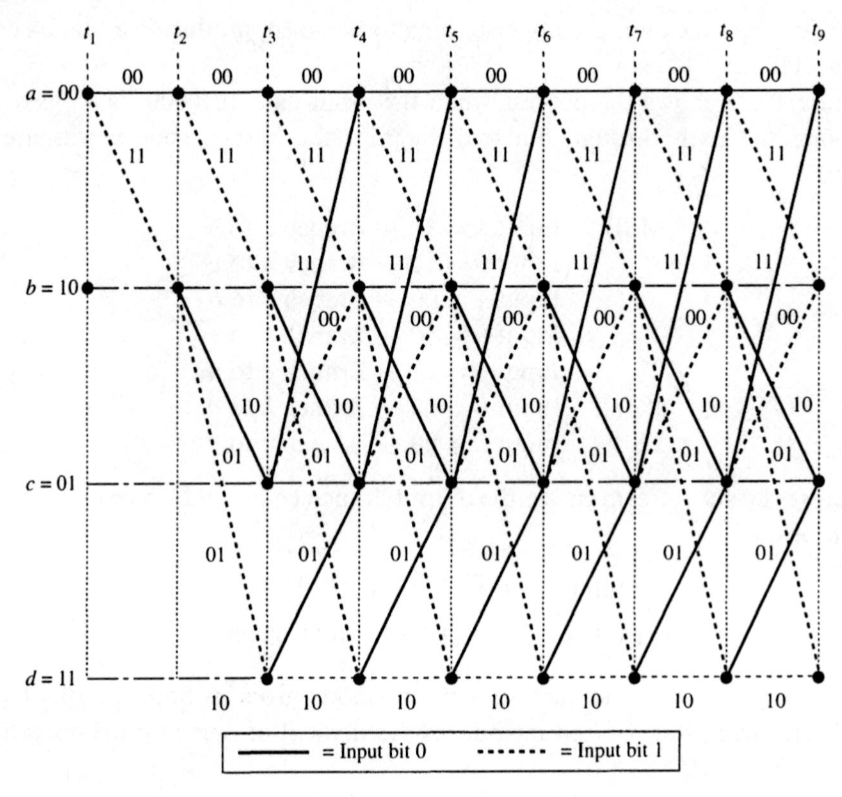

Fig. 10.14 Trellis diagram

- Bit 4 is a logic 0 and causes the output code to be 10. The path is from *b* to *c*.
- Bit 5 is a logic 0 and causes the output code to be 11. The path is from *c* to *a*.
- Bit 6 is a logic 0 and causes the output code to be 00. The path is from *a* to *a*.
- Bit 7 is a logic 0 and causes the output code to be 00. The path is from *a* to *a*.

Output code = 11 10 00 10 11 00 00

Output branching = a–b b–c c–b b–c c–a a–a a–a

This is the same as the code produced by the state diagram and the tree diagram.

If the code 11 10 00 10 11 00 00_2 is received at a receiver then the convolutional decoder will produce the output data as follows:

Output code = 11 10 00 10 11 00 00

Output branching = a–b b–c c–b b–c c–a a–a a–a

Data = 1 0 1 0 0 0 0_2

The important question is how will the convolutional decoder react when the input data is errored. The decoder has no idea that the data it receives is errored. The errors will only become apparent as the data is applied to the trellis algorithm. In practice this is a computer program.

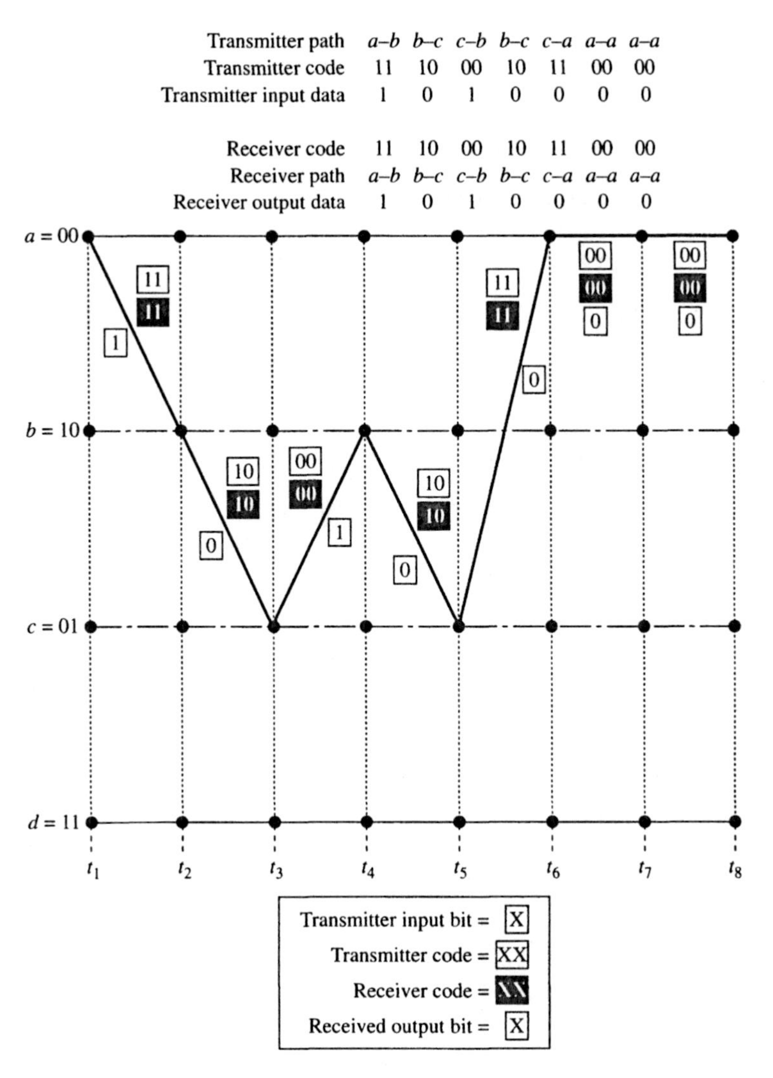

Fig. 10.15 Trellis diagram for Example 10.10

Example 10.11

The code received by a convolutional decoder is as follows:

$$\text{Code} = 11\ 10\ 00\ 10\ 11\ 00\ 00_2$$

$$\text{Symbol no.} = \ 1\ \ 2\ \ 3\ \ 4\ \ 5\ \ 6\ \ 7$$

Determine the most probable transmitted data that has been errored by the medium.

Solution

Taking path 1, the encoding process is shown in Fig. 10.16.

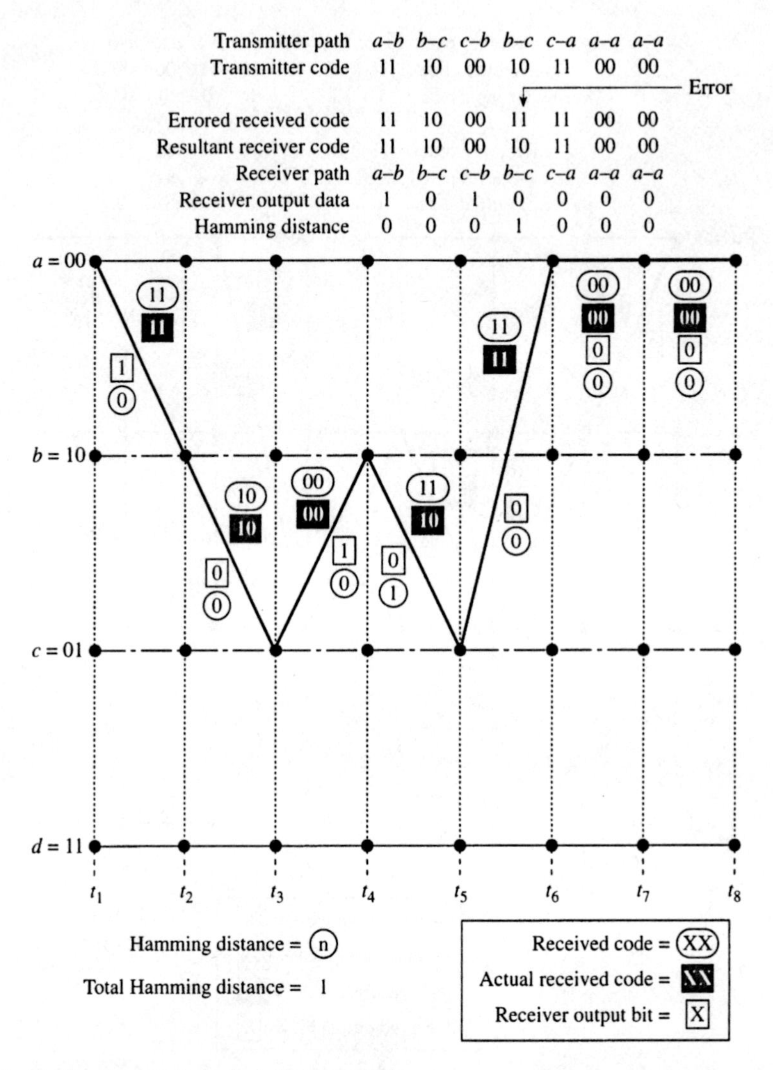

Fig. 10.16 Trellis diagram for path 1, Example 10.11

Symbol 1:			
Branching =	*a* to *b*		Output data = 1
Input code	=	11	
Branching code	=	11	
Hamming distance	=	0	
Symbol 2:			
Branching =	*b* to *c*		Output data = 0
Input code	=	10	
Branching code	=	10	
Hamming distance	=	0	

Symbol 3:

Branching =	*c* to *b*		Output data = 1
Input code	=	00	
Branching code	=	00	
Hamming distance	=	0	

Symbol 4:

Branching =	*b* to *c*		Output data = 0
Input code	=	11	
Branching code	=	10	
Hamming distance	=	1	

Symbol 5:

Branching =	*c* to *a*		Output data = 0
Input code	=	11	
Branching code	=	11	
Hamming distance	=	0	

Symbol 6:

Branching =	*a* to *a*		Output data = 0
Input code	=	00	
Branching code	=	00	
Hamming distance	=	0	

Symbol 7:

Branching =	*a* to *a*		Output data = 0
Input code	=	00	
Branching code	=	00	
Hamming distance	=	0	

Total Hamming distance = 1

However, this is not the only possible path. There are 16 different paths. Figures 10.17 to 10.31 shows paths 2 to 16 respectively.

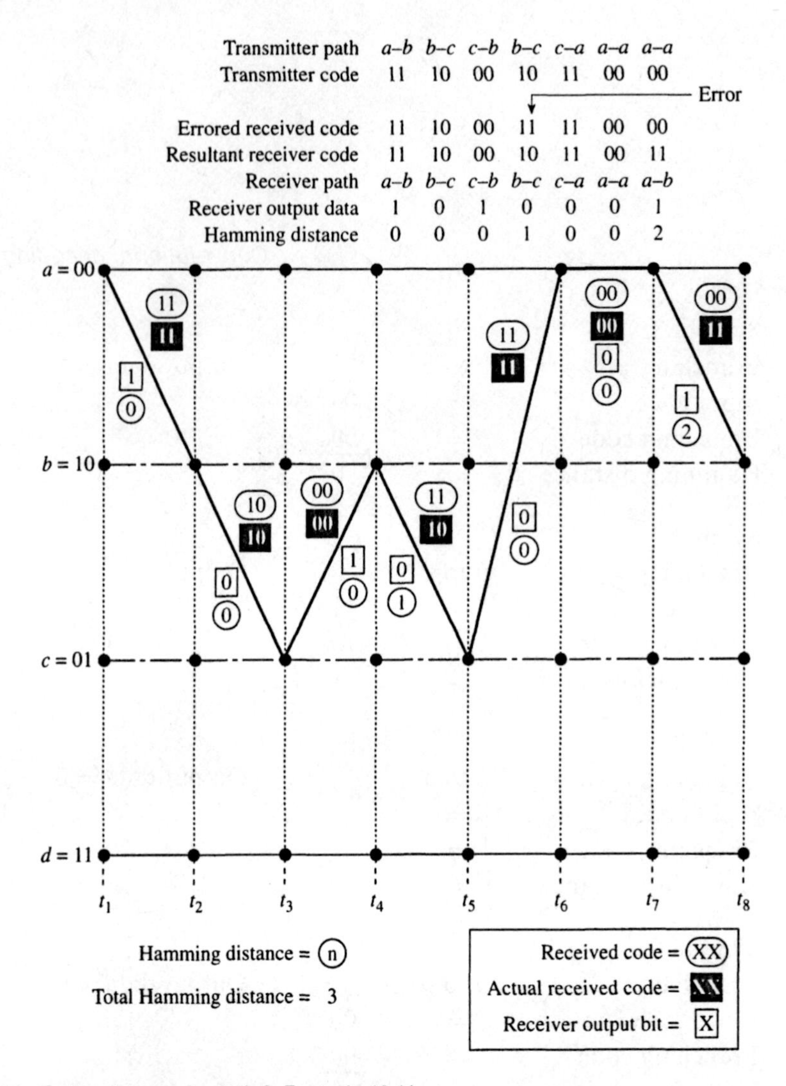

Fig. 10.17 Trellis diagram for path 2, Example 10.11

Consider path 2 (Fig. 10.17):

Tx data	=	1	0	1	0	0	0	0
Tx path	=	*a–b*	*b–c*	*c–b*	*b–c*	*c–a*	*a–a*	*a–a*
Tx code output	=	11	10	00	10	11	00	00
Rx output data	=	1	0	1	0	0	0	1
Rx path	=	*a–b*	*b–c*	*c–b*	*b–c*	*c–a*	*a–a*	*a–b*
Rx code input	=	11	10	00	11	11	00	00
Rx code actual	=	11	10	00	10	11	00	11
Hamming distance	=	0	0	0	1	0	0	2

Total Hamming distance = 3

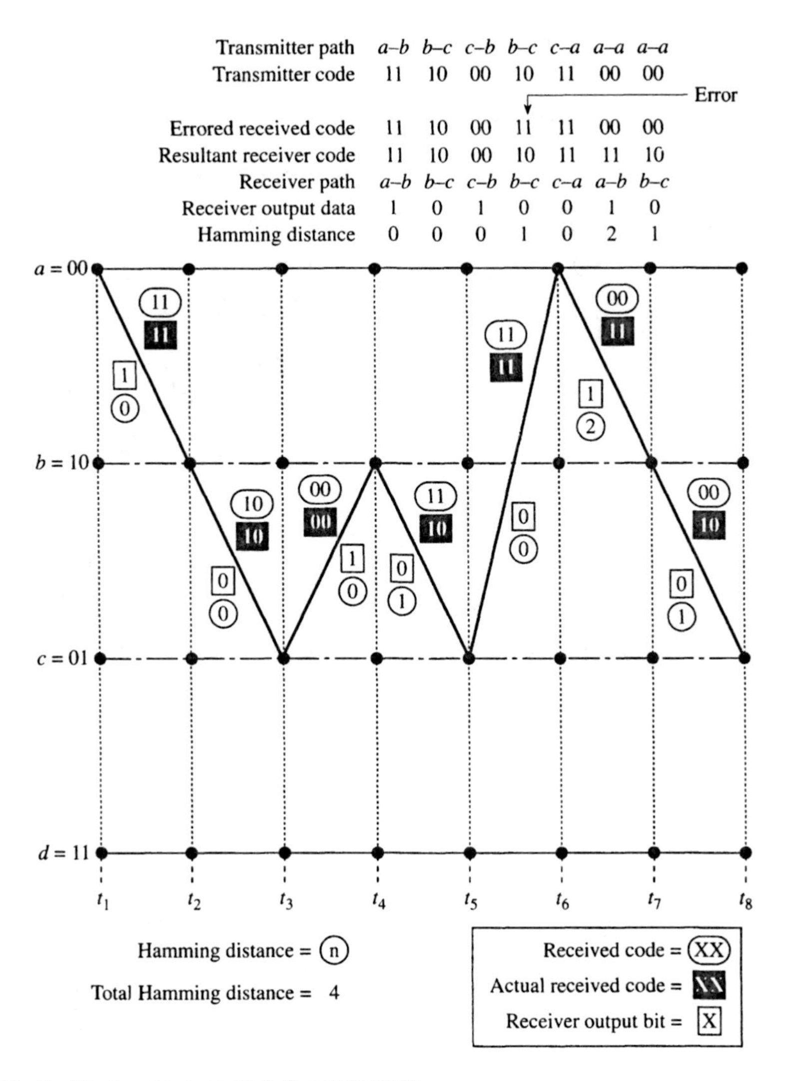

Fig. 10.18 Trellis diagram for path 3, Example 10.11

Consider path 3 (Fig. 10.18):

Tx data	=	1	0	1	0	0	0	0
Tx path	=	*a–b*	*b–c*	*c–b*	*b–c*	*c–a*	*a–a*	*a–a*
Tx code output	=	11	10	00	10	11	00	00
Rx output data	=	1	0	1	0	0	1	0
Rx path	=	*a–b*	*b–c*	*c–b*	*b–c*	*c–a*	*a–b*	*b–c*
Rx code input	=	11	10	00	11	11	00	00
Rx code actual	=	11	10	00	10	11	11	10
Hamming distance	=	0	0	0	1	0	2	1

Total Hamming distance = 4

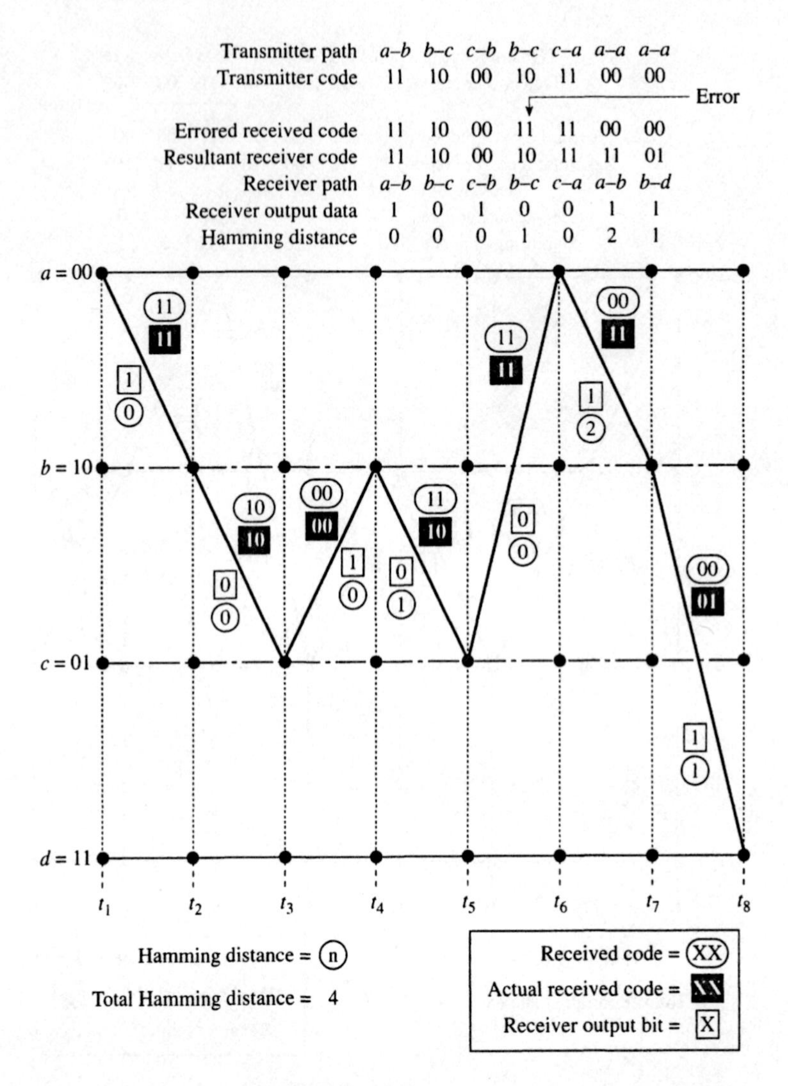

Fig. 10.19 Trellis diagram for path 4, Example 10.11

Consider path 4 (Fig. 10.19):

Tx data	= 1	0	1	0	0	0	0
Tx path	= *a–b*	*b–c*	*c–b*	*b–c*	*c–a*	*a–a*	*a–a*
Tx code output	= 11	10	00	10	11	00	00
Rx output data	= 1	0	1	0	0	1	1
Rx path	= *a–b*	*b–c*	*c–b*	*b–c*	*c–a*	*a–b*	*b–d*
Rx code input	= 11	10	00	11	11	00	00
Rx code actual	= 11	10	00	10	11	11	01
Hamming distance	= 0	0	0	1	0	2	1

Total Hamming distance = 4

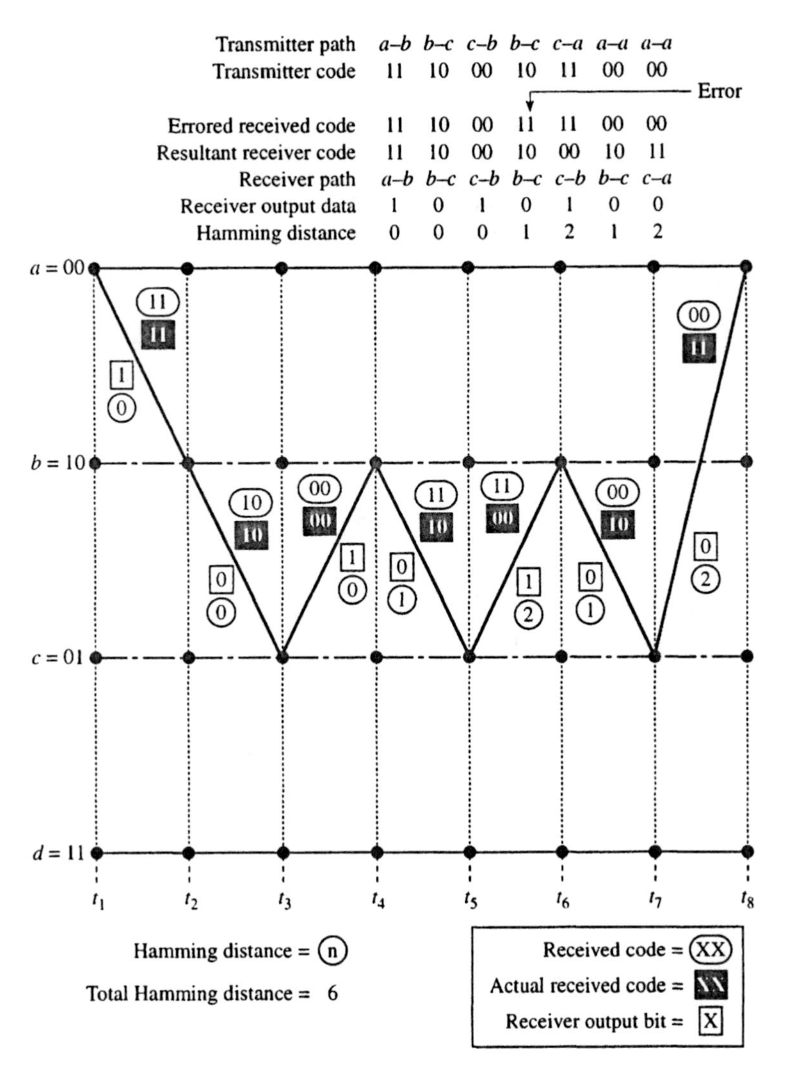

Fig. 10.20 Trellis diagram for path 5, Example 10.11

Consider path 5 (Fig. 10.20):

Tx data	=	1	0	1	0	0	0	0
Tx path	=	*a–b*	*b–c*	*c–b*	*b–c*	*c–a*	*a–a*	*a–a*
Tx code output	=	11	10	00	10	11	00	00
Rx output data	=	1	0	1	0	1	0	0
Rx path	=	*a–b*	*b–c*	*c–b*	*b–c*	*c–b*	*b–c*	*c–a*
Rx code input	=	11	10	00	11	11	00	00
Rx code actual	=	11	10	00	10	00	10	11
Hamming distance	=	0	0	0	1	2	1	2

Total Hamming distance = 6

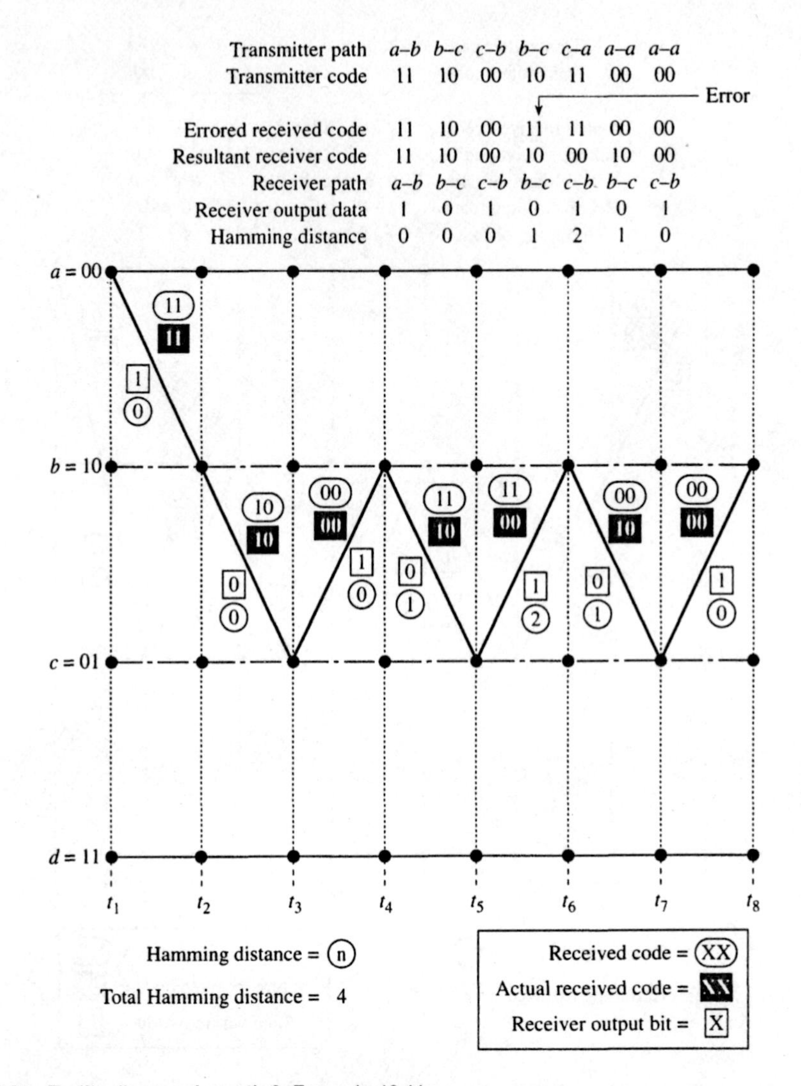

Fig. 10.21 Trellis diagram for path 6, Example 10.11

Consider path 6 (Fig. 10.21):

Tx data	= 1	0	1	0	0	0	0
Tx path	= *a–b*	*b–c*	*c–b*	*b–c*	*c–a*	*a–a*	*a–a*
Tx code output	= 11	10	00	10	11	00	00
Rx output data	= 1	0	1	0	1	0	1
Rx path	= *a–b*	*b–c*	*c–b*	*b–c*	*c–b*	*b–c*	*c–d*
Rx code input	= 11	10	00	11	11	00	00
Rx code actual	= 11	10	00	10	00	10	00
Hamming distance	= 0	0	0	1	2	1	0

Total Hamming distance = 4

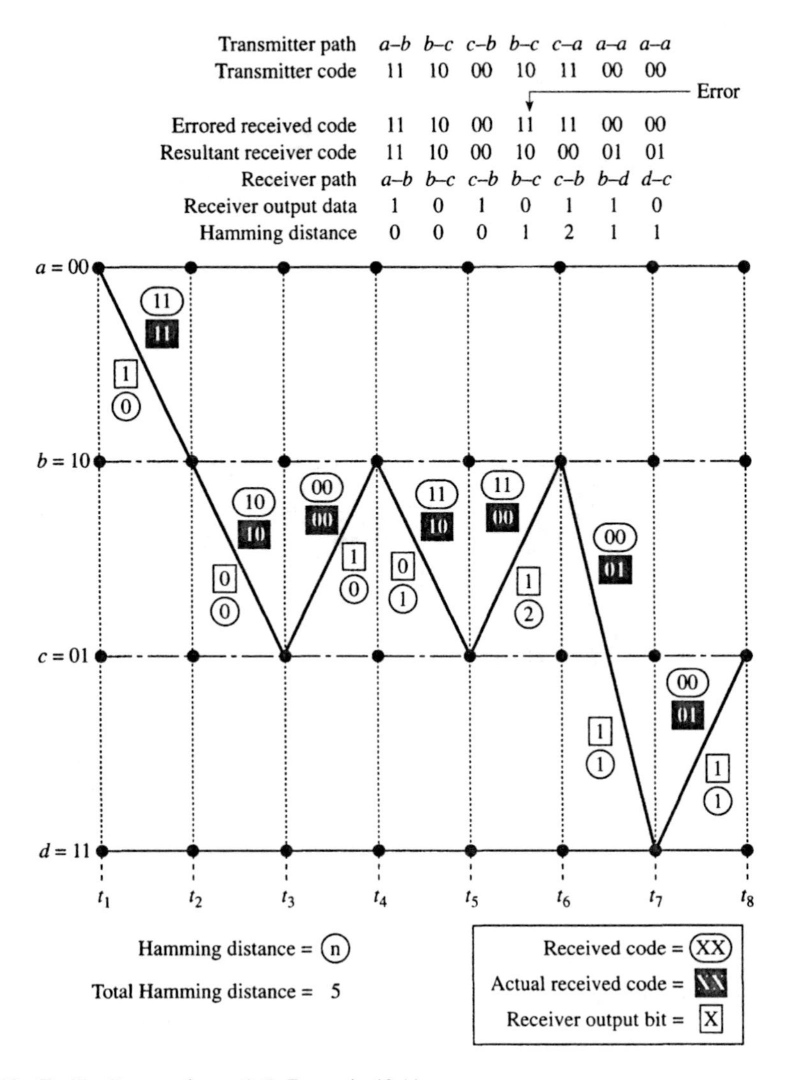

Fig. 10.22 Trellis diagram for path 7, Example 10.11

Consider path 7 (Fig. 10.22):

Tx data	=	1	0	1	0	0	0	0
Tx path	=	*a–b*	*b–c*	*c–b*	*b–c*	*c–a*	*a–a*	*a–a*
Tx code output	=	11	10	00	10	11	00	00
Rx output data	=	1	0	1	0	1	1	0
Rx path	=	*a–b*	*b–c*	*c–b*	*b–c*	*c–b*	*b–d*	*d–c*
Rx code input	=	11	10	00	11	11	00	00
Rx code actual	=	11	10	00	10	00	01	01
Hamming distance	=	0	0	0	1	2	1	1

Total Hamming distance = 5

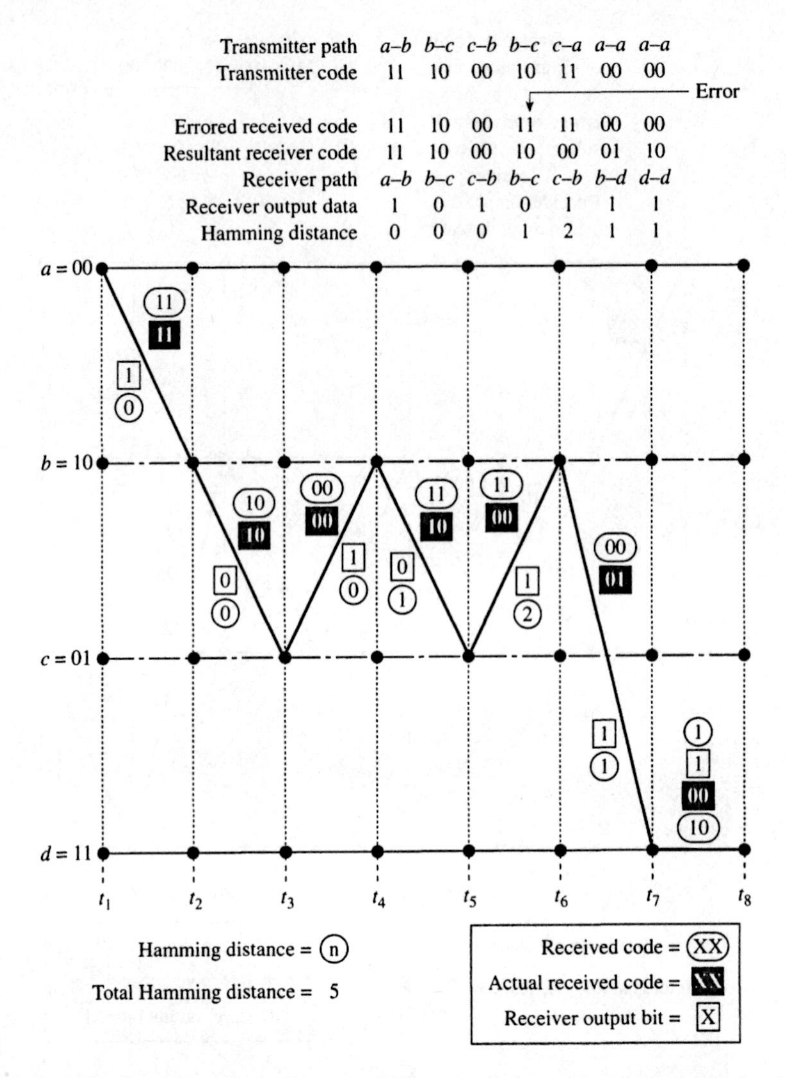

Fig. 10.23 Trellis diagram for path 8, Example 10.11

Consider path 8 (Fig. 10.23):

Tx data	= 1	0	1	0	0	0	0
Tx path	= *a–b*	*b–c*	*c–b*	*b–c*	*c–a*	*a–a*	*a–a*
Tx code output	= 11	10	00	10	11	00	00
Rx output data	= 1	0	1	0	1	1	1
Rx path	= *a–b*	*b–c*	*c–b*	*b–c*	*c–b*	*b–d*	*d–d*
Rx code input	= 11	10	00	11	11	00	00
Rx code actual	= 11	10	00	10	00	01	10
Hamming distance	= 0	0	0	1	2	1	1

Total Hamming distance = 5

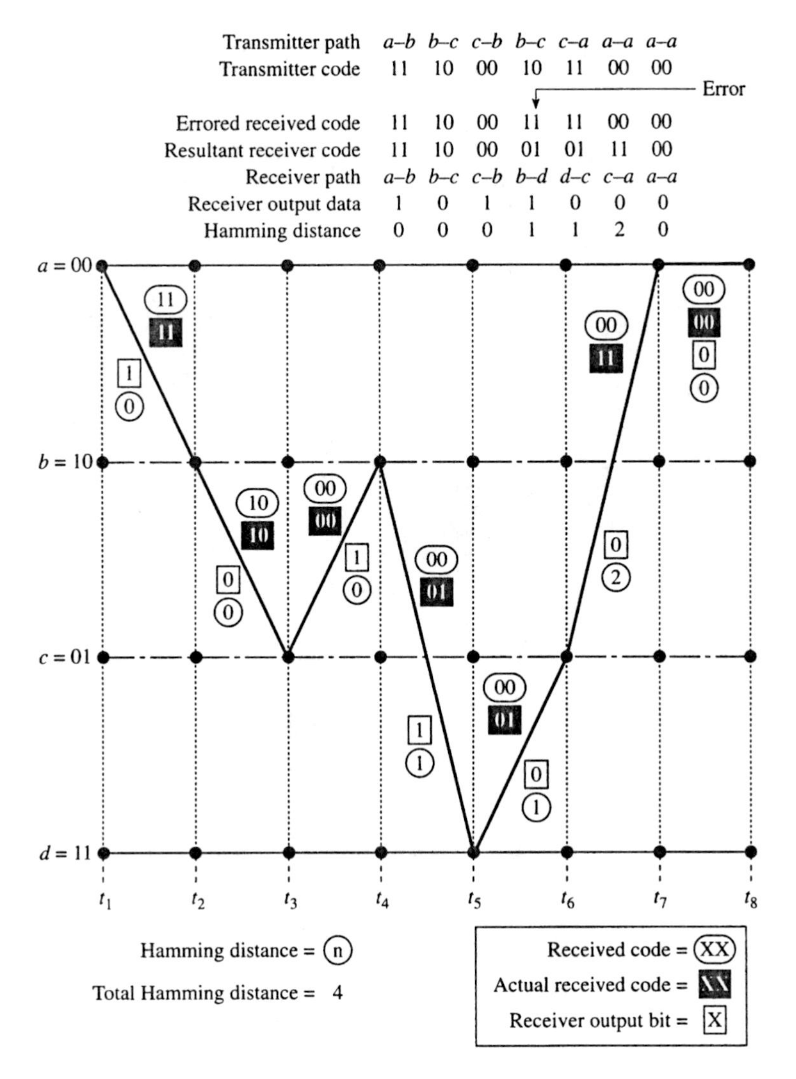

Fig. 10.24 Trellis diagram for path 9, Example 10.11

Consider path 9 (Fig. 10.24):

Tx data	=	1	0	1	0	0	0	0
Tx path	=	*a–b*	*b–c*	*c–b*	*b–c*	*c–a*	*a–a*	*a–a*
Tx code output	=	11	10	00	10	11	00	00
Rx output data	=	1	0	1	1	0	0	0
Rx path	=	*a–b*	*b–c*	*c–b*	*b–d*	*d–c*	*c–a*	*a–a*
Rx code input	=	11	10	00	11	11	00	00
Rx code actual	=	11	10	00	01	01	11	00
Hamming distance	=	0	0	0	1	1	2	0

Total Hamming distance = 4

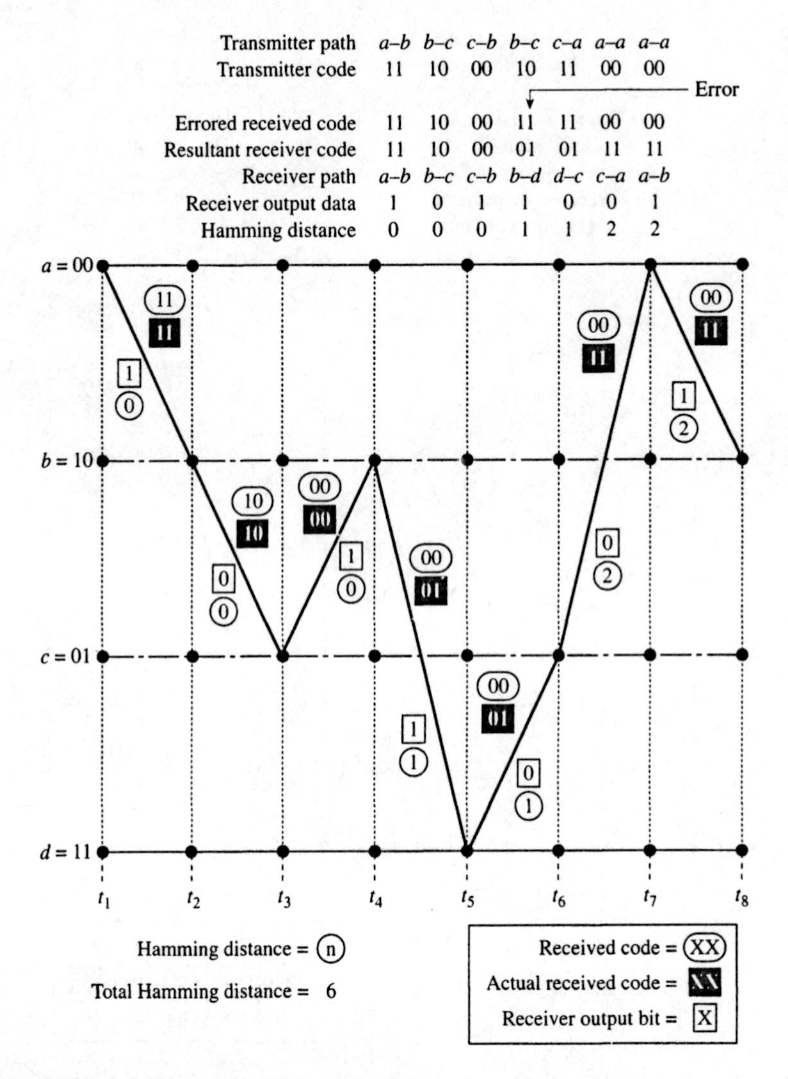

Fig. 10.25 Trellis diagram for path 10, Example 10.11

Consider path 10 (Fig. 10.25):

Tx data	=	1	0	1	0	0	0	0
Tx path	=	*a–b*	*b–c*	*c–b*	*b–c*	*c–a*	*a–a*	*a–a*
Tx code output	=	11	10	00	10	11	00	00
Rx output data	=	1	0	1	1	0	0	1
Rx path	=	*a–b*	*b–c*	*c–b*	*b–d*	*d–c*	*c–a*	*a–b*
Rx code input	=	11	10	00	11	11	00	00
Rx code actual	=	11	10	00	01	01	11	11
Hamming distance	=	0	0	0	1	1	2	2

Total Hamming distance = 6

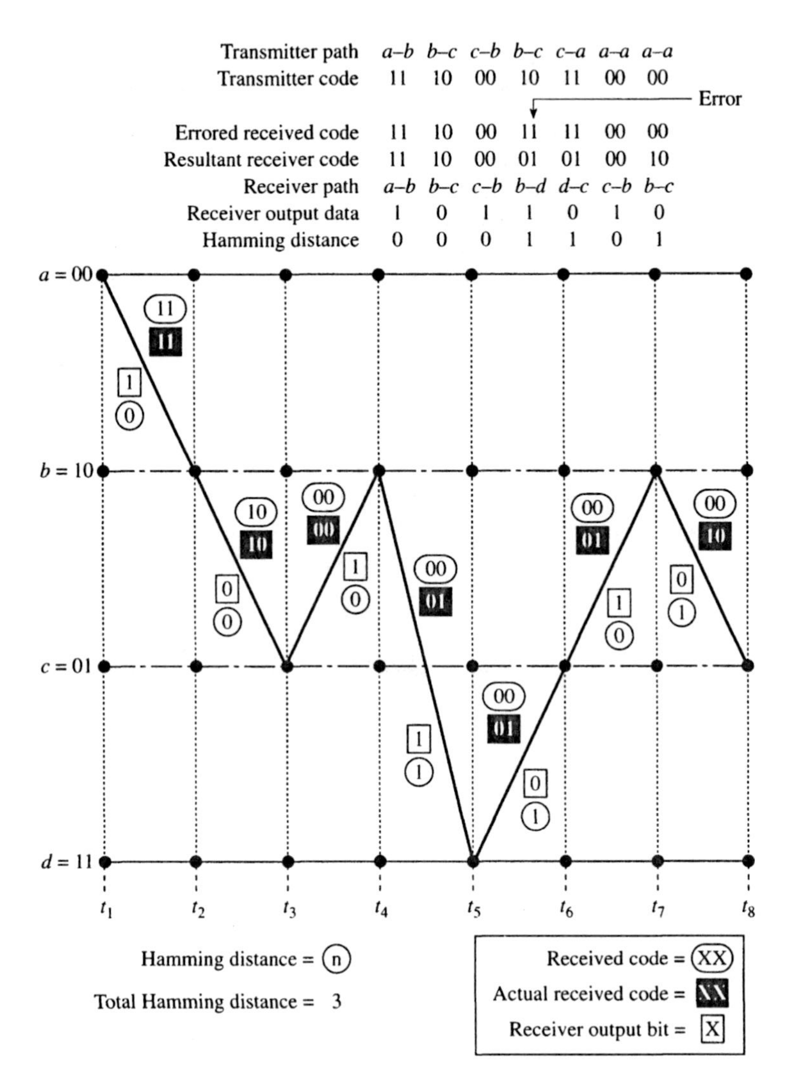

Fig. 10.26 Trellis diagram for path 11, Example 10.11

Consider path 11 (Fig. 10.26):

Tx data	= 1	0	1	0	0	0	0
Tx path	= *a–b*	*b–c*	*c–b*	*b–c*	*c–a*	*a–a*	*a–a*
Tx code output	= 11	10	00	10	11	00	00
Rx output data	= 1	0	1	1	0	1	0
Rx path	= *a–b*	*b–c*	*c–b*	*b–d*	*d–c*	*c–b*	*b–c*
Rx code input	= 11	10	00	11	11	00	00
Rx code actual	= 11	10	00	01	01	00	10
Hamming distance	= 0	0	0	1	1	0	1

Total Hamming distance = 3

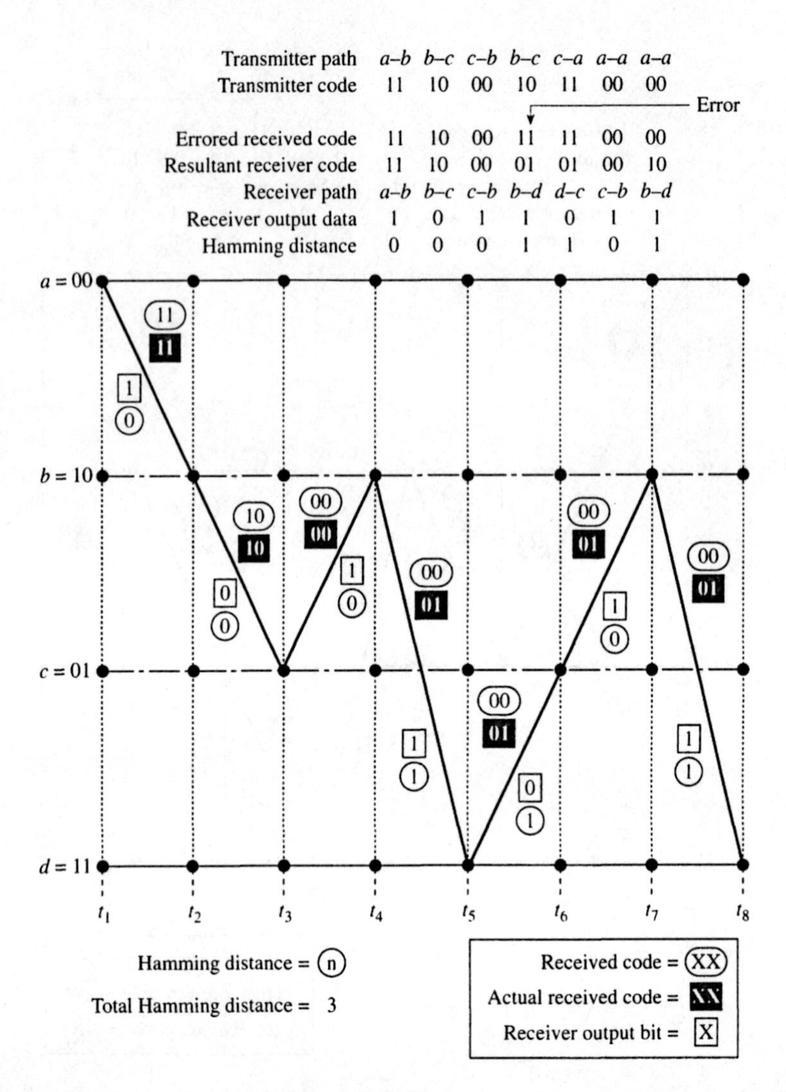

Fig. 10.27 Trellis diagram for path 12, Example 10.11

Consider path 12 (Fig. 10.27):

Tx data	=	1	0	1	0	0	0	0
Tx path	=	*a–b*	*b–c*	*c–b*	*b–c*	*c–a*	*a–a*	*a–a*
Tx code output	=	11	10	00	10	11	00	00
Rx output data	=	1	0	1	1	0	1	1
Rx path	=	*a–b*	*b–c*	*c–b*	*b–d*	*d–c*	*c–b*	*b–d*
Rx code input	=	11	10	00	11	11	00	00
Rx code actual	=	11	10	00	01	01	00	01
Hamming distance	=	0	0	0	1	1	0	1

Total Hamming distance = 3

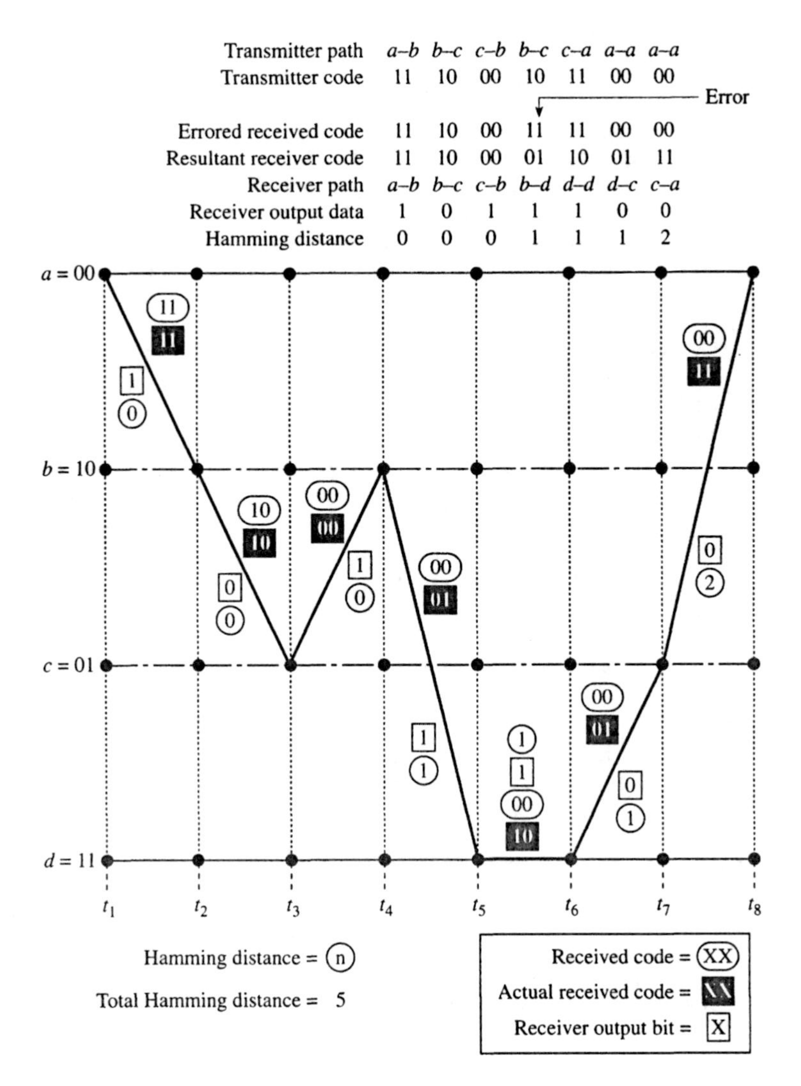

Fig. 10.28 Trellis diagram for path 13, Example 10.11

Consider path 13 (Fig. 10.28):

Tx data	=	1	0	1	0	0	0	0
Tx path	=	*a–b*	*b–c*	*c–b*	*b–c*	*c–a*	*a–a*	*a–a*
Tx code output	=	11	10	00	10	11	00	00
Rx output data	=	1	0	1	1	1	0	0
Rx path	=	*a–b*	*b–c*	*c–b*	*b–d*	*d–d*	*d–c*	*c–a*
Rx code input	=	11	10	00	11	11	00	00
Rx code actual	=	11	10	00	01	10	01	11
Hamming distance	=	0	0	0	1	1	1	2

Total Hamming distance = 5

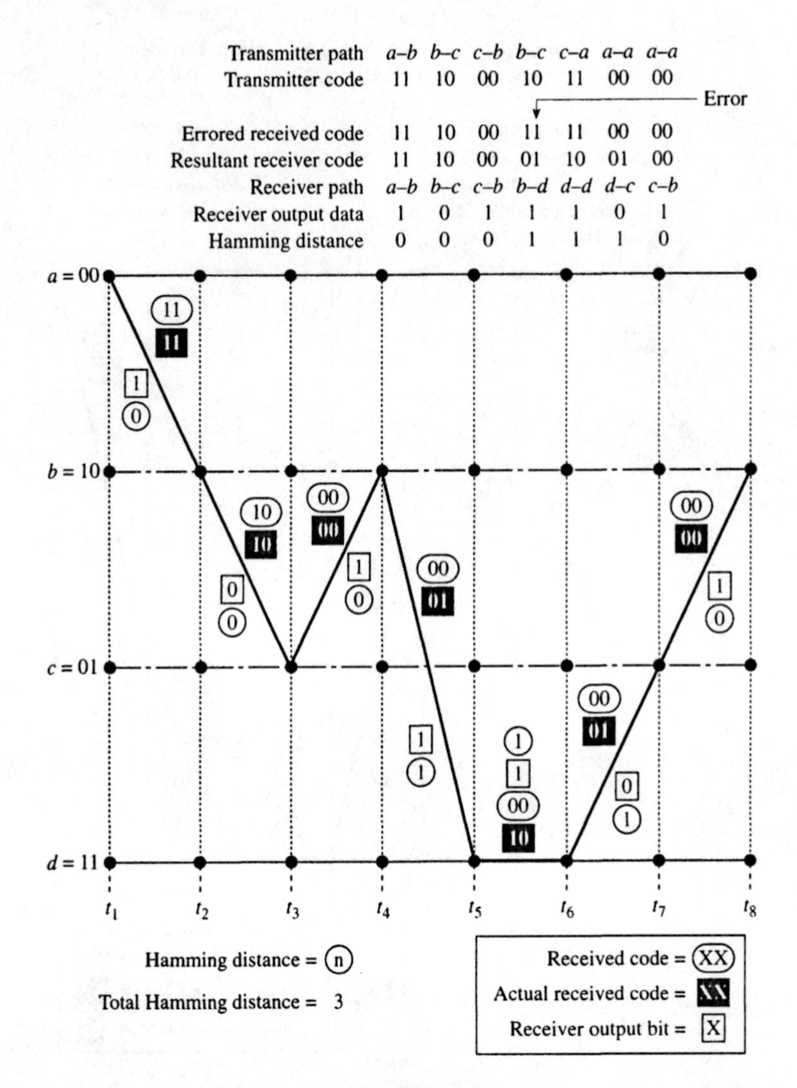

Fig. 10.29 Trellis diagram for path 14, Example 10.11

Consider path 14 (Fig. 10.29):

Tx data	= 1	0	1	0	0	0	0
Tx path	= a–b	b–c	c–b	b–c	c–a	a–a	a–a
Tx code output	= 11	10	00	10	11	00	00
Rx output data	= 1	0	1	1	1	0	1
Rx path	= a–b	b–c	c–b	b–d	d–d	d–c	c–b
Rx code input	= 11	10	00	11	11	00	00
Rx code actual	= 11	10	00	01	10	01	00
Hamming distance	= 0	0	0	1	1	1	0

Total Hamming distance = 3

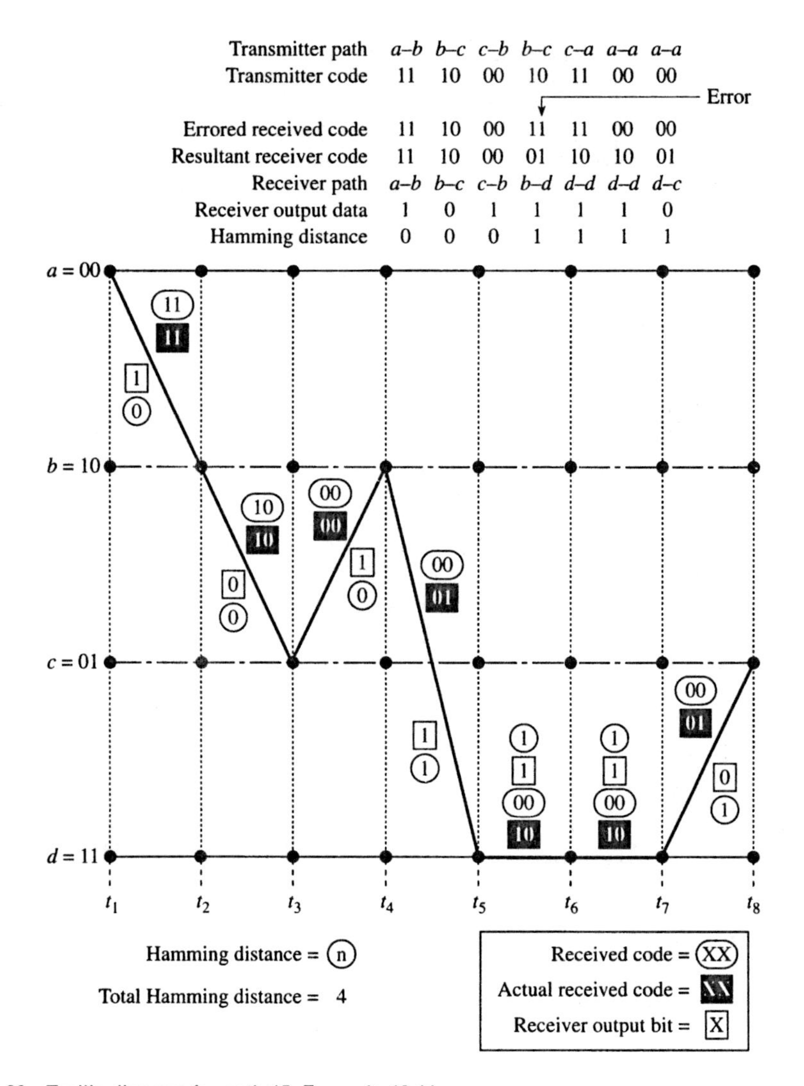

Fig. 10.30 Trellis diagram for path 15, Example 10.11

Consider path 15 (Fig. 10.30):

Tx data	= 1	0	1	0	0	0	0
Tx path	= *a–b*	*b–c*	*c–b*	*b–c*	*c–a*	*a–a*	*a–a*
Tx code output	= 11	10	00	10	11	00	00
Rx output data	= 1	0	1	1	1	1	0
Rx path	= *a–b*	*b–c*	*c–b*	*b–d*	*d–d*	*d–d*	*d–c*
Rx code input	= 11	10	00	11	11	00	00
Rx code actual	= 11	10	00	01	10	10	01
Hamming distance	= 0	0	0	1	1	1	1

Total Hamming distance = 4

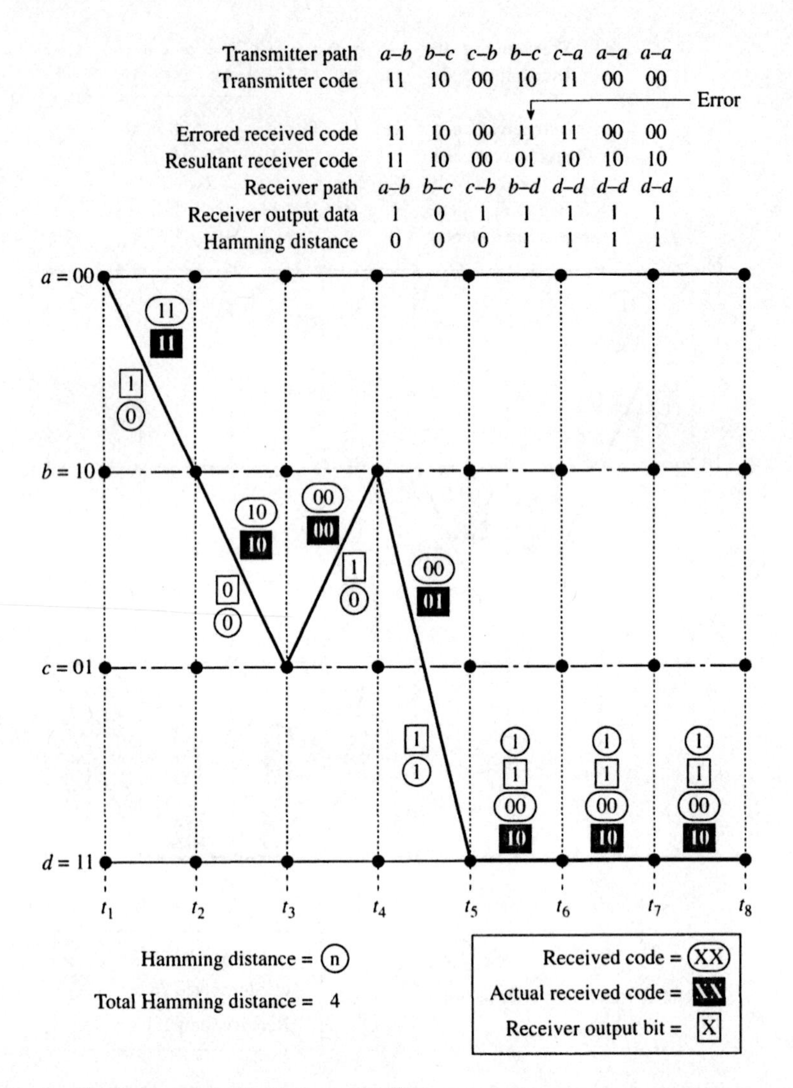

Fig. 10.31 Trellis diagram for path 16, Example 10.11

Consider path 16 (Fig. 10.31):

Tx data	= 1	0	1	0	0	0	0
Tx path	= *a–b*	*b–c*	*c–b*	*b–c*	*c–a*	*a–a*	*a–a*
Tx code output	= 11	10	00	10	11	00	00
Rx output data	= 1	0	1	1	1	1	1
Rx path	= *a–b*	*b–c*	*c–b*	*b–d*	*d–d*	*d–d*	*d–d*
Rx code input	= 11	10	00	11	11	00	00
Rx code actual	= 11	10	00	01	10	10	10
Hamming distance	= 0	0	0	1	1	1	1

Total Hamming distance = 4

As can be seen from the above there is only one correct path, which is path 1. This path yields a total Hamming distance of 1 whereas all the other paths yield a total

Table 10.6 Hamming distance results for paths 1 to 16

	Transmitted code	Tx data	
	11 10 00 10 11 00 00	101 0000	
	Received code		
	11 10 00 11 11 00 00		
Path number	Receiver codes	O/P data	Hamming distance
1	11 10 00 10 11 00 00	101 0000	1
2	11 10 00 10 11 00 11	101 0001	3
3	11 10 00 10 11 11 10	101 0010	4
4	11 10 00 10 11 11 01	101 0011	4
5	11 10 00 10 00 10 11	101 0100	6
6	11 10 00 10 00 10 00	101 0101	4
7	11 10 00 10 00 01 01	101 0110	5
8	11 10 00 10 00 01 10	101 0111	5
9	11 10 00 01 01 11 00	101 1000	4
10	11 10 00 01 01 11 11	101 1001	6
11	11 10 00 01 01 00 10	101 1010	3
12	11 10 00 01 01 00 01	101 1011	3
13	11 10 00 01 10 01 11	101 1100	5
14	11 10 00 01 10 01 11	101 1101	5
15	11 10 00 01 10 10 01	101 1110	4
16	11 10 00 01 10 10 10	101 1111	4

Hamming distance greater than 1. In practice the path that yields the lowest Hamming distance is the preferred path and the data that results due to that path is the data that is output by the receiver.

From the above it is also clear that the receiver will determine the total Hamming distance for all the code combinations. In the above example the first three codes, 11 10 00, due to the first three data, 101_2, are the same; however, the next received code is errored and hence the last four codes, 11 11 00 00, will cause the receiver to examine all the paths for the output data 0000_2 to 1111_2. Therefore there are 16 different paths that must be examined.

The results are shown in Table 10.6.

Figure 10.32 shows all the 16 possible paths for Example 10.11. In practice, if there are two paths that yield the same Hamming distance due to received data that is errored then the path that is closest to *a* is chosen as the correct path.

This coding method is very powerful as it can correct errors through the decoder and as a result is often referred to as forward error correction.

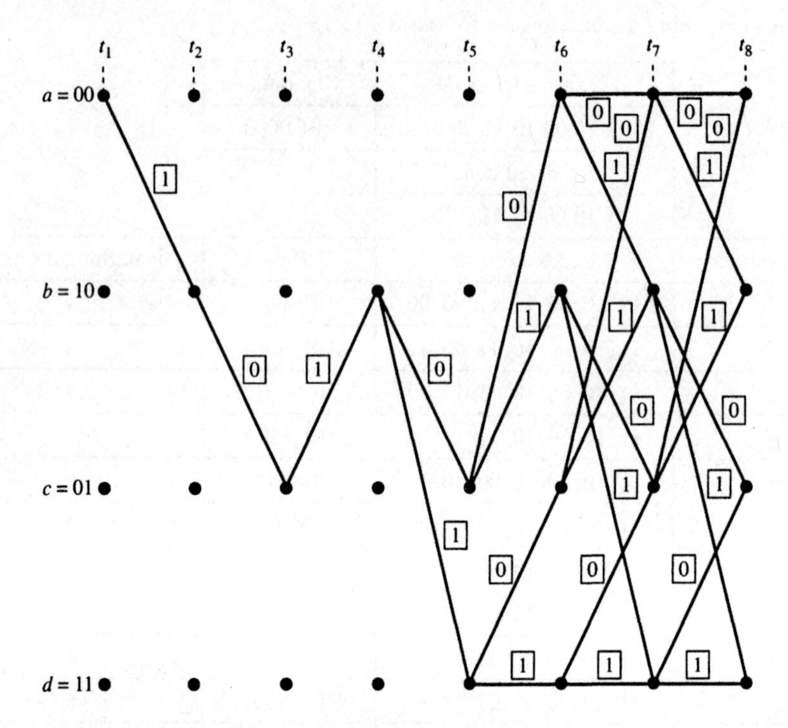

Fig. 10.32 Trellis diagram for all 16 paths, Example 10.11

10.7 REVIEW QUESTIONS

10.1 State the disadvantages of using a single parity bit for the detection of errors at a receiver.

10.2 Determine the asynchronous serial output data stream for the ASCII character P, where one start bit, a parity bit and two stop bits are used.

10.3 Code each ASCII character in the word 'Goal' for transmission over an asynchronous link where one start bit, a parity bit and a single stop bit is employed.

10.4 Describe the disadvantages of using block check codes for error detection and correction.

10.5 Determine the CRC-4 word for the following 8-bit data stream. The generator polynomial is 19_{16}.

$$\text{Data} = \text{C A}_{16}$$

10.6 The following data stream is received by a circuit that employs CRC-4. Determine whether the data is error free or not.

$$\text{Received data} = 10110010111_2$$

10.7 Determine the coding for the following data stream that is input to a transmitter that employs trellis coding:

$$\text{Data} = 0110_2$$

11 Line and interface coding

For mankind to respond in situations the available information must enable rational and logical decisions to be made.

11.1 INTRODUCTION

Interface coding and line coding is the subject of this chapter. All digital transmission systems must transmit the data in such a way that the clock frequency can be extracted from the received signal. This prevents having to supply a separate path for the clock frequency.

Some codes are designed to be transmitted as frequency-modulated signals, while others are designed to be conveyed over metallic pairs. All the codes have advantages and disadvantages. In practice the code that is best suited to the application is chosen.

There are numerous different transmission codes; only a few have been selected and are described here.

11.2 REQUIREMENTS OF LINE AND INTERFACE CODES

Interface codes are used when data at the output of one piece of equipment must be connected to the input of another piece, where the two pieces of equipment could be up to 100 m apart. As will be shown later, ordinary binary data cannot be used and as a result the data must be changed into some other format.

Line codes are used when the data is transmitted down some medium to a distant receiver. In this case the distance between the transmitter and the receiver could vary quite considerably, from a few kilometres, as used on cable systems, to thousands of kilometres, as with satellite systems.

For a code to be used as an interface code or line code, the requirements that the code must satisfy are dependent on the type of medium and the format of the data transmission. However, some general requirements are:

- *D.c. component*. For transmission lines, the d.c. component should be removed to enable correct detection of the individual bits at the receiver. The elimination of the d.c. component also enables a.c. coupling to take place. By eliminating the varying d.c., the resultant low frequency is eliminated which could otherwise affect adjacent pairs through crosstalk.
- *Self-clocking*. Symbol or bit synchronisation is required for digital communication. In many applications the original clock frequency must be extracted from the incoming signal. Hence the incoming signal must have the clock frequency embedded in it for all bit stream sequences.
- *Error detection*. Some schemes have the means of detecting data errors without introducing additional error detection bits into the data sequence.
- *Bandwidth compression*. Some schemes increase the efficiency of the available bandwidth by enabling more information to be conveyed when compared to other schemes using the same type of medium.
- *Noise immunity*. Some schemes display a lower bit error rate than others even when they are transmitted over similar media having similar signal-to-noise ratios.

All the different types of interface codes and line codes can be categorised under one of the following:

- Non-return to zero (NRZ).
- Return to zero (RZ).
- Phase encoded.
- Pulse modulation.

11.3 NON-RETURN TO ZERO CODES

These are the most commonly used codes, and can be partitioned into the following subgroups:

- NRZ-L (L for level).
- NRZ-S (S for space).
- NRZ-M (M for mark).

All these codes are shown in Fig. 11.1.

11.3.1 Non-return to zero – level (NRZ-L)

This code is widely used in digital logic circuits. It is often referred to as NRZ data and is shown in Fig. 11.1(a).

Coding

Logic 1 = No transitions during the bit period.
Remains high for the duration of the bit period

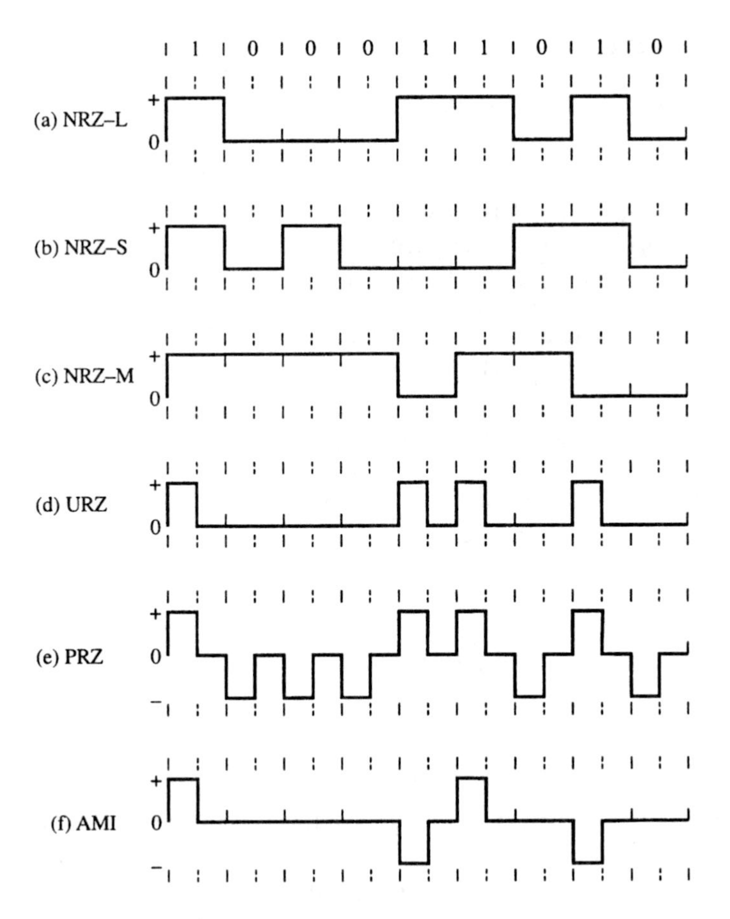

Fig. 11.1 Non-return to zero codes

Logic 0 = No transitions during the bit period.
Remains low for the duration of the bit period

The logic state only changes when the following logic state is different to the preceding logic state.

Disadvantages

- High d.c. component.
- Loss of clock for long strings of logic 1s and logic 0s.

If this code is used to transfer data from one circuit to another then a separate lead carrying the clock signal must also be used to enable the data to be correctly clocked in the second circuit.

If a long string consisting of, predominantly, logic 0 states or logic 1 states is transmitted then the frequency content drops towards 0 Hz. Also if there are a large number of logic 1 states in the bit stream then the average d.c. rises towards the voltage of the logic 1 state.

11.3.2 Non-return to zero – space (NRZ-S)

This code is shown in Fig. 11.1(b).

Coding

Logic 1 = No transitions during the bit period.
There is no change in level from the previous state for the duration of the bit period

Logic 0 = Single transition at the beginning of the bit period.
If the previous state was low then the transition is from low to high. If the previous state was high then the transition is from high to low

Every time a logic 0 appears in the bit stream the logic level changes. A logic 1 causes no change in the logic state.

Disadvantages

- High d.c. component.
- Loss of clock for long strings of logic 1s.

In this case if the input bit stream consists of a long string of logic 1 states then the frequency content drops towards 0 Hz. The d.c. content can either drop towards 0 V, if the previous state was at 0 V prior to the first logic 1 in the sequence, or the d.c. content can rise towards the voltage of a logic 1, if the previous state was at a positive voltage prior to the first logic 1 in the sequence. However, if the input bit stream consists of a large number of logic 0 states then the output bit stream becomes almost a 101010_2 sequence which has a fundamental frequency and harmonic frequencies.

11.3.3 Non-return to zero – mark (NRZ-M)

This is often referred to as differential coding. It is sometimes referred to as non-return to zero – inverted (NRZ-I). This code is shown in Fig. 11.1(c).

Coding

Logic 1 = Single transition at the beginning of the bit period.
If the previous state was low then the transition is from low to high. If the previous state was high then the transition is from high to low

Logic 0 = No transitions during the bit period.
There is no change in level from the previous state for the duration of the bit period

Every time a logic 1 appears in the bit stream the logic level changes. A logic 0 causes no change in the logic state.

Disadvantages
- High d.c. component.
- Loss of clock for long strings of logic 0s.

In this case if the input bit stream consists of a long string of logic 0 states then the frequency content drops towards 0 Hz and the d.c. content can either drop towards 0 V or rise towards the voltage of a logic 1. However, if the input bit stream consists of a large number of logic 1 states then the output bit stream becomes almost a 101010_2 sequence which has a fundamental frequency and harmonic frequencies.

11.4 RETURN TO ZERO (RZ)

This code can be subdivided into the following subgroups.

- Unipolar return to zero (URZ).
- Polar return to zero (PRZ).
- Return to zero – alternate mark inversion (RZ-AMI).
- High-density bipolar format *n* (HDB-*n*).

These codes find applications in digital baseband transmission and magnetic recording. They are shown in Fig. 11.1.

11.4.1 Unipolar return to zero (URZ)

This code is shown in Fig. 11.1(d).

Coding

Logic 1 = Two transitions during the bit period.
The first transition is from a low to a high at the beginning of the bit period. The second transition is from a high to low halfway through the bit period

Logic 0 = No transitions during the bit period.
The state remains low for the duration of the bit period

An input logic 1 causes the output to remain at a logic 1 for the first half of the bit period and return to a logic 0 for the second half of the bit period. The code has a 50% duty cycle for the logic 1 states.

Disadvantages
- Reduced d.c. component.
- Twice the bandwidth of NRZ codes.

Because the high states in the output bit stream have a shorter duration than that of the full bit period, the average d.c. is reduced. This reduction in the time period of the output high durations causes the fundamental frequency to be doubled relative to an NRZ-L data stream.

11.4.2 Polar return to zero (PRZ)

This code is shown in Fig. 11.1(e).

Coding

Logic 1 = Two transitions during the bit period.
The first transition is from 0 V to a positive voltage at the beginning of the bit period. The second transition is from a positive voltage to 0 V halfway through the bit period

Logic 0 = Two transitions during the bit period.
The first transition is from 0 V to a negative voltage at the beginning of the bit period. The second transition is from a negative voltage to 0 V halfway through the bit period

This code is referred to as a bipolar code or ternary code as there are three distinct voltages, i.e. a positive voltage, zero volts or earth potential and a negative voltage. The code has a 50% duty cycle for the logic 0 and logic 1 states.

Disadvantage

- Reduced d.c. component.

Advantage

- Clock frequency is embedded in the code for all bit sequences.

If the input bit stream consists of a long string of logic 1 states then only positive-going pulses will be produced, whereas if the input bit stream consists of a long string of logic 0 states then the output would consist of only negative-going pulses. As a result the average d.c. would either rise to a positive voltage or drop to a negative voltage respectively. Because of the reduction in the length of the pulse the average d.c. is reduced.

11.4.3 Return to zero – alternate mark inversion (RZ-AMI)

This code is usually referred to simply as alternate mark inversion or AMI. This code is shown in Fig. 11.1(f).

Coding

Logic 1 = Two transitions during the bit period.
The first transition is from 0 V to either a positive or negative voltage at the beginning of the bit period. The second transition is from the positive or negative voltage to 0 V halfway through the bit period

Logic 0 = No transitions during the bit period.
The output remains a low for the duration of the bit period

The first logic 1 (mark) in the input bit stream will cause a positive voltage output pulse, the second will cause the output pulse to go negative, and so forth. In other words each successive mark causes the output pulse to be in the opposite polarity to the previous mark. The code has a 50% duty cycle for the logic 1 states.

Disadvantage
- Loss of clock for long strings of logic 0s.

Advantages
- No d.c. component.
- Same bandwidth as the NRZ codes.

Because the output pulses are alternately positive and negative, the average d.c. remains very close to 0 V for a succession of logic 1 states at the input. Also because of the alternating nature of the signal the bandwidth is similar to that of NRZ-L data streams.

11.4.4 High-density bipolar format – *n* code (HDB-*n*)

This code is a derivative of RZ-AMI described in Section 11.4.3. Each successive mark reverses the output polarity of the RZ mark so that the resultant bit stream consists of pulses following an alternating polarity sequence.

For the 30/32-channel CEPT PCM system the CCITT recommendation is the HDB-3 code. The '3' indicates that the maximum number of consecutive logic 0 states may not exceed three. When there are more than three logic 0 states, a violation bit must be used. Sometimes it is also necessary to use a parity bit in order to cause the violation bit to assume the correct polarity. The term violation means 'break the alternate mark inversion law'.

The CCITT recommendation for line coding on a PCM system consists of the following codes:

- Alternate digit inversion (ADI).
- Return to zero (RZ).
- High density bipolar format 3 (HDB-3).

Coding

Alternate digit inversion (ADI)
Each sample on each of the 30 speech channels is quantised and encoded into an 8-bit word. The structure of the 8-bit word is such that bit 1 represents the polarity and bits 2 to 8 indicate the amplitude of the sample. This coding is shown in Fig. 11.2.

Odd bits (1, 3, 5, 7):

Logic 1 = Logic 1 for the full bit period
Logic 0 = Logic 0 for the full bit period

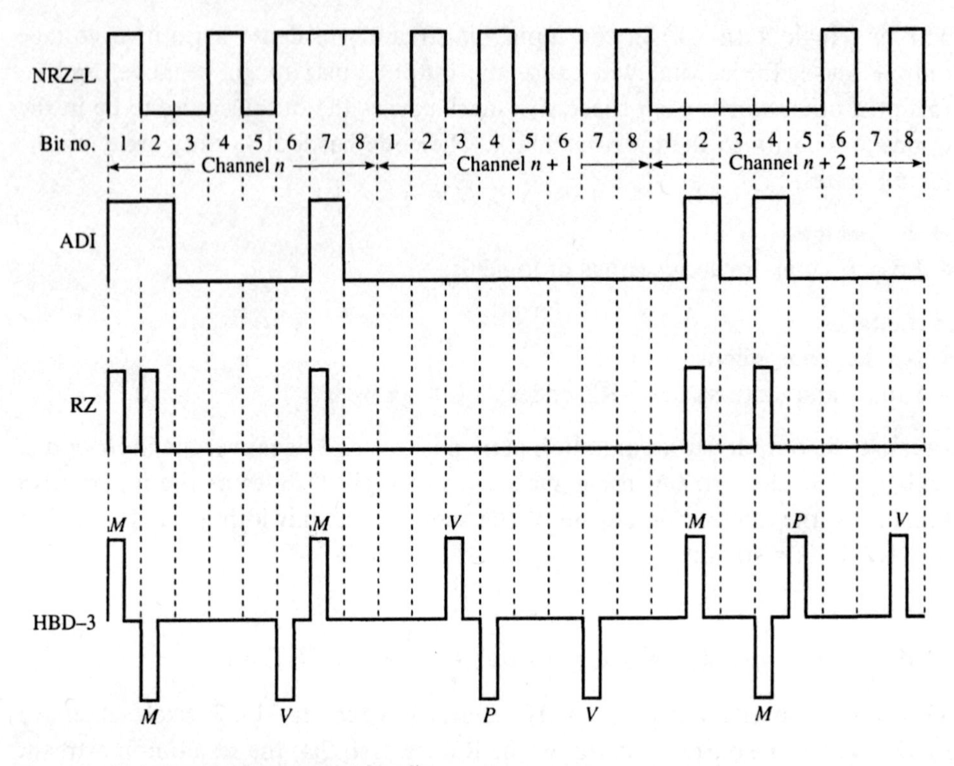

Fig. 11.2 High-density bipolar format 3 coding

Even bits (2, 4, 6, 8):

Logic 1 = Logic 0 for the full bit period
Logic 0 = Logic 1 for the full bit period

Consider the following bit stream:

Bit numbers	1 2 3 4 5 6 7 8
Input bit stream	1 0 0 1 1 0 0 0
ADI code	1 **1** 0 **0** 1 **1** 0 **1**

As shown above, the emboldened bits in the ADI code are in the opposite logic state relative to the input state. Notice that only the bits in the even bit positions have changed state.

The main problem with ADI is that for the following bit sequences there will be a loss of clock if this code was used as the interface or line code.

Bit numbers	1 2 3 4 5 6 7 8
Input bit stream	0 1 0 1 0 1 0 1
ADI code	0 **0** 0 **0** 0 **0** 0 **0**

Bit numbers	1 2 3 4 5 6 7 8
Input bit stream	1 0 1 0 1 0 1 0
ADI code	1 **1** 1 **1** 1 **1** 1 **1**

In the second example given above not only is there no clock but the average d.c. rises to the voltage of a logic 1.

Return to zero (RZ)

The RZ code used is the URZ code. The coding technique is described in Section 11.4.1 and is shown in Fig. 11.1(d) as well as Fig. 11.2.

High-density bipolar format 3 (HDB-3)

The code follows AMI for all bit sequences where there are three or fewer consecutive logic 0 states (spaces) between two valid logic 1 states (marks). Figure 11.2 shows the typical conversion process from binary to HDB-3.

The *violation* bit is a bit that violates the AMI principle. The *parity* bit is not a valid mark but is a bit that is inserted into the bit stream to force the violation bit to assume a particular polarity.

When there are more than three consecutive logic 0 states then the fourth consecutive space is made into a *violation* bit.

Rule 1. The violation bit must assume the polarity of the valid mark (i.e. mark or parity bit) that immediately precedes it.

Rule 2. The successive violation must assume a polarity that is opposite to the violation that immediately precedes it.

Rule 3. If there is no valid mark to force a violation bit to have the opposite polarity to the violation that immediately precedes it then a parity bit is used to force the violation to assume the opposite polarity.

Rule 4. The number of spaces between a valid mark and a violation bit is 3.

Rule 5. The number of spaces between a parity bit and a violation bit is 2.

Figure 11.2 shows the conversion process from NRZ-L to HDB-3 when violation and parity bits are required. Notice that the HDB-3 output is a RZ code.

Referring to Fig. 11.2, the input NRZ-L bit stream is as follows:

	Channel n								Channel $n+1$								Channel $n+2$							
Bit No.	1	2	3	4	5	6	7	8	1	2	3	4	5	6	7	8	1	2	3	4	5	6	7	8
Data	1	0	0	1	0	1	1	1	0	1	0	1	0	1	0	1	0	0	0	0	0	1	0	1_2
ADI	1	1	0	0	0	0	1	0	0	0	0	0	0	0	0	0	0	1	0	1	0	0	0	0
HDB-3	$+M$	$-M$	0	0	0	$-V$	$+M$	0	0	0	$+V$	$-P$	0	0	$-V$	0	0	$+M$	0	$-M$	$+P$	0	0	$+V$

where $+M$ = positive valid mark
$-M$ = negative valid mark
$-V$ = negative violation bit
$+V$ = positive violation bit
$+P$ = positive parity bit
$-P$ = negative parity bit

This code is a bipolar or ternary code as there are three states, namely a positive voltage, 0 V or earth potential and a negative voltage. The code has a 50% duty cycle for the logic 1 states.

Advantages of HDB-3

- The clock frequency is embedded in the code.
- There is only a very small d.c. component.
- It has a bandwidth that is smaller than RZ codes.

11.5 THE PHASE-ENCODED GROUP

This group consists of the following:

- Biphase – mark (BIΦ-M) or Manchester I code.
- Biphase – level (BIΦ-L) or Manchester II code.
- Biphase – space (BIΦ-S).
- Delay modulation (DM).

These codes are used in magnetic recording, optical communications and some satellite systems. All these codes are shown in Fig. 11.3.

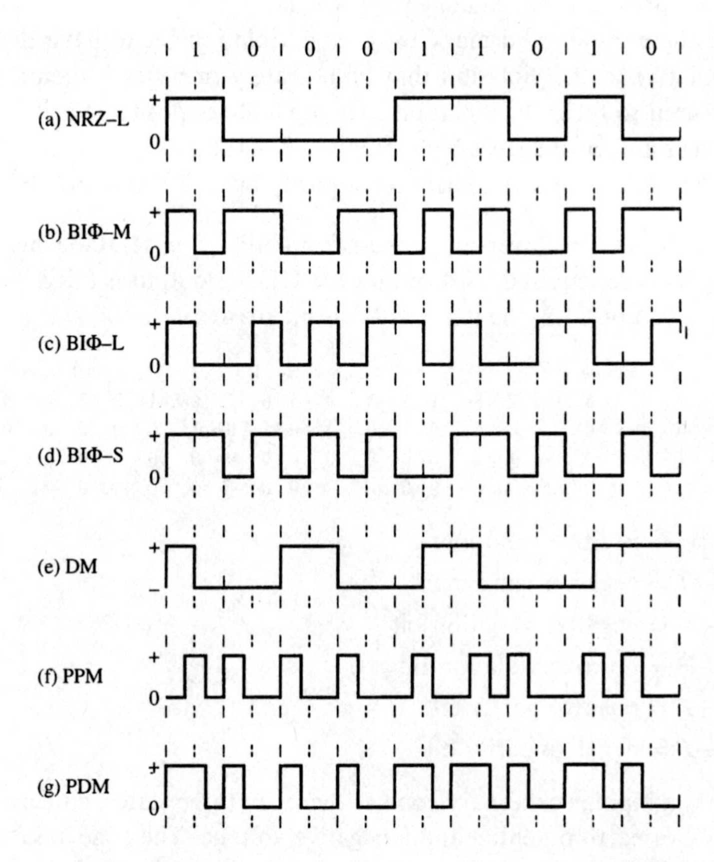

Fig. 11.3 The phase-encoded group

11.5.1 Biphase – mark (BIΦ-M) or Manchester I code

The biphase – mark (BIΦ-M) code was developed by Manchester University and hence is also known as the Manchester I code. It is a split-phase code. The code is shown in Fig. 11.3(b).

Coding

Logic 1 = Two transitions during the bit period. The first transition is at the beginning of the bit period. If the previous state was low then the first transition will be from low to high and if the previous state was high then the first transition will be from high to low. The second transition is halfway through the bit period in the opposite direction relative to the first transition

Logic 0 = One transition during bit period, at the beginning of the bit period. If the previous state is low then the transition is from low to high. If the previous state was high then the transition is from high to low

Disadvantages

- High average d.c. component.
- Twice the bandwidth of NRZ codes.

Advantage

- Clock frequency is embedded in the code for all bit sequences.

The code has a 50% duty cycle for the logic 1 states.

11.5.2 Biphase – level (BIΦ-L) or Manchester II code

The biphase level (BIΦ-L) code was also developed by Manchester University and hence is also known as the Manchester II code. It is a split-phase code. This code is shown in Fig. 11.3(c).

Coding

Logic 1 = One transition during the bit period, halfway through the bit period from high to low if the previous state is high. Two transitions during the bit period if the previous state is low. The first transition will be from low to high at the beginning of the bit period and the second transition will be from high to low halfway through the bit period

Logic 0 = One transition during the bit period, halfway through the bit period from low to high if the previous state is low. Two transitions during the bit period if the previous state is high. The first transition will be from high to low at the beginning of the bit period and the second transition will be from low to high halfway through the bit period

Disadvantages
- Reduced d.c. component.
- Twice the bandwidth of NRZ codes.

Advantage
- Clock frequency is embedded in the code for all bit sequences.

11.5.3 Biphase – space (BIΦ-S)

This code is a split-phase code. It is shown in Fig. 11.3(d).

Coding

Logic 1 = One transition during the bit period at the beginning of the bit period. If the previous state was low then the transition is from low to high. If the previous state was high then the transition will be from high to low

Logic 0 = Two transitions during the bit period. The first transition is at the beginning of the bit period and the second transition is halfway through the bit period. For the first transition, if the previous state was low then the transition will be from low to high and if the previous state was high then the transition will be from high to low. The second transition will be in the opposite direction to the first transition

Disadvantages
- Very high d.c. component.
- Twice the bandwidth of NRZ codes.

Advantage
- Clock frequency is embedded in the code for all bit sequences.

The code has a 50% duty cycle for the logic 0 states.

11.5.4 Delay modulation (DM) or Miller code

This code is a split-phase code and is also known as the Miller code. This code is shown in Fig. 11.3(e)

Coding

Logic 1 = One transition halfway through the bit period. If the previous state was low then the transition will be from low to high. If the previous state was high then the transition will be from high to low

Logic 0 = No transitions during the bit period unless followed by a second logic 0. The second logic 0 has one transition at the beginning of the bit period in the opposite direction to the previous state

Disadvantage

- Reduced d.c. component.

Advantages

- Clock frequency is embedded in the code for all bit sequences.
- The same bandwidth as NRZ codes.

11.6 PULSE MODULATION

This group consists of the following:

- Pulse position modulation (PPM).
- Pulse duration modulation (PDM).

These codes are shown in Fig. 11.3.

11.6.1 Pulse position modulation (PPM)

This coding method was briefly considered in Chapter 2 when spread spectrum systems were covered. The theoretical definition for the code is given here. This code is shown in Fig. 11.3(f).

Coding

Logic 1 = Two transitions during the bit period. The first transition is one third of the period from the start of the bit period from low to high. The second transition is two-thirds of the period from the start of the bit period from high to low

Logic 0 = Two transitions during the bit period. The first transition is at the beginning of the bit period from low to high. The second transition is one-third of the period from the start of the bit period from high to low

An input logic 1 forces the output to remain at a logic 0 for one-third of the bit period, a logic 1 for one-third of the bit period and then a logic 0 for one-third of the bit period.

An input logic 0 causes the output to change to a logic 1 for one-third of the bit period and then to a logic 0 for two-thirds of the bit period.

Disadvantage

- Large bandwidth compared to NRZ codes.

Advantages

- Reduced d.c. component.
- Clock frequency is embedded in the code.

11.6.2 Pulse duration modulation (PDM)

This coding method was briefly considered in Chapter 2 when spread spectrum systems were covered. The theoretical definition for the code is given here. This code is shown in Fig. 11.3(g).

Coding

Logic 1 = Two transitions during the bit period. The first transition is at the beginning of the bit period from low to high. The second transition is two-thirds of the period from the start of the bit period from high to low

Logic 0 = Two transitions during the bit period. The first transition is at the beginning of the bit period from low to high. The second transition is one-third of the bit period from the start of the bit period from high to low

An input logic 1 forces the output to a logic 1 for two-thirds of the bit period and then to a logic 0 for one-third of the bit period.

An input logic 0 causes the output to change to a logic 1 for one-third of the bit period and then to a logic 0 for two-thirds of the bit period.

Disadvantages

- Large d.c. component for long strings of logic 1s.
- Large bandwidth compared to NRZ codes.

Advantage

- Clock frequency is embedded in the code for all bit sequences.

11.7 FREQUENCY DISTRIBUTION

Figure 11.4 shows the spectral density against normalised bandwidth and Fig. 11.5 shows the frequency distribution for the different groups. Figure 11.5 shows that the RZ and biphase codes have a larger bandwidth requirement than the NRZ code. Figure 11.4 shows that the bipolar codes have a smaller bandwidth requirement than the NRZ code. It can also be seen that the biphase codes require larger bandwidths than any of the other coding methods.

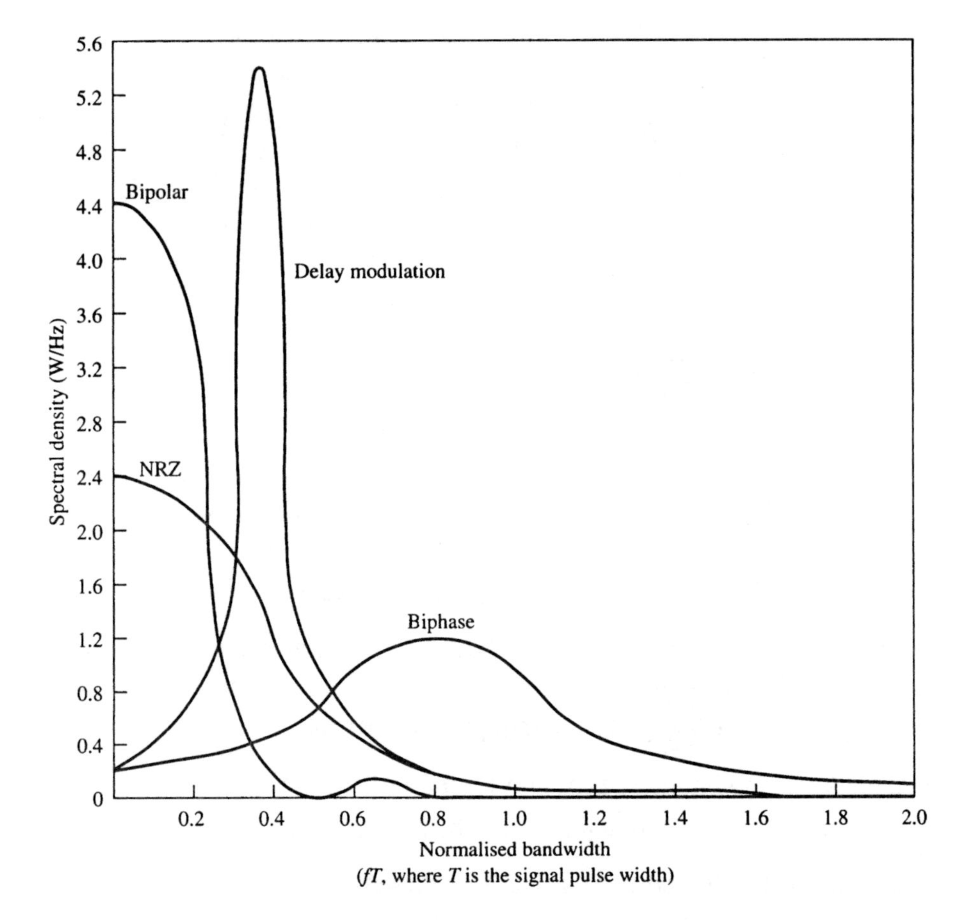

Fig. 11.4 Spectral densities of various PCM waveforms

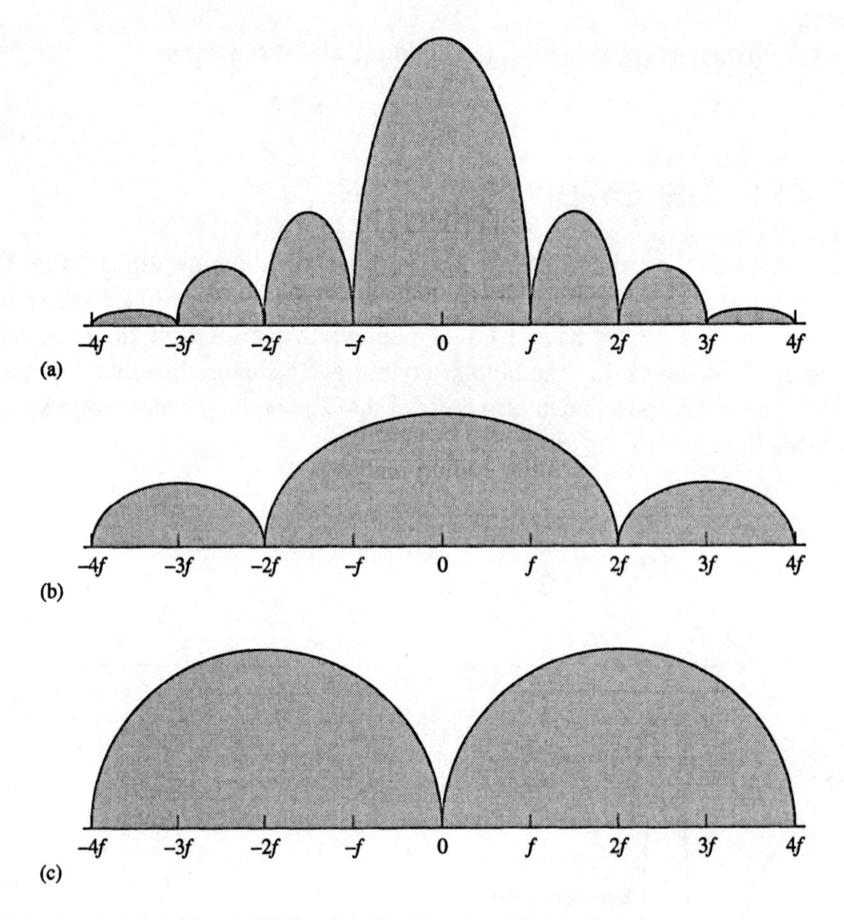

Fig. 11.5 Spectral densities: (a) NRZ codes; (b) RZ codes; (c) biphase codes

11.8 REVIEW QUESTIONS

11.1 State the requirements of a line or interface code.

11.2 State the four different categories into which all the codes can be classed.

11.3 Code the following bit stream into AMI. The coding must show the ADI, RZ and AMI steps.

$$\text{Data stream} = 10011101010111010000_2$$

11.4 Code the following bit stream into HDB-3. The coding must show the ADI, RZ and HDB-3 steps.

$$\text{Data stream} = 10011101010111010000_2$$

11.5 Code the following bit stream into Manchester I code.

$$\text{Data stream} = 10011101010111010000_2$$

11.6 Code the following bit stream into Manchester II code.

$$\text{Data stream} = 10011101010111010000_2$$

12 ISO open systems interconnect seven-layer model

> Quality of information not quantity of information makes decision making easier.

12.1 INTRODUCTION

Data transmission is becoming more and more widespread. An understanding of the different protocols is important in the modern world. This is especially true with the advent of the worldwide net. This chapter discusses the International Standards Organisation (ISO) seven-layer model and the principle of operation of data being transmitted via the ISO model.

Initially a discussion on the different types of protocol is given. This leads to a general discussion of the ISO model. A discussion of the different interfaces used at the physical layer is then given.

The chapter also deals with the data format and the structures of the other layers within the model and the importance of each layer.

12.2 MESSAGE TRANSFER AND MESSAGE SWITCHING

All modern data transfer systems use a system referred to as automatic repeat request (ARQ). This method is used to ensure that unerrored data is transferred to the end user. If errored data is received then the system asks for the data to be retransmitted. The ARQ instructions are written into a protocol, which is a computer program that is loaded into a switching computer. The protocol enables the switching computer to correctly process the data that it receives from a terminal so that it can transmit it to another terminal.

12.3 PROTOCOL CATEGORIES AND OPERATION

The ARQ protocols are divided into two categories:

- Stop and wait.
- Continuous.

12.3.1 Stop and wait protocol

This is a message-oriented protocol. This means that the complete message must be transmitted to a switching computer and stored in the computer memory. Only once the complete message is stored in the memory is it then sent to the next computer.

This implies that the switching computer must have a very large memory as it must serve several users simultaneously. Also the message transfer is slow through this system in that while the computer is busy with the data from one user it cannot service the other users. The emphasis in this system is on storage of the message and not on message transfer. For large messages the other users are forced to wait and hence the users are not fairly served.

12.3.2 Continuous protocol

In this system the message is broken up into small packets. The packets are then continuously transferred to the switching computer. The switching computer now only has to store the packet before transmitting it to the next computer. The emphasis here is on forwarding the data and not on storage. This has the added advantage that the memory of the switching computer is smaller and also the system is a lot faster so that the users are more efficiently served.

12.3.3 Protocol operation

There are three different ways in which these protocols can operate, these are:

- Simplex operation.
- Half-duplex operation.
- Full-duplex operation.

Simplex operation is where the communication is only in one direction. This is similar to commercial radio and TV.

Half-duplex operation is where both terminals can transmit and receive but not at the same time. Data can only be transferred in one direction at a time. This is similar to the communication using walkie-talkies.

Full-duplex operation is where both terminals can transmit and receive at the same time.

12.4 TYPES OF PROTOCOLS

There are five different types of protocols namely:

- Asynchronous.
- Protected asynchronous.
- Character oriented.
- Byte oriented.
- Bit oriented.

12.4.1 Asynchronous protocol

These protocols are extremely simplistic. They provide no error correction. Each character is packaged between a start bit and stop bits. If a parity bit is included and an error occurs, the errored data is still accepted as the receiver ignores the status of the parity bit.

12.4.2 Protected asynchronous protocol

These protocols use the parity bit for error detection and inform the computer that the data is errored. The protocol does not ask for retransmission of the errored data. These protocols are also simplistic in nature. Each character is packaged between a start bit and stop bits with a parity bit included just before the stop bits.

12.4.3 Character-oriented protocols

A typical example of this type of protocol is the binary synchronous protocol (Bisync or BSC) developed by IBM. The binary synchronous protocol format is as follows:

SYN	SYN	SOH	⟨HEADER⟩	STX	⟨TEXT⟩	ETX	⟨bcc⟩

where SYN = synchronous character

SOH = start of header (marks the beginning of the message block)

STX = start of text (marks the beginning of the text block)

ETX = end of text (marks the end of the text block)

bcc = block check character (used for error detection).

The SOH and ⟨Header⟩ are optional and are dependent on the user and the application.

A data link escape (DLE) command is incorporated when the data is to be transparent to BSC. Consider the following data string that is to be transmitted to a distant terminal.

Data string = ABC STX DEF ETX 123 DLE 456

In this string the character strings 'STX', 'ETX' and 'DLE' are commands that the BSC protocol will respond to. In order for these text strings to be treated as text strings and not commands, they must be made transparent to the protocol. This is achieved as follows:

SYN	SYN	DLE	STX	ABC STX DEF ETX 123	DLE	DLE 456	DLE	ETX

The first DLE command makes the text block transparent. The second DLE command is ignored, the third DLE command is accepted as data and the last DLE command indicates the end of the transparency.

Bisync transmission method

The protocol is an automatic repeat request (ARQ) protocol. Each block must be acknowledged prior to the next block being transmitted.

The protocol is a half-duplex protocol. Each block must be acknowledged prior to the next block being sent. The system uses two different acknowledgement signals called ACK 0 and ACK 1. This is used to clearly identify which block of data has been correctly received by the distant receiver. Figure 12.1(a) shows

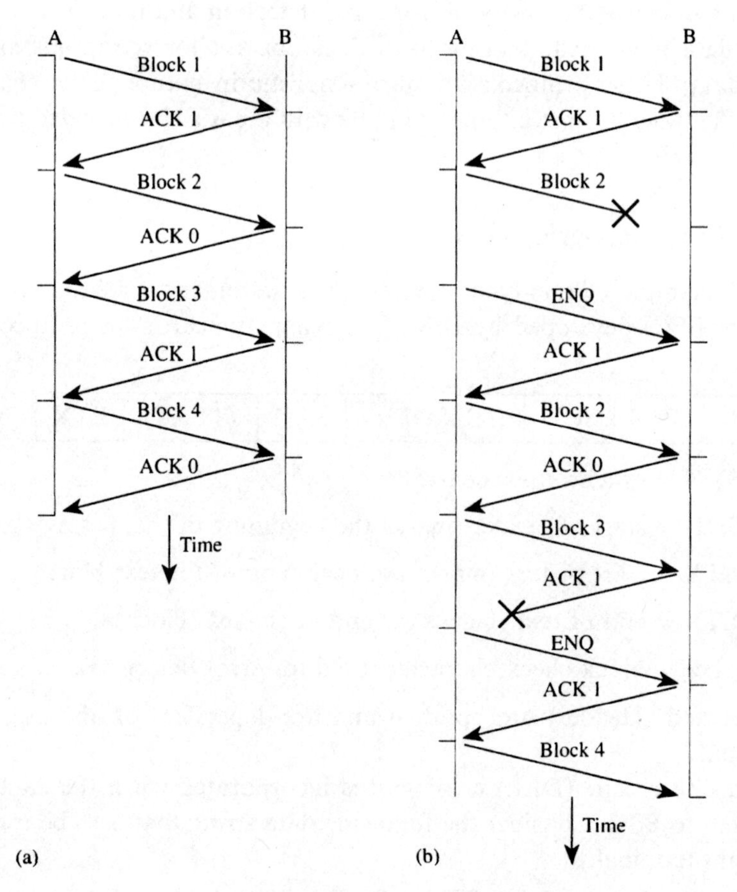

Fig. 12.1 Bisync call procedure

the normal transmission method of the bisync protocol. As shown, when data block 1 is transmitted from A to B, before data block 2 can be transmitted, the acknowledgement signal (ACK 1) must be sent from B to A. When data block 2 from A arrives at B then B sends the ACK 0 signal to A. The same applies when data blocks are sent from B to A.

If, as shown in Fig. 12.1(b), A does not receive an acknowledgement from B then after a timeout period A will send an enquiry signal (ENQ) to B. On receipt of this signal, B sends the same acknowledgement signal that it sent out previously, in this case ACK 1. However, if A is expecting the ACK 0 signal then it knows that data block 2 did not arrive at B and hence this block must be retransmitted.

The same procedure applies when data blocks are transferred from B to A.

Limitations of the bisync protocol

This protocol has three major limitations:

- Inappropriate over satellite links as the delays become too long.
- Each block of data must be acknowledged before the next block is sent.
- Data is only presented in one direction at a time although the circuit could possibly support full-duplex operation.

12.4.4 Byte-oriented protocols

The length of each frame (i.e. number of bytes) is contained in a count field which is part of the frame header. This provides built-in transparency as the receiver knows how many bytes to receive. This means that the receiver can uniquely identify the data between the SOH and ETX commands.

12.4.5 Bit-oriented protocols

A typical example of this type of protocol is the high-level data link control protocol (HDLC). This protocol is described in detail when the data link layer in the open systems interconnect seven-layer model is discussed in Section 12.8.2.

12.5 THE INTERNATIONAL STANDARDS ORGANISATION SEVEN-LAYER MODEL

The International Standards Organisation (ISO) developed the recommendations for the open systems interconnect (OSI) seven-layer model. This model is recommended to be implemented in switching computers to enable different computers to communicate with one another. The whole idea behind the model is to enable data terminals to be interconnected by the network in an unrestricted way. Figure 12.2 shows the model. Each one of the layers contains a software

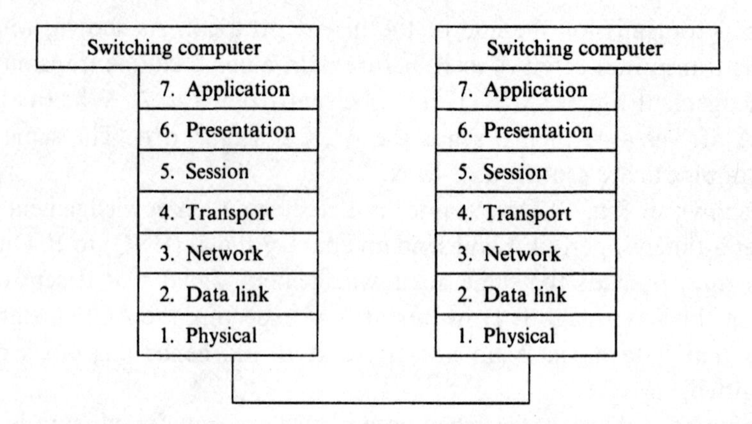

Fig. 12.2 Open systems interconnect seven-layer model

package. The physical layer contains both a software package and hardware such as an RS232C or RS449 port.

12.6 FUNCTIONS OF THE DIFFERENT LAYERS

12.6.1 Physical layer

The physical layer protocol is responsible for moving data over the transmission medium. It does not supply error correction or data sequencing. It is the physical interface between the open system and the transmission medium. This is not to be confused with the error detection and correction schemes that are used within the transmission system, which ensure error-free performance from the input to the system at the transmitter to the output of the system at the receiver. If the data applied to the input to the transmitter is already errored, the transmission system cannot correct it.

The physical layer deals with the transfer of the data between the users or end terminals.

12.6.2 Data link layer

The data link layer protocol provides error detection and recovery. It is responsible for transferring user data (information) and control data between the end terminals. The control data is used to initiate the call, maintain the integrity of the call and terminate the call that is transferred over the link.

12.6.3 Network layer

The network layer protocol is the interface between the open system and the communication network. It selects routes to establish logical connections. It relays protocol data units (PDUs) for other open systems.

12.6.4 Transport layer

The transport layer protocol provides higher layers with reliable connections over routes and networks that have different levels of quality. Some routes provide high-quality connections where the error rate of the transferred data is exceptionally good. However, other routes result in a large error rate. The transport layer must ensure that irrespective of the route, the quality of the received data is the same.

It correctly sequences the PDUs if the network layer does not provide this service. It shields higher layers from the realities of the network and route.

12.6.5 Session layer

The session layer protocol establishes a communication session, maintains the session for the duration of the call and then terminates the session at the end of the call. The transport layer does this on behalf of the application on two open systems.

12.6.6 Presentation layer

The presentation layer protocol is concerned with the transformations that are required to get the data from the format acceptable to one application into the presentation format acceptable to another application. An example is ASCII presentation on one terminal to EBCDIC presentation on another terminal and vice versa.

12.6.7 Application layer

The application layer protocol is responsible for exchanging user data or information. Examples of this are: bank account details, flight bookings, hotel reservations etc.

12.6.8 Null layers and sublayers

Every layer in the OSI model must be present; null layers, or layers that are totally empty, are not permitted. Each layer must contain a software package.

A layer may be sublayered if the functions of a layer can be more precisely defined by breaking it into smaller pieces. This is shown in Fig. 12.3.

12.6.9 Headers, data blocks and layers

Figure 12.4 shows where headers and footers are added to the data. The data from the user's computer is passed to the application layer, where a header is added to the data. The data and this header form the data that is passed to the presentation layer, which adds its header to the data. This is then transferred to the session

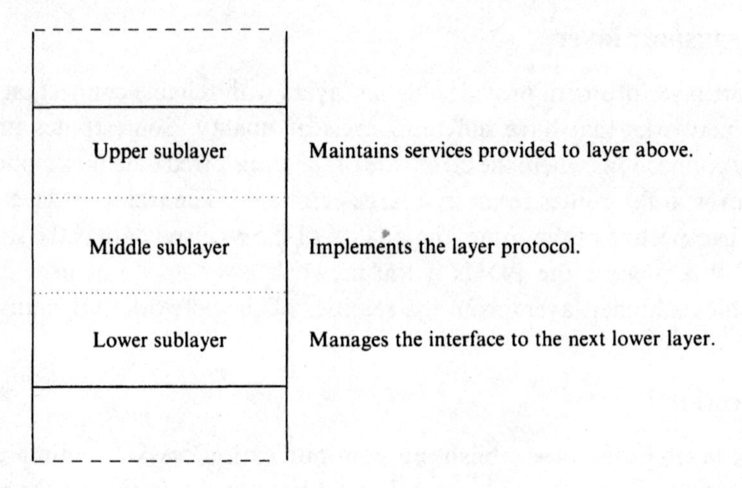

Fig. 12.3 Sublayers

layer, which adds its header. The transport layer adds a header, the network layer adds a header. The data link layer adds a header as well as a footer. This is then transferred over the physical connection via the physical layer. This means that the data transferred over the physical layer has a high redundancy and hence it is important to ensure that the raw data has no or very little redundancy.

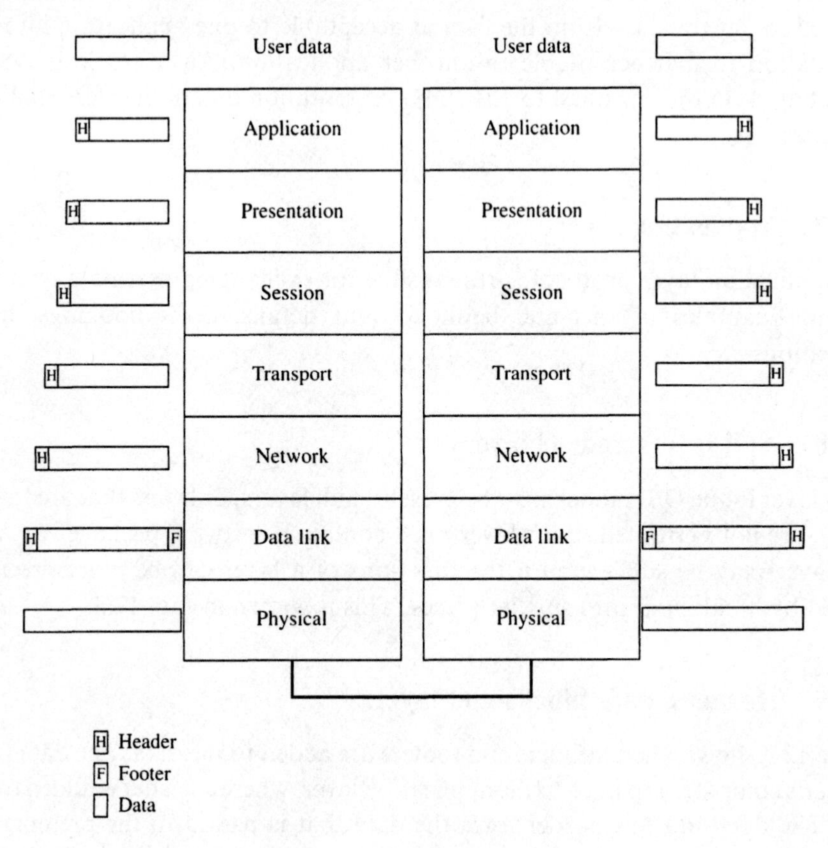

Fig. 12.4 Headers and footers in the seven-layer model

12.6.10 General

The physical, data link and network layers are the responsibility of the transmission engineer. The application, presentation and session layers are the responsibility of the switching engineer. The transport layer is the glue that holds the model together and is the responsibility of both the transmission and switching engineers.

12.7 THE ITU, ISO AND CCITT

The International Telecommunications Union (ITU) consists of several forums such as ITU-T (formerly CCITT), ITU-R (formerly CCIR) and ITU-D. CCITT stands for the Consultative Committee of International Telegraphy and Telephony. Many of the ITU recommendations still use the CCITT designated numbers.

- ITU-T determines the recommendations for telecommunication systems such as cable systems.
- ITU-R determines the recommendations for radio, microwave and satellite systems.
- ITU-D determines recommendations for data communication over local area networks (LANs) and wide area networks (WANs).

The main function of the International Standards Organisation (ISO) is to recommend standards that will be accepted by the whole world to enable the data equipment made by different manufacturers to be able to communicate properly.

There are many different organisations that recommend standards. Some of these organisations are vendor oriented, some are nationally oriented and some are internationally oriented. Many of the standards recommended by these various organisations are very similar. The only differences are in terms of implementation.

12.8 RECOMMENDED INTERFACES

12.8.1 Physical layer

The ISO recommendations are the 'X.21' and 'X.21 bis' protocols. The X.21 bis recommendations cover the RS232C port and the V.24 interface, which is used when the connection to the public switched network is an analogue connection and the medium connecting the data terminal equipment (DTE) to the analogue exchange is a metallic cable pair and the data rate is relatively slow.

The X.21 bis recommendation also covers the unbalanced and balanced RS232C ports which are used when the connection from the DTE to a digital exchange is by means of a digital link.

The X.21 recommendations cover the RS449 protocol and the V.11 and V.35 interfaces, which are used for higher data rates than the RS232C port. The V.35 interface is used when the link to the analogue exchange is an analogue link by means of metallic cable pair. The V.11 interface is used when the link to a digital exchange is a digital link.

The X.21 bis protocol

The V.24 interface

Figure 12.5(a) shows the position of the modems in relationship to the DTE and data circuit equipment (DCE). The link connecting the two modems together is now totally analogue.

Figure 12.5(b) shows the actual V.24 interface. In its simplest form the V.24 interface consists of 13 different circuits. The modem was originally designed to

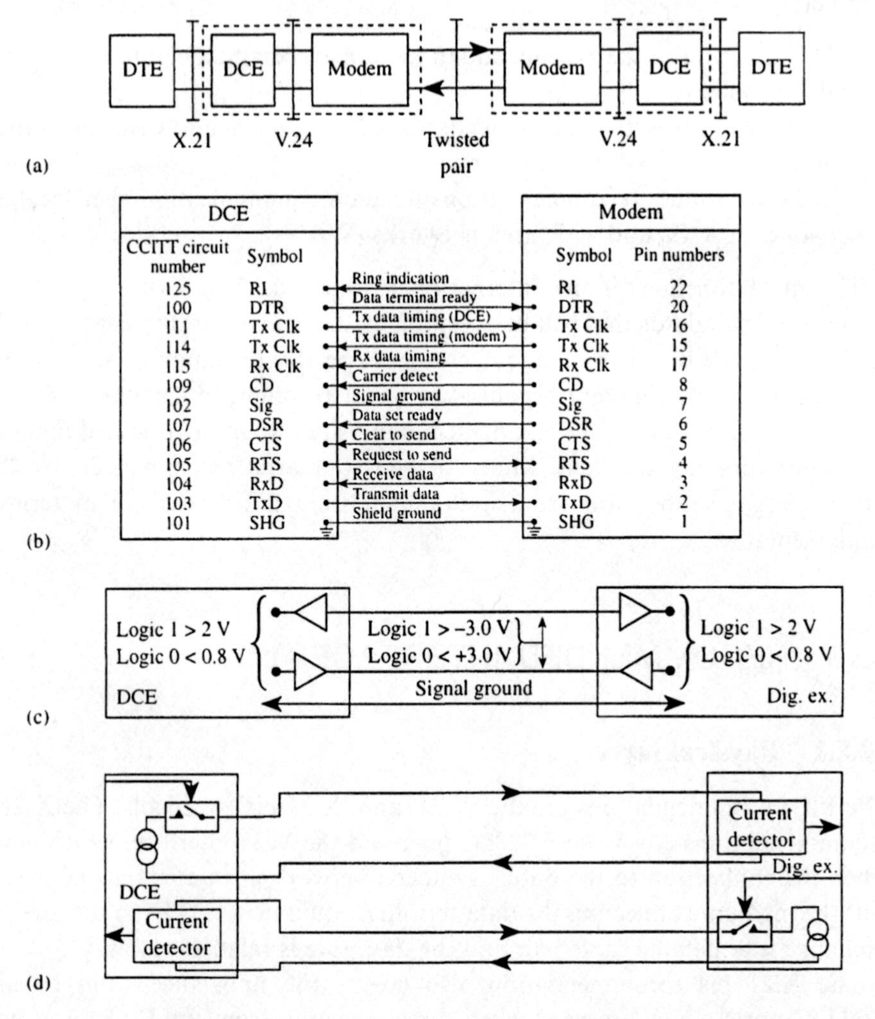

Fig. 12.5 (a) Signal definitions; (b) V.24 interface; (c) unbalanced RS232C interface; (d) 20 mA loop current RS232C interface

connect a data rate of 9600 baud (9600 b/s) to the exchange in the public switched telephone network (PSTN). The V.24 interface consists of a 25-way D connector that connects the RS232C port to the modem. The transmit clock frequency can be either generated on the RS232C port or on the modem. It can thus be fed from the RS232C port to the modem on the Tx timing lead from the DTE to the DCE or it can be fed from the modem on the Tx timing lead on the DCE to the DTE. The receiver clock is always fed from the modem to the RS232C port on the Rx timing lead. The functions of the leads are shown in Fig. 12.5(b).

Ideally the modem should be placed very close to the DCE and the DCE should be placed into one of the slots in the DTE. The analogue exchange can be situated quite a few kilometres from the DTE. The modem normally forms part of the DCE whereas the DTE is the user terminal or computer.

RS232C/V.24 signal definitions

Digital links

For short distances between the DTE and a digital exchange (less than 100 m), the RS232C interface, as shown in Fig. 12.5(c), is used. In this instance the link between the DCE and the exchange is a digital link. Figure 12.5(c) shows an unbalanced link using ECL (emitter-coupled logic) where a logic 1 is allocated a negative voltage greater than −3 V and a logic 0 is allocated a positive voltage greater than +3 V.

When the distance between the DTE and the digital exchange is longer than 100 m and the data is to be transferred in a digital format, the 20 mA loop interface can be used. This interface is a balanced interface and is shown in Fig. 12.5(d). The data is changed into pulses of current or no current at the transmit end and the receiver reproduces the data by detecting the current/no-current conditions. This is similar to the transfer of dial pulses from a rotary dial on a telephone and is referred to as loop disconnect signalling as the receiver senses when the loop is made and when the loop is broken.

Analogue links

Figure 12.6 shows a typical call procedure over an RS232C port connected to a modem. When a call is to be made, the first step is to establish a link and this is done by means of dial pulses that must be transmitted first. The dial pulses are used in the exchange to route the call to the correct destination.

If the distant terminal is free then the DCE will cause the data set ready (DSR) lead to go high, which indicates to the receive modem that the DTE at the distant end is ready to accept a call. The modem on receipt of the call initialisation signal from the exchange will cause the ring indication (RI) to go high. The DCE responds by causing the transmit data (TxD) lead to go high, which causes the modem to transmit a constant tone to the exchange on the cable pair.

This tone is then received by the calling modem which causes the carrier detect (CD) lead to go high. This informs the calling DCE that the called DCE is ready for the call to proceed.

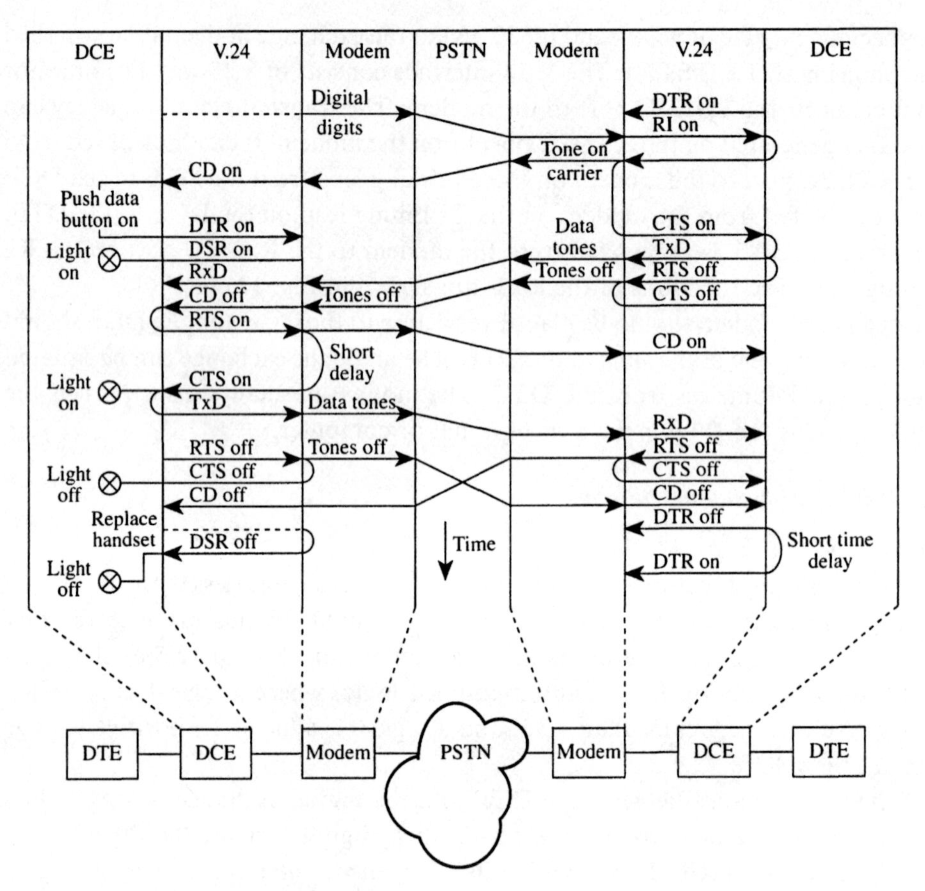

Fig. 12.6 RS232C/V.24 call procedure

At the called modem, after a timeout period, the modem causes the clear to send (CTS) lead to go high. This causes the DCE to place a data signal on the TxD lead. After a short duration the called DCE causes the request to send (RTS) lead to go low, which causes the called modem to stop transmitting the data tones back to the calling end. On receipt of the low on the RTS lead the called modem causes the CTS lead to go low.

At the calling end the DCE causes the data terminal ready (DTR) lead to go high. The modem responds by causing the DSR lead to go high. The modem, on receipt of the data tones, causes the data to be output to the DTE on the receive data (RxD) lead. When the modem senses that the data tones have been removed it causes the CD lead to go low. This in turn results in the DTE causing the RST lead to go high, which causes the calling modem to transmit data tones to the called modem.

On receipt of the data tones by the called modem, the CD lead goes high. At the calling modem after a short delay, the CTS lead goes high. The calling DCE now responds by transmitting the data on the TxD lead to the modem. The calling modem now transmits the data tones to the called modem. The called modem converts the received data tones back to data and sends it to the DCE on the RxD lead.

Once all the data has been received the calling DCE causes the RTS lead to go low, which causes the CTS to go low. The modem stops sending data tones. The same takes place in the opposite direction. Both modems then detect that the data tones have been removed and cause the CD lead to go low, which terminates the call.

The X.21 protocol

This protocol describes the RS449 interface. The RS449 interface is microprocessor based. The X.21 interface describes the link between the DTE and the DCE, which is shown in Fig. 12.7(a). The V.35 interface describes the link between the DCE and the modem, which is shown in Fig. 12.7(a) and (b).

The X.21 interface consists of five separate leads or five pairs of wires. These are the transmit data lead or pair, receive data lead or pair, control lead or pair, the

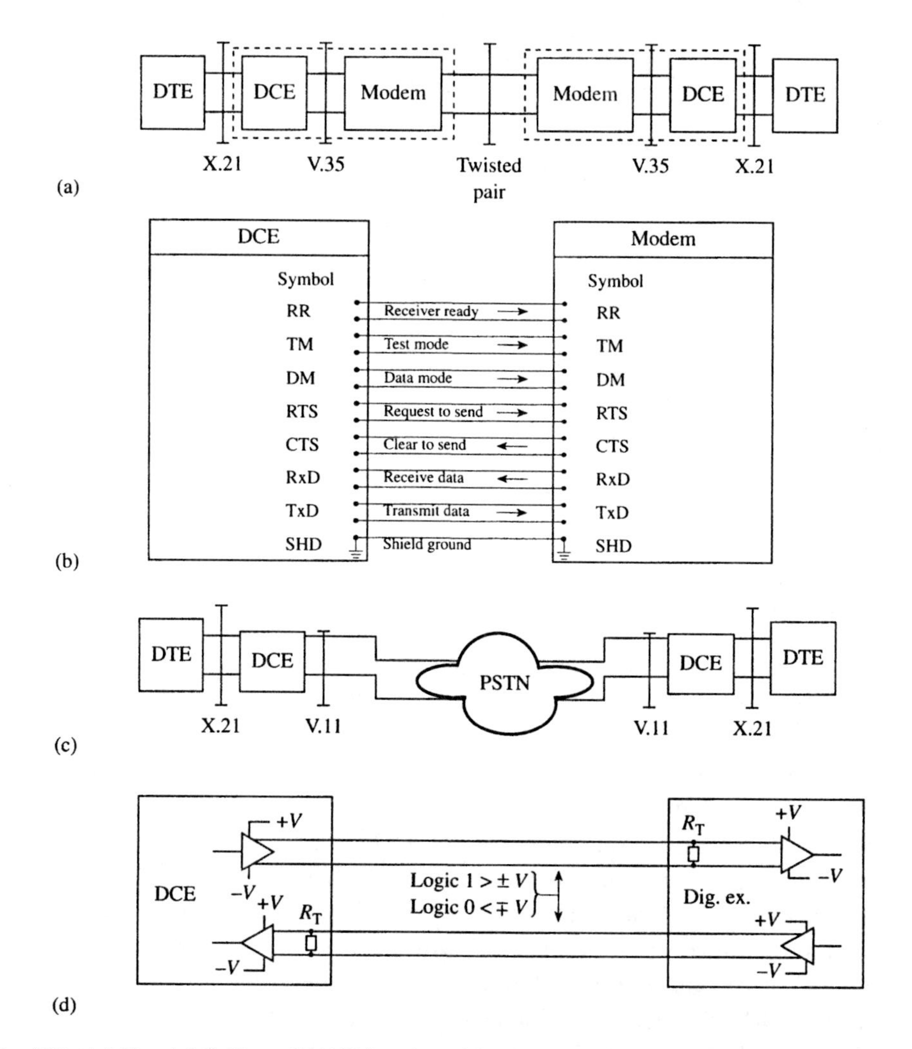

Fig. 12.7 (a) Signal definitions; (b) V.35 interface; (c) unbalanced V.11 interface; (d) balanced V.11 interfaced

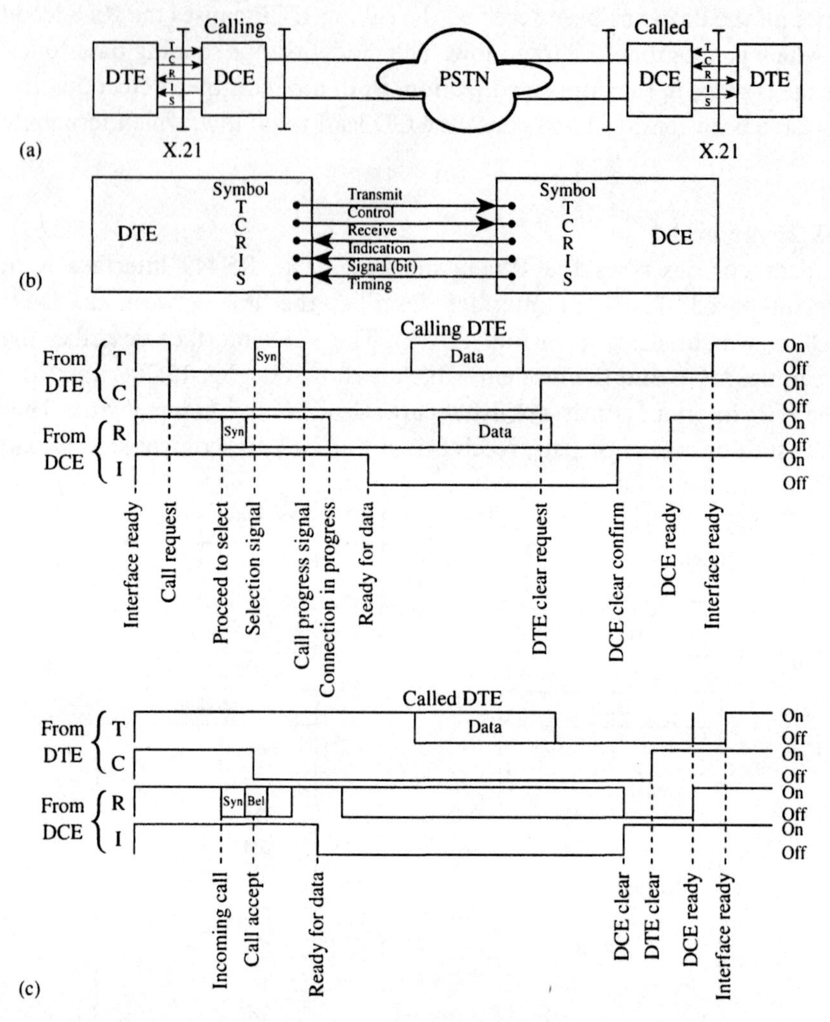

Fig. 12.8 (a) Signal definitions; (b) X.21 interface; (c) call procedure

indication lead or pair and signal element lead or pair which carries the timing information. The unbalanced interface is shown in Fig. 12.7(a) as well as in Fig. 12.8(a) and (b).

A typical call procedure is shown in Fig. 12.8(c) which has been included for reference purposes only.

The RS449 recommendation covers the RS422 interface, which is a balanced interface and uses a pair of wires for each control signal, and the RS423 which is an unbalanced interface and uses a single wire for each control signal. The maximum separation in metres together with the maximum bit rates are indicated in Table 12.1 for both the RS422 and the RS423 interfaces.

When the RS449 interface is connected to a modem for use over an analogue link then the V.35 interface is used. When the RS449 is used on a digital link then the V.11 interface is used.

Table 12.1 Maximum separation and bit rates RS422 and RS423

Circuit type	Maximum separation (m)	Maximum bit rate
RS422	10	100 kb/s
	100	10 kb/s
	1000	1 kb/s
RS423	10	10 Mb/s
	100	1 Mb/s
	1000	100 kb/s

The V.35 interface
The V.35 interface is used to connect the DTE via the DCE to a wideband (analogue) high-bit-rate modem. A 34-pin connector is used and data rates of between 48 kb/s and 168 kb/s are possible. Figure 12.7(b) shows the balanced V.35 interface and its control signals. When the interface is in operation only two of the controls are active at a time.

The V.11 interface
This interface is only used when the connection between the DTE and the exchange is a digital connection. Figure 12.7(c) and (d) shows the interface. In this case the ECL is used which is the same as the unbalanced RS232C shown in Fig. 12.5(c).

12.8.2 Data link layer

The protocol that is recommended by the ITU is the CCITT-recommended high-level data link control (HDLC). This protocol is a derivative of the serial data link control (SDLC) protocol.

High-level data link control protocol
The frame format is shown below:

Flag	Address	Control	Data	FCS	Flag
8 bits	8/16 bits	8/16 bits	*n* bytes	16 bits	8 bits

where Flag = 01111110
FCS = frame check sequence employing CRC-16 or CCITT-CRC
CRC = cyclic redundancy check

The HDLC protocol supports three different types of frames; these are the unnumbered frame (U-frame), the supervisory frame (S-frame) and the information frame (I-frame).

The information frame

This frame is used to control the flow of data. It has a frame format as shown below. The control field can be either 8 bits or 16 bits wide.

Flag	Address	Control	Data	FCS	Flag
8 bits	8/16 bits	8/16 bits	*n* bytes	16 bits	8 bits

Control field and address field (8-bit)

For a frame having an 8-bit control field, the control field is as follows:

Bit no.	7	6	5	4	3	2	1	0
Usage	0		*N*(s)		P/F		*N*(r)	

- Bit 7 A logic 0 in this position indicates that this frame is an information frame.
- *N*(s) Contains the frame number of the frame presently being sent by the transmitter. The frame number is from 000_2 to 111_2, i.e. frames 0_{10} to $7_{10} \rightarrow$ modulo 8.
- *N*(r) Contains the number of the next expected frame from the distant transmitter. The frame number is from 000_2 to 111_2, i.e. frames 0_{10} to $7_{10} \rightarrow$ modulo 8.
- P/F Poll or final bit. This bit is used for control purposes. It is used by one terminal to see if the other terminal is still there. If it is set to a logic 1 then the receiver must respond immediately by setting this bit to a logic 1. The receiver sets this bit to a logic 1 when it transmits its final frame.

An 8-bit address field enables one of 256_{10} ports to be addressed.

Control field and address field (16-bit)

For a frame having a 16-bit control field, the control field is as follows:

15	14	13	12	11	10	9	8	7	6	5	4	3	2	1	0
0				*N*(s)				P/F				*N*(r)			

- Bit 15 A logic 0 in this position indicates that this frame is an information frame.
- *N*(s) Contains the frame number of the frame presently being sent by the transmitter. The frame number is from 0000000_2 to 1111111_2, i.e. frames 0_{10} to $127_{10} \rightarrow$ modulo 128.
- *N*(r) Contains the number of the next expected frame from the distant transmitter. The frame number is from 0000000_2 to 1111111_2, i.e. 0_{10} to $127_{10} \rightarrow$ modulo 128.
- P/F Poll or final bit. This bit is used for control purposes. It is used by one terminal to see if the other terminal is still there. If it is set to a logic 1

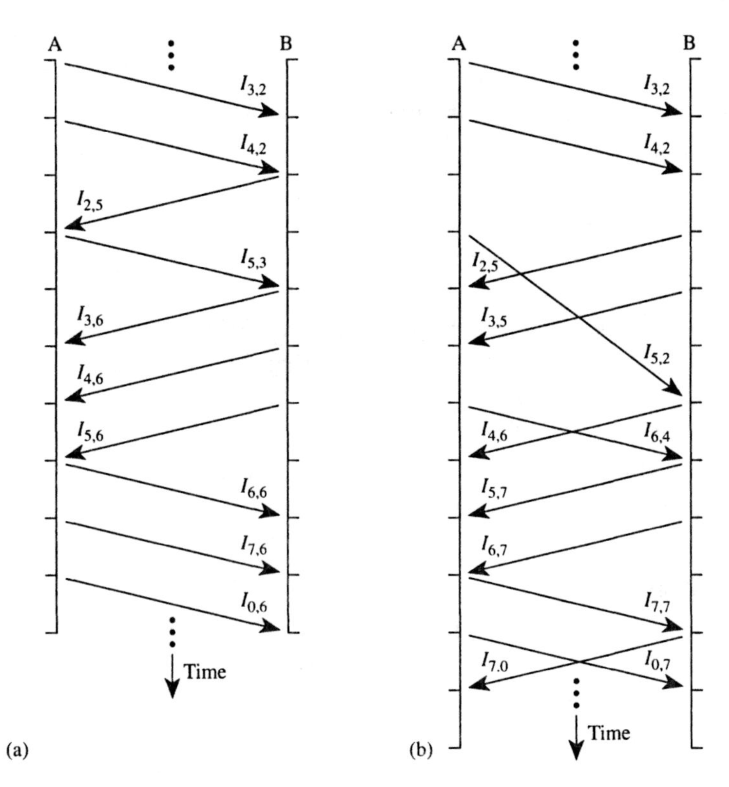

Note. The above diagrams indicate the information frames having an 8-bit control field. As a result of this acknowledgement must take place after every eight frames. However, acknowledgement can take place at any point during the transmission. If a 16-bit control field is used then acknowledgement must take place after 128 frames.

Fig. 12.9 HDLC information frame transfer: (a) half-duplex working; (b) full-duplex working

then the receiver must respond immediately by setting this bit to a logic 1. The receiver sets this bit to a logic 1 when it transmits its final frame.

A 16-bit address field enables one of $65\,536_{10}$ ports to be addressed.

The HDLC information frame data transfer is shown in Fig. 12.9. Figure 12.9(a) shows half-duplex working, whereas Fig. 12.9(b) shows full-duplex working, which is purely for reference purposes. Both figures show the procedure using a modulo 8 count for the N(s) and N(r) fields. The format used in these figures for the transfer of information frames is $I_{N(s),N(r)}$.

Consider a call that has already been established, shown in Fig. 12.9(a), and the point is reached where A has transmitted information frame 3 to B and is expecting information frame 2 from B ($I_{3,2}$). Then A transmits information frame 4 to B and is still expecting information frame 2 from B ($I_{4,2}$). B responds by transmitting information frame 2 to A and is now expecting information frame 5 from A ($I_{2,5}$). On receipt of this frame at A the N(r) is 5, which informs A that information frames 3 and 4 were received and accepted as error free by B and thus these frames are acknowledged. Then A transmits information frame 5, expecting information frame 3 from B ($I_{5,3}$). On receipt of frame $I_{5,3}$ at B, N(r) is 3, indicating that A is now expecting information frame 3 from B and acknowledging that A received information frame 2 correctly and without errors from B.

When A has transmitted $I_{7,6}$, the next information frame that A transmits must be $I_{0,6}$ because of the modulo 8 count. The same applies for the transmitted frames from B to A.

If a modulo 8 count is used for the N(s) and N(r) fields and if A transmits $I_{0,0}$ to $I_{7,0}$ and B does not transmit any frames during this time, then before A can transmit $I_{0,0}$ again B must respond by acknowledging that it has received information frames 0 to 7 correctly.

The procedure that takes place when data is received or the distant terminal does not receive a frame is described in the next section.

The supervisory frame

This frame is used to control the flow of data. The frame structure is shown below:

Flag	Address	Control	FCS	Flag
8 bits	8/16 bits	8/16 bits	16 bits	8 bits

Eight-bit control field

The control field is as follows:

Bit no.	8	7	6	5	4	3	2	1
Usage	1	0	Supy		P/F	N(r)		

- Bits 8 and 7 The code 10_2 identifies this frame as a supervisory frame.
- Supy These are supervisory codes for the control of the flow of data.
- N(r) Contains the number of the next expected frame from the distant transmitter. The frame number is from 000_2 to 111_2, i.e. 0_{10} to $7_{10} \rightarrow$ modulo 8.
- P/F Poll or final bit. This bit is used for control purposes. It is used by one terminal to see if the other terminal is still there. If it is set to a logic 1 then the receiver must respond immediately by setting this bit to a logic 1. The receiver sets this bit to a logic 1 when it transmits its final frame.

Sixteen-bit control field

The control field is as follows:

15	14	13	12	11	10	9	8	7	6	5	4	3	2	1	0
0	1	Supy		Reserve				P/F	N(r)						

- Bit 15 and 14 The code 01_2 indicates that this is a supervisory frame.

- Supy — These are supervisory bits which are used to control the flow of data.
- Reserve — These are unallocated bits for future expansion.
- N(r) — Contains the number of the next expected frame from the distant transmitter. The frame number is from 0000000_2 to 1111111_2, i.e. 0_{10} to 127_{10} → modulo 128.
- P/F — Poll or final bit. This bit is used for control purposes. It is used by one terminal to see if the other terminal is still there. If it is set to a logic 1 then the receiver must respond immediately by setting this bit to a logic 1. The receiver sets this bit to a logic 1 when it transmits its final frame.

The HDLC supervisory frame transfer is shown in Fig. 12.10.

Types of supervisory frames

There are four different types of supervisory frame. This means there are four different control functions that can be carried out by the supervisory frame. These are shown in Table 12.2.

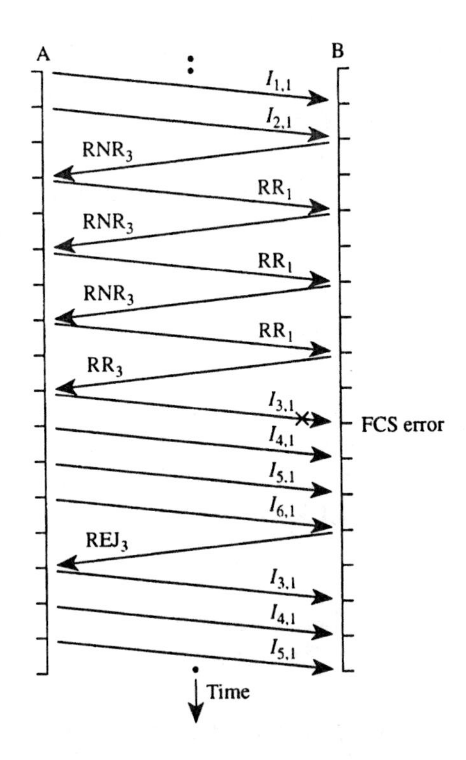

Note. The above diagram indicates the supervisory frames having an 8-bit control field. As a result of this, acknowledgement must take place every eight frames. However, acknowlegement can take place at any point during the transmission. If a 16-bit control field is used then acknowledgement must take place after 128 frames.

Fig. 12.10 HDLC supervisory frame transfer – half-duplex working

Table 12.2 Supervisory frames

Type	Meaning and usage
RR	Receiver ready. Explicit acknowledgement of all frames prior to the specified frame. Receiver ready to receive data in the specified frame
RNR	Receiver not ready. Explicit acknowledgement of all frames prior to the specified frame. Receiver not ready to receive additional frames
REJ	Reject. Explicit acknowledgement of all frames prior to the specified frame. Request retransmission of all frames starting with the specified frame number
SREJ	Selective reject. Requests retransmission of the specified frame only

With reference to Fig. 12.10, a call has been established and A has transmitted $I_{1,1}$ to B and then $I_{2,1}$ to B. At this point B becomes busy and cannot accept any more data from A, so it transmits a receiver-not-ready (RNR) frame to A. The format of this is $\text{RNR}_{N(\text{r})}$. In this case $N(\text{r})$ is 3, hence B transmits RNR_3 to A. This informs A that B has received information frame 2 correctly but that B cannot accept information frame 3 from A as it is busy. This indicates to A that information frames 1 and 2 have been acknowledged by B.

If B does not transmit any frames, then after a timeout period A transmits a receiver ready supervisory frame (RR_1) to B which asks B whether it is ready to receive information frame 3 from A and also tells B that A is expecting information frame 1. On receipt of the RR_1 frame B must respond immediately. If B is still not ready then it will transmit RNR_3 to A but if it is ready B will transmit RR_3 to A.

On receipt of the supervisory frame RR_3 from B, A then responds by transmitting $I_{3,1}$ to B. Assume that there is an FCS error in this frame when it is received at B but B does not respond immediately. Instead A transmits $I_{4,1}$ to $I_{6,1}$ to B. Because of the FCS error B does not accept $I_{3,1}$ from A and at this point transmits a reject supervisory frame (REJ_3) to A rejecting information frame 3 but acknowledging all previous information frames up to and including frame 2. On receipt of the REJ_3 A responds by retransmitting $I_{3,1}$ to $I_{6,1}$ to B.

Unnumbered frames

The function of the unnumbered frame is to set up the connection and then to disconnect the connection. The unnumbered frame is also used to set up the mode of operation. There are three different modes of operation. Each different mode applies to a specific type of circuit. There are three different types of circuits. These are given below:

- Unbalanced point to point.
- Unbalanced multipoint.
- Balanced point to point.

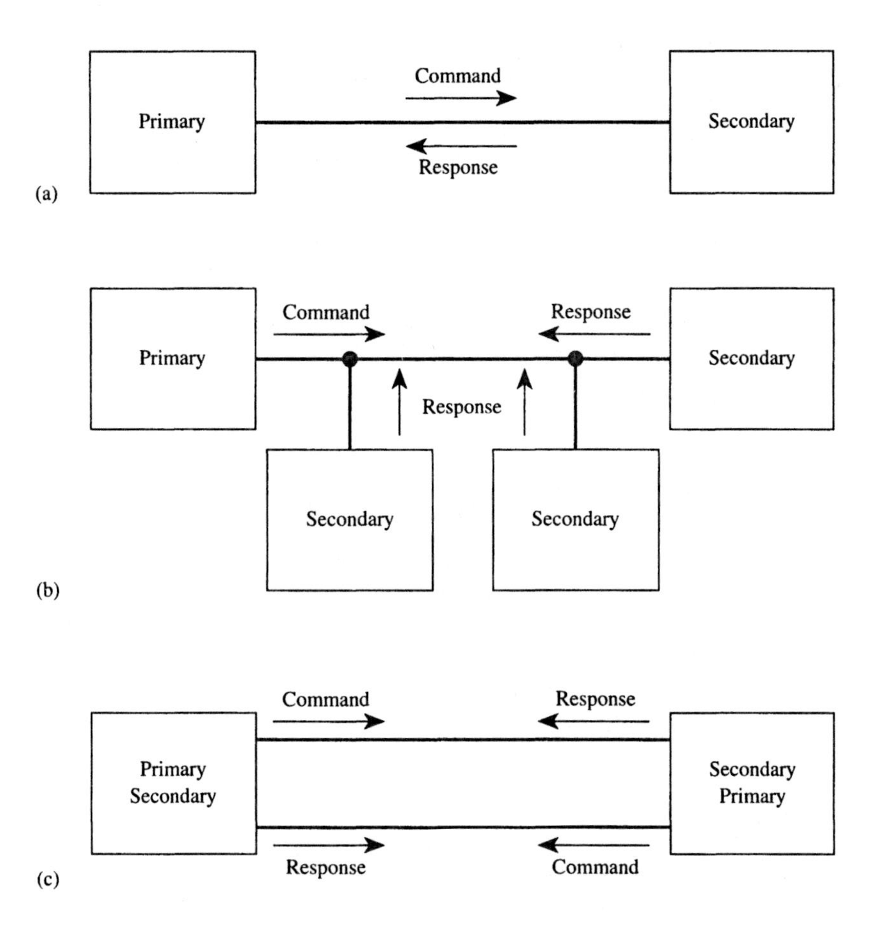

Fig. 12.11 (a) Unbalanced point to point; (b) unbalanced multipoint; (c) balanced point to point

Figure 12.11 shows these three basic configurations and the abbreviations are given in Table 12.3.

Unnumbered frame commands are given in Table 12.4 and unnumbered frame responses are given in Table 12.5.

Table 12.3 Unnumbered frame configurations

Mode	Type	Circuit type	Meaning and usage
1	NRM	Unbalanced point to point Unbalanced multipoint	Normal response mode Unbalanced configuration Secondary only transmits when requested to by the primary
2	ARM	Unbalanced point to point	Asynchronous response mode Secondary allowed to transmit to primary without first being requested to do so
3	ABM	Balanced point to point	Asynchronous balanced mode Both stations have equal status

Table 12.4 Unnumbered frame commands

Command	Meaning
SARM	Set asynchronous response mode 8-bit address/control field
SARME	Set expanded asynchronous response mode 16-bit address/control field
SNRM	Set normal response mode 8-bit address/control field
SNRME	Set expanded normal response mode 16-bit address/control field
SABM	Set asynchronous balanced mode 8-bit address/control field
SABME	Set expanded asynchronous balanced mode 16-bit address/control field
RSET	Reset
FRMR	Frame reject
DISC	Disconnect

Table 12.5 Unnumbered frame responses

Response	Meaning
UA	Unnumbered frame
CMDR	Command reject
FRMR	Frame reject
DM	Disconnect

Figure 12.12 shows the call procedure for the unnumbered frame. To initialise a call, A transmits an unnumbered frame, which sets the mode of operation. In this case the command is SABM, which sets the circuit to asynchronous balanced mode. If B accepts that this mode is correct for that particular circuit then it responds by transmitting a UA frame to A. The call now proceeds with the transfer of information and supervisory frames. At the end of the call A transmits a disconnect (DISC) unnumbered frame to B. B responds by transmitting a DM unnumbered frame to A which acknowledges the disconnect command.

12.8.3 Network layer

This layer is sublayered into two sections. The recommendation for the top sublayer is the internet protocol (IP), which interfaces with the transport layer.

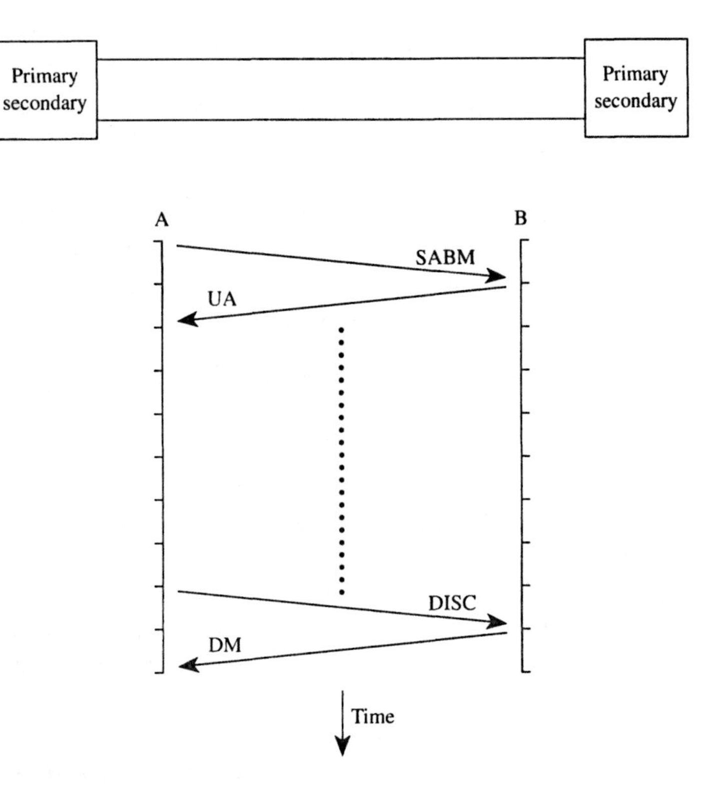

Note. The above diagrams indicate the supervisory frames having an 8-bit control field. As a result of this, acknowledgement must take place every eight frames. However, acknowlegement can take place at any point during the transmission. If a 16-bit control field is used then acknowledgement must take place after 128 frames.

Fig. 12.12 HDLC unnumbered frame transfer – balanced point-to-point circuit

The recommendation for the bottom layer is the X.213 protocol. This protocol implements the layer protocol and interfaces with the data link layer. The reason for this is that the X.213 recommendation does not include a specification for how the network layer will interface with the transport layer.

X.213 protocol

This protocol provides for the establishment of virtual circuits. There are two types of virtual circuit:

- *Permanent virtual circuit (PVC).* This circuit is similar to a leased line in that when the two DTEs want to communicate with each other they do not have to set up a connection first. The connection already exists. This means that they have permanent access to one another, through the packet-switched network, all the time. This is similar to a private wire circuit.
- *Switched virtual circuit (SVC).* This is a dial-up service where calls are established on demand through the packet-switched network. This is similar to a normal telephone call.

The network layer also allows two types of service:

- *Connection-oriented service.* Most PVCs and SVCs operate on this service. The PDUs are correctly sequenced by the network and hence arrive in the correct sequence.
- *Connectionless-oriented service.* In this service the PDUs may or may not arrive in the correct sequence. The recommendation does not guarantee correct sequencing of PDUs using this service. The user might have to rearrange the received data so that it follows the correct order.

The network layer also allows several different data streams to be multiplexed together.

Multiplexing

Several data streams can be multiplexed together and sent over a single X.25 link. Each data stream is allocated a unique logical channel number (LCN). The protocol allows for a maximum of 4096_{10} LCNs between each DTE–DTE pair.

Table 12.6 shows the arrangement. As can be seen, channel number 0 is allocated to the network layer protocol for recovery from fatal failures. The next group of channel numbers, 1_{10} to X_1, are allocated to the permanent virtual circuits. The next group of channel numbers, X_2 to X_3, are allocated to incoming calls only. This means that calls that are initialised at some distant terminals are allocated these channel numbers when the network is accessed. The next group of channel numbers, X_4 to X_5, are allocated to bothway circuits which can be accessed by either local terminals or by distant terminals if there are no free channel numbers in the outgoing or incoming group, respectively. The next group of channel numbers, X_6 to 4095_{10}, are allocated to outgoing calls which means that local terminals only have access to these numbers.

From the above it is obvious that the LCN assigned to the DTE–DCE pair at the local exchange has nothing to do with the LCN assigned at the DCE–DTE pair at the distant exchange for a particular DTE–DTE pair. This is shown in Fig. 12.13. The LCN for the DTE–DCE pair at the calling exchange will fall in either the outgoing or bothway group, whereas at the called end the LCN for the DCE–DTE pair will fall in either the incoming or bothway group.

Table 12.6 Assignment of logical channel numbers

Channel number	Purpose
0_{10}	X.25 network layer protocol
$1_{10} \ldots X_1$	PVCs
$X_2 \ldots X_3$	Incoming calls
$X_4 \ldots X_5$	Bothway calls
$X_6 \ldots 4095_{10}$	Outgoing calls

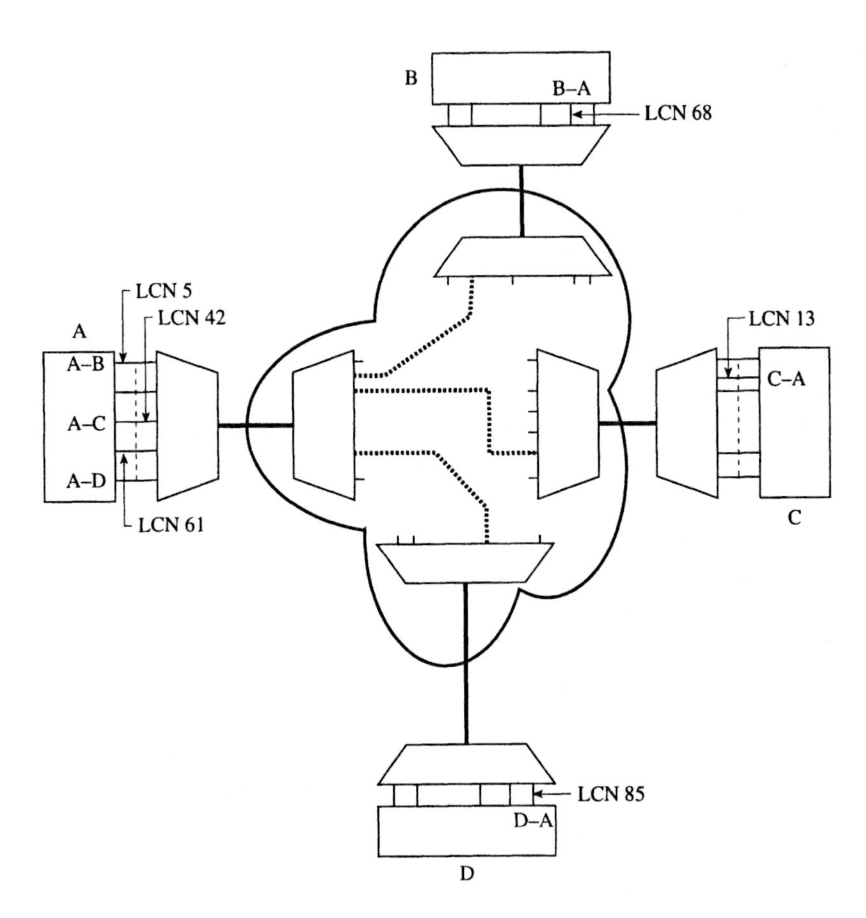

Fig. 12.13 Logical channel number assignment

Packet header

The format of the frame is shown in Table 12.7.

Table 12.7 Packet header frame format

Octet	4 bits	4 bits
1	General format ID	Logical channel group no.
2	Logical channel number	
3	Packet type ID	
4	Optional octets based on packet type	
⋮		

General format ID: Used to select certain packet layer options.
Logical group number: Selects one of 16 groups.
Logical channel number: Selects one of 256 outlets.
Packet type ID: Distinguishes packet type.

Packet types

The different packet types are shown in Table 12.8.

Table 12.8 Packet types

From DTE to DCE	From DCE to DTE	Purpose
Call request Call accept	Incoming call Call connect	Call set-up
Data	Data	Data transfer
RR	RR	Receiver ready
RNR	RNR	Receiver not ready
Clear request Clear confirm	Clear indication Clear confirm	Call clearing
Reset request Reset confirm	Reset indication Reset confirm	Reset a VC
Restart request	Restart indication	Restart the entire packet layer

Packet assembler/dissembler (PAD)

Packet assemblers/dissemblers are used on non-X.25 or packet-switched terminals and computers that require communication with packet-switched or X.25 terminals or computers. Essentially they are used on terminals, such as character-mode terminals, that are not capable of packeting the data according to the X.25 protocol.

Standards relating to PADS

- X.3 PAD in the public switched network.
- X.28 Protocol for DTE, DCE interface for start–stop-mode terminals.
- X.29 Protocol for the exchange of control data between PADs.
- X.75 Terminal and transit call procedures and data transfer systems on international circuits between packet-switched data networks.

X.3 protocol

The X.3 protocol is designed to enable the normal X.25 functions to be carried out by the PAD. These functions include things such as call establishment, flow control etc. The X.3 protocol offers other facilities which are controllable by the terminal or the remote packet-mode DTE. These facilities are as follows:

- Is local echo checking required or not?
- Are alternative control characters required such as line feed and carriage return?

When a PAD is accessed by a non-packet-mode terminal the software will start with the default parameters. However, some of these parameters will have to be different from the default values for the specific terminal.

X.28 protocol

The X.28 protocol is used to enable only the specific values to be changed for the particular asynchronous character-mode terminal. The X.28 recommendation covers the following functions:

- Call establishment between the character-mode terminal and the PAD.
- Setting of the parameters of the particular terminal.
- Control of the data exchange process between the terminal and the PAD.
- Call disconnection at the end of the data transfer.

The call can be established on a permanent virtual circuit basis (PVC) or switched virtual circuit basis (SVC). The main stipulation is that it is a switched circuit set-up using a PSTN. The PAD is capable of determining the data rate used by the terminal. The terminal will initially send a number of service request characters. This enables the PAD to initialise the standard values for the terminal as well as the data rate used by the terminal. Once this is done, then the PAD can establish the necessary virtual circuit through the PSTN.

Figure 12.14(a) shows where the PADs are situated relative to the different terminals connected to the PSTN. Figure 12.14(b) shows the situation where

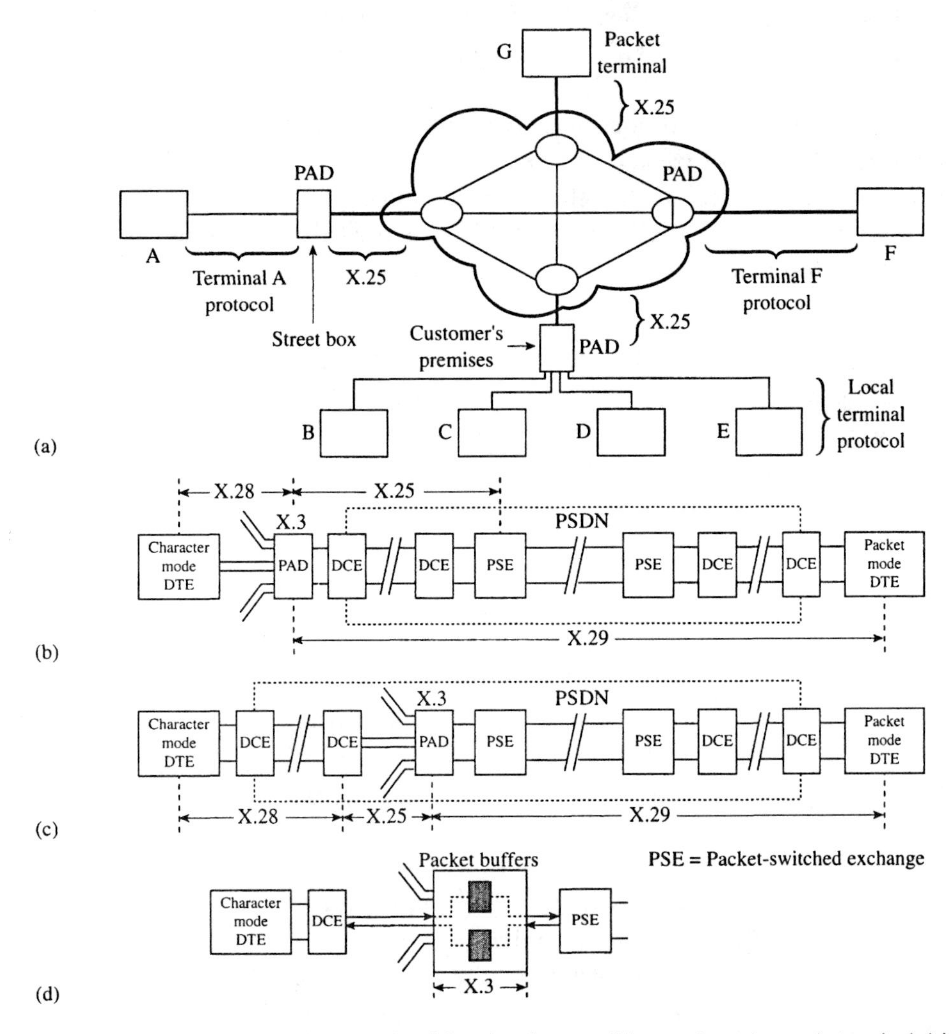

Fig. 12.14 Packet assembler/disassembler: (a) national usage; (b) near character-mode terminal; (c) remote from terminal; (d) internal structure

the PAD is situated relatively close to a character-mode terminal, i.e. in the same room. In Fig. 12.14(c) the PAD is situated some distance away from the character-mode terminal, i.e. in a street box.

Figure 12.14(d) shows that the PAD buffers the data that it receives from the character-mode terminal. Because the data is stored in the PAD, it can then packet the data for transmission over the X.25 network. At the receiver the PAD receives the packets and stores them so that it can transmit the complete message back to the character-mode terminal.

X.29 protocol

The X.29 protocol lays down the recommendations for the communication between a PAD and a distant packet-switched terminal. The call request, call connect, data transfer and call disconnect are the same as the procedures used in X.25. In order for the distant packet-switched terminal to be able to identify the type of start–stop terminal, the PAD uses the first four octets of the identifier field. The distant packet-switched terminal can then use alternative protocols should they be required.

Also the Q bit in the header of each packet is used to inform the PAD that the data from the distant packet-switched terminal is either for the PAD or must be disassembled and sent to the start–stop terminal. If the Q bit is set then the data is meant for the PAD and if the Q bit is reset then the data is meant for the start–stop terminal.

X.75 protocol

This recommendation is for the establishment of virtual circuits between two signalling terminal exchanges (STE), which is normally required for international calls where one STE is in one country and the other STE is in a different country. The packet types are given in Table 12.9.

Table 12.9 Packet types for X.75

Packet type	Protocol usage
Call request	Call set-up
Call connect	
Clear request	Call clearing
Clear confirm	
Data	Data transfer
Interrupt	
Interrupt confirm	
Receiver ready	Flow control
Receiver not ready	
Reset confirm	
Restart	Restart
Restart confirm	

The format of the call request frame is shown in Table 12.10.

Table 12.10 Call request frame for X.75

8	7	6	5	4	3	2	1
0	0	Modulo		Group			
Channel							
Type							*I*
Source address length				Destination address length			
Destination address Source address							
0	0	Network utilities length					
Network utilities							
0	0	Facilities length					
Facilities							
User data							

The format of the control frame is shown in Table 12.11.

Table 12.11 Control frame for X.75

8	7	6	5	4	3	2	1
0	0	Modulo		Group			
Channel							
Type							*I*
Additional fields							

The format of the data transfer frame is shown in Table 12.12.

Table 12.12 Data transfer frame for X.75

8	7	6	5	4	3	2	1
Q	*D*	Modulo		Group			
Channel							
P(*R*)			*M*	*P*(*S*)			0
Data fields							

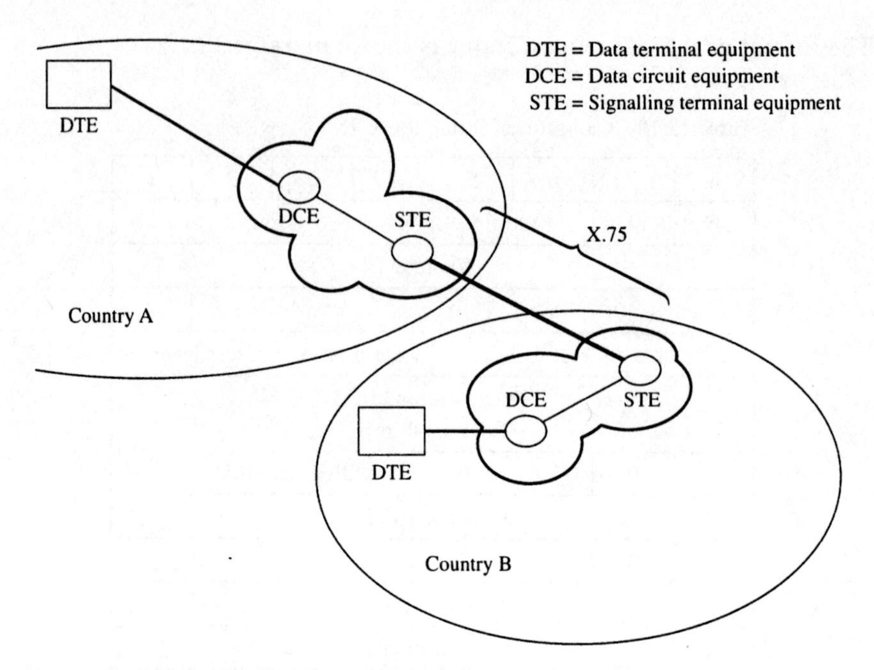

Fig. 12.15 ISO and CCITT recommended protocols

Figure 12.15 shows where this standard is applied.

12.8.4 Transport layer

The transport layer can supply three types of service over five classes of circuit protocol. The types of service are as follows:

- Type A: Reliable service (simple protocol).
- Type B: Moderate service (more robust protocol).
- Type C: Poor service (complex protocol).

The five classes of protocol are as follows:

- TP0: class 0 – used on type A services. Flow control provided by the network.
- TP1: class 1 – used on type B services. Flow control provided by the network. Basic error recovery.
- TP2: class 2 – used on type A services. Multiplexes several transport connections over a single network connection. Provides flow control as this must be done on a per transport connection basis.
- TP3: class 3 – used on type B services. Error detection and recovery. Multiplexes several transport connections over a single network connection. Flow control provided as this must be done on a per transport connection basis.
- TP4: class 4 – used on type C services. Full error detection and recovery. Employs check sums and timeouts for error detection and recovery. Flow control provided due to poor quality link.

12.8.5 Session layer

The session layer regulates the data transfer and control for both half-duplex and full-duplex working. It inserts synchronisation points in the session for easy recovery if there is a failure.

12.8.6 Presentation layer

The protocol in the presentation layer is responsible for the establishment of presentation connections and the release of these connections. It is also responsible for the transfer of data.

This layer also changes the format of the data to the required format of the terminal/computer.

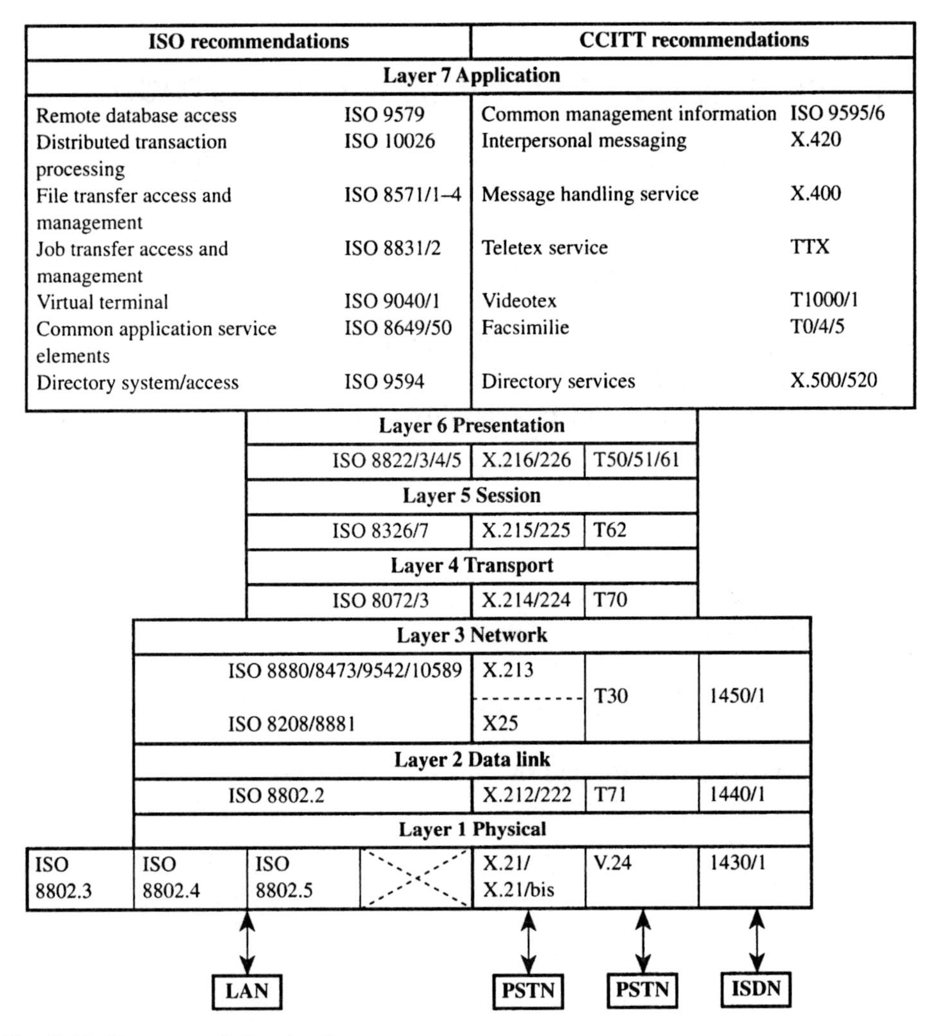

Fig. 12.16 Recommended protocols

12.8.7 Application layer

This protocol implements the actual application as required by the user.

12.9 ISO AND CCITT RECOMMENDED PROTOCOLS

Figure 12.16 shows the recommended protocols for each layer in the model.

12.10 REVIEW QUESTIONS

12.1 Describe the various categories into which data protocols fall.
12.2 Describe, using a diagram, the ISO seven-layer model.
12.3 Describe the functions and formats of the different types of frames produced by the data link layer.
12.4 Using a diagram, describe how a layer can be sublayered.
12.5 Describe each of the following:
  (a) Switched virtual circuit.
  (b) Permanent virtual circuit.
  (c) Connection-oriented services.
  (d) Connectionless-oriented services.
12.6 Describe the different types and classes of service provided by the transport layer.
12.7 Describe the reasons for and the functions of PADs.

Answers to review questions

Chapter 1

1.1: A square wave consists of a fundamental frequency and all the odd harmonic frequencies. The initial phase of all these frequencies must be 0° and they must all initially move in either a positive direction or a negative direction. The amplitude of each harmonic frequency must be the amplitude of the fundamental frequency divided by that harmonic number.

1.2: A sawtooth waveform consists of a fundamental frequency and all the harmonic frequencies. The initial phase relationship must be the same as that described for the square wave. The amplitude relationship must be the same as that described for a square wave. See Fig. 1.4 for the construction of the waveform.

1.3: (a) A sine wave is a continuous wave that has an infinite level range and consists of a single frequency. (b) An analogue waveform is a complex signal consisting of two or more frequencies that are added together. The waveform is continuous and has an infinite range of levels. (c) A true digital signal is a complex waveform that is a discontinuous signal and has a finite range of levels.

1.4: A true noise spike consists of a fundamental frequency and all the harmonic frequencies. However, each frequency in the noise spike has the same amplitude and hence because of the large amplitudes and large bandwidth of the signal they cause interference with telecommunication equipment or in some cases damage the equipment.

1.5: (a) Gain = 48.15 dB. (b) Loss = 10 dB. (c) Gain = 3.365 dB.

1.6: (a) Level = +7.4 dBm. (b) Level = +18.34 dBm. (c) Level = −6.5 dBm.

Chapter 2

2.1: For an unbalanced amplitude modulator the products of modulation are the USB, LSB and the carrier frequency. For a single balanced amplitude modulator the products of modulation are the USB, LSB and the modulating frequency. For a double balanced amplitude modulator the products of modulation are the USB and LSB only.

2.2: f_a = 60 kHz to 108 kHz. LSB = 232 kHz to 280 kHz. USB = 400 kHz to 448 kHz.

2.3: LSB = 12 kHz to 252 kHz. USB = 612 kHz to 852 kHz.

2.4: Modulation depth = 65.62%.
2.5: (a) $P_c = 0.51\,\text{mW}$. (b) $P_{USB} = P_{LSB} = 92.19\,\mu\text{W}$. (c) $P_{SB} = 184.39\,\mu\text{W}$. (d) Total power = 694.39 μW. (e) Efficiency = 26.55%.
2.6: A system that uses a modulation index of less than 0.5 is considered as a narrowband FM system and only transmits the carrier frequency together with the 1st USB and 1st LSB. A system that uses a modulation index greater than 0.5 is considered to be a wideband FM system and transmits multiple sidebands together with the carrier frequency. The relative power in each sideband and the carrier frequency is dependent on the actual modulation index.
2.7: Advantage: less affected by noise than AM. Disadvantage: very wide bandwidth.

Chapter 3

3.1: The transmitted signal must occupy a bandwidth much greater than the bandwidth of the modulating frequency. The bandwidth occupied by the transmitted signal must be determined by a prescribed waveform and not by the modulating frequency.
3.2: They aid privacy of the transmission since the spectral density of the spread spectrum may be less than the noise spectral density of the receiver. The de-spreading process in the receiver will spread the spectra of unwanted narrow band signals thus improving interference rejection. The effect on a spread spectrum receiver that receives a spread spectrum from a different spread spectrum system that uses the same frequency bands but implementing a different spreading pattern approximates to noise in the receiver.
3.3: See Section 3.6.1.
3.4: Hop set – this is the number of channels that are used by the system. Dwell time – this is the length of time that the system transmits on an individual channel. Hop rate – this is the rate at which the hopping takes place.
3.5: See Section 3.6.2, 'System operation'.
3.6: See Section 3.6.2, 'Synchronisation'.
3.7: Method 1: with this technique each binary bit is transmitted as a short pulse or chirp. The actual interval between each transmitted chirp is varied but each chirp has the same duration. Method 2: with this technique each chirp has a different duration. The start of each chirp takes place at the same point during the bit period.
3.8: See Section 3.7.

Chapter 4

4.1: Figure 4.3(a) is an ASK modulator. When the carrier frequency is such that the two diodes are forward biased, the full amplitude of the data signal appears on the output at the frequency of the carrier. When the carrier frequency is such that the two diodes are reverse biased, only a small amplitude carrier due to carrier leak appears on the output. Figure 4.3(b) is also an ASK modulator. When the carrier is such that the diodes are reverse biased, the peak output signal appears on the output at the frequency of the carrier frequency. When the diodes are forward biased by the carrier, only a small peak amplitude signal appears on the output due to carrier leak.

4.2: Refer to Fig. 4.5(a). The input data stream is applied to a digital to analogue converter. The analogue voltage is summed with the control voltage from a reference oscillator. This resultant control voltage is used to cause the VCO to change frequency such that when a logic 1 is applied the frequency increases to a maximum and when a logic 0 is applied the frequency drops to a minimum. The deviation is with reference to the reference frequency supplied by the reference oscillator.

4.3: Refer to Fig. 4.5(b). The input analogue signal is applied to the system by means of a filter which allows a frequency range equivalent to the frequency swing to be applied to the phase comparator. The resultant control voltage from the phase comparator is amplified by an amplifier. This control voltage varies in accordance with the frequency swing. The d.c. control voltage is applied to an analogue-to-digital converter which reproduces the digital signal. The control voltage is also applied to the VCO which changes frequency in accordance with the control voltage and the VCO tries to track the varying frequency input.

4.4: Refer to Fig. 4.3. In this figure exchange the carrier frequency input with the data signal and the data input with the carrier frequency.

4.5: See Fig. 4.8(b).

4.6: See Fig. 4.9(b). 011_2 appears in the fourth quadrant. 101_2 appears in the first quadrant. 110_2 appears in the third quadrant. 001_2 appears in the first quadrant.

4.7:

Present				Quad	Next				Quad	Phase	Amplitude						Output			
A	*B*	*C*	*D*		*A*	*B*	*C*	*D*		change	*a*	*c*	*B*	*D*	*b*	*d*	*a*	*b*	*c*	*d*
1	0	0	1	4	0	1	1	1	2	180	1	1	1	1	0	0	1	0	1	0
1	0	1	0	1	0	0	1	1	2	+90	1	0	0	1	1	0	1	1	0	0
1	1	0	0	4	0	0	0	0	3	−90	0	1	0	0	1	1	0	1	1	1
0	1	1	1	2	0	1	1	0	2	0	0	0	1	0	0	1	0	0	0	1
0	0	0	1	3	1	1	1	1	1	180	1	1	1	1	0	0	1	0	1	0

Data = (1001) 1010 1100 0111 0001 1010_2

4.8:

Present				Quad	Next						Phase	Quad	Amplitude						Output			
a	*b*	*c*	*d*		*a*	*b*	*c*	*d*	*a*	*c*	change		*A*	*C*	*b*	*d*	*B*	*D*	*A*	*B*	*C*	*D*
1	0	0	1	4	1	0	1	0	1	1	180	3	0	1	0	0	1	1	0	1	1	1
1	0	1	0	1	1	1	0	0	1	0	+90	2	0	1	1	0	0	1	0	0	1	1
1	1	0	0	4	0	1	1	1	0	1	−90	3	0	0	1	1	0	0	0	0	0	0
0	1	1	1	2	0	0	0	1	0	0	0	2	0	1	0	1	1	0	0	1	1	0
0	0	0	1	3	1	0	1	0	1	1	180	1	1	1	0	0	1	1	1	1	1	1

Data = 0111 0011 0000 0110 1111_2

Chapter 5

5.1: See Section 5.7.1.
5.2: See Fig. 5.4.
5.3: (a) Bit duration = 152.53 ns. (b) Time slot duration = 1.22 μs. (c) Frame duration = 48.81 μs. (d) Multiframe duration = 976.2 μs.
5.4: Sampling frequency of signalling channel = 500 Hz.
5.5: Alarms at town A: FAL, MFAL. Alarms at town B: RFAL, RMFAL.
5.6: When, say, terminal A loses multiframe alignment then the MFAL alarm at that terminal becomes active and the receive signalling circuits are disabled. Bit 2 in the MFSW is set by the multiframe alarm detector circuit which is then transmitted to terminal B. At terminal B, bit 2 in the MFSW is detected as being set and causes the RMFAL alarm to become active. This inhibits the transmit signalling circuits.

Chapter 6

6.1: Noise voltage = 395.38 nV.
6.2: Electrostatic and electromagnetic coupling can be minimised by proper screening of the IF, RF and power supply circuits. Crosstalk can be reduced by separating the transmit cable pairs from the receive cable pairs. Thermal noise can be reduced by cooling the active components.
6.3: SNR = 54.15 dB.
6.4: SNR = 42.76 dB.
6.5: (a) Order = amplifier C first, amplifier B second, amplifier A last. Amplifier C has lowest noise figure and gain, amplifier A has the highest noise figure and gain. (b) $F_{(3)} = 1.57$. (c) $T_A = 435K$, $T_B = 203K$, $T_C = 58K$. (d) $T_{(3)} = 165.3K$.
6.6: (a) $F_B = 1.15$. (b) Order = amplifier B first, amplifier C second and amplifier A last. (c) $F_3 = 1.237$. (d) $T_A = 435K$, $T_B = 43.5K$, $T_C = 58K$. (e) $T_{(3)} = 68.73K$.

Chapter 7

7.1: Amplitude distortion occurs when the attenuation of a medium is not constant for all frequencies. On metallic pairs the higher frequencies are attenuated more than the lower frequencies. This results in a different amplitude relationship between the fundamental frequency and the harmonic frequencies at the receiver relative to the transmitter, and causes the output signal at the receiver to be distorted relative to the input signal to the transmitter.
7.2: The fundamental frequency and the different harmonic frequencies travel at different velocities down a transmission medium other than free space. This means that the phase relationship between the fundamental frequency and the harmonic frequencies at the input to the receiver is different to that at the output of the transmitter. This distortion causes the signal to contain harmonic frequencies that were not present in the original signal.

7.3: Refer to Fig. 7.4. Along each section noise from external sources is induced into the transmission medium. The signal is attenuated as it propagates down the transmission medium. At the first repeater both the signal and the noise are amplified. Hence at the output of the first repeater the signal-to-noise ratio is worse than it was at the output of the terminal repeater. The same effect takes place in all subsequent sections, resulting in a degraded signal-to-noise ratio at the input to the distant receiver. If too many repeaters are used between the transmitter and receiver then the noise could swamp the signal, making the system unusable.
7.4: In a digital system the signal is regenerated at each repeater. This means that the noise in the previous section is left behind in that section and is not carried forward to the next section. The digital signal is recreated by the regenerator and hence the signal-to-noise ratio at the distant receiver is the same as that at the output of the transmitter.
7.5: Refer to Fig. 7.7. Because all frequencies do not travel at the same velocity down the transmission medium this has the effect of causing the digital pulse to spread out. This spreading out can cause interference with the previous and post digital pulses and can cause a logic 0 to be decoded as a logic 1 instead.

Chapter 8

8.1: Refer to Fig. 8.3. Because the transmission media such as metallic pairs attenuate the higher frequencies more than the lower frequencies, the Gaussian white noise which has a constant amplitude across the full frequency range, affects the higher frequencies more than the lower frequencies and as a result is sometimes referred to as triangular noise.
8.2: Total number of characters = 155_{10}.
$P(x_i)$ for A $= 25/155 = 0.161$; for B $= 32/155 = 0.206$; for C $= 12/155 = 0.078$; for D $= 4/155 = 0.026$; for E $= 64/155 = 0.413$; for F $= 18/155 = 0.116$. $S(x) = 2.205_{10}$.
8.3: (a) $C = 46.508$ kb/s. (b) $C = 20.111$ kb/s. (c) $C = 11.479$ kb/s.
8.4: $x = 3.5857$. $P(b) = 179.647 \times 10^{-6}$.
8.5: $x = 5.071$. $P(b) = 205.053 \times 10^{-9}$
8.6: $C = 50.49$ kb/s.

Chapter 9

9.1: ASCII = American Standards Institute Code for Information Interchange. EBCDIC = Extended Binary Coded Decimal Code.
9.2: (a) $S(x) = 2.33_{10}$. (b) Huffmann codes are determined using the tree and state diagrams in Figure A.1.

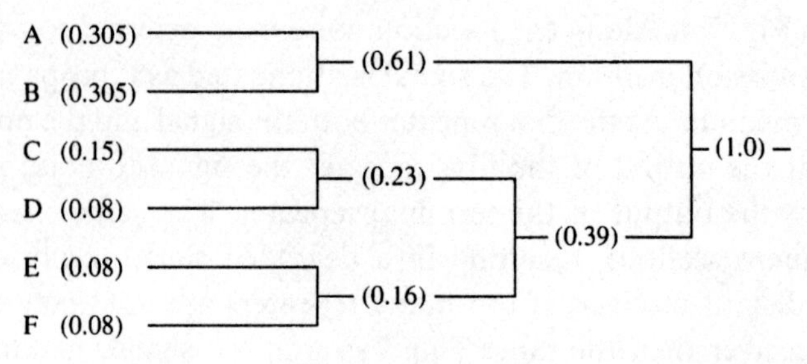

Tree diagram

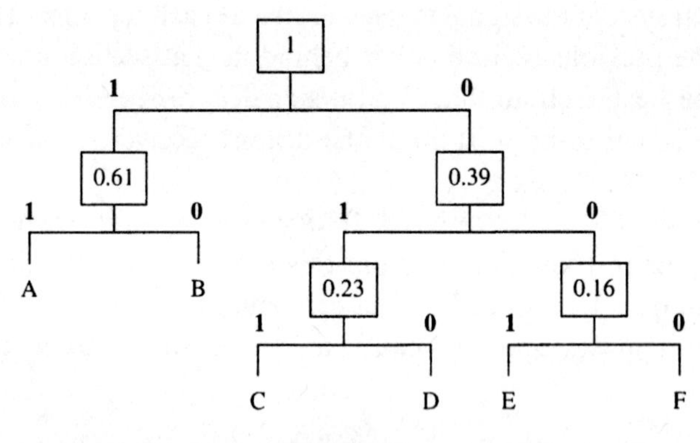

State diagram

x_i	$P(x_i)$	Code	n_i	$n_i \cdot P(x_i)$
A	0.305	11	2	0.610
B	0.305	10	2	0.610
C	0.15	011	3	0.450
D	0.08	010	3	0.240
E	0.08	001	3	0.240
F	0.08	000	3	0.240
Sum $n_i \cdot P(x_i)$				2.390

Fig. A.1

(c) Compression rate = 1.255. (d) Efficiency = 97.49%
9.3: Total number of characters = 1050_{10}. $P(x_i)$ for A = 0.195; for B = 0.174; for C = 0.17; for D = 0.17; for E = 0.08; for F = 0.08; for G = 0.05; for H = 0.05; for I = 0.031. (a) $S(x) = 2.9386_{10}$. (b) Huffmann codes for A = 11; for B = 101; for C = 100; for D = 011; for E = 010; for F = 0011; for G = 0010; for H = 0001; for I = 0000. (c) Compression = 1.326. (d) Efficiency = 97.43%.
9.4: (a) $r = 4$. (b) $(n, k) = (16, 5)$. (c) Hamming code = 01110_2. Transmitted data = 0100110010101010_2. (d) Efficiency = 75%. (e) Redundancy = 25%.
9.5: Received Hamming code = 01011_2. Receiver-determined Hamming code = 01110_2. The received data has been errored.

Chapter 10

10.1: If an even number of bits in the bit stream are errored by the medium then the receiver will determine the status of the receiver parity bit to be the same as the receiver determined status of the parity bit and hence the data will be accepted as being error free.

10.2: ASCII character P = 1011000_2. For even parity the parity bit = 1. Transmitted data = 00001101111_2. For odd parity the parity bit = 0. Transmitted data = 00001101011_2.

10.3: For even parity,
G = 1001110_2, even parity bit = 0
o = 1101111_2, even parity bit = 0
a = 1101000_2, even parity bit = 1
l = 1100011_2, even parity bit = 0
Transmitted data:
G = 00111001011_2
o = 01111011011_2
a = 00001011111_2
l = 01100011011_2
Tx data = $00111001011011110110110000101111101100011011_2$
For odd parity,
G = 1001110_2, odd parity bit = 1
o = 1101111_2, odd parity bit = 1
a = 1101000_2, odd parity bit = 0
l = 1100011_2, odd parity bit = 1
Transmitted data:
G = 00111001111_2
o = 01111011111_2
a = 00001011011_2
l = 01100011111_2
Tx data = $00111001111011110111110000101101101100011111_2$

10.4: If two or more bits are errored then although the VRC and HRC at the receiver would detect that there are errors in a word or words and a column or columns, the receiver would be unable to uniquely identify the errored bits and hence correct the errors.

10.5: CRC-4 = 1100_2

10.6: Received CRC-4 = 0111_2. Receiver-generated CRC-4 = 1101_2. Hence the received bit stream has been errored.

10.7: Data = $0110\ 00_2$ (two zeros must be added to clear the registers).

Data	=	0	1	1	0	0	0
Path	=	*a–a*	*a–b*	*b–d*	*d–c*	*c–a*	*a–a*
Code	=	00	11	01	01	11	00

Trellis code = 001101011100_2.

Chapter 11

11.1: *D.c. component:* For transmission lines the d.c. component should be removed to enable correct detection of the individual bits at the receiver. The elimination of the d.c. component enables a.c. coupling to take place and eliminates the resultant low frequency which could affect adjacent pairs through crosstalk.

Self-clocking: Symbol or bit synchronisation is required for digital communication and is achieved by having the clock frequency embedded in the bit stream.

Error detection: Some schemes have the means of detecting data errors without introducing additional error detection bits into the data sequence.

Bandwidth compression: Some schemes increase the efficiency of the available bandwidth by enabling more information to be conveyed when compared to other schemes using the same type of medium.

Noise immunity: Some schemes display a bit error rate that is better than other schemes, when they are transmitted over similar media having similar signal-to-noise ratios.

11.2: Non-return to zero codes; return to zero codes; phase-encoded codes; multi-level binary codes.

11.3:

Data stream	=	1	0	0	1	1	1	0	1	0	1	1	1	0	1	0	0	0	0_2
ADI	=	1	1	0	0	1	0	0	0	0	0	1	0	0	0	0	1	0	1_2
AMI	=	$+M$	$-M$	0	0	$+M$	0	0	0	0	0	$-M$	0	0	0	0	$+M$	0	$-M$

The code must be shown as an RZ code.

11.4:

Data	=	1	0	0	1	1	1	0	1	0	1	1	1	0	1	0	0	0	0_2
ADI	=	1	1	0	0	1	0	0	0	0	0	1	0	0	0	0	1	0	1
HDB-3	=	$+M$	$-M$	0	0	$+M$	0	0	0	$+V$	0	$-M$	0	0	0	$-V$	$+M$	0	$-M$

The code must be shown as an RZ code.

11.5:

Data	=	1	0	0	1	1	1	0	1	0	1	1	1	0	1	0	0	0	0_2
Code	=	10	1	0	10	10	10	1	01	0	10	10	10	1	01	0	1	0	1_2

11.6:

Data	=	1	0	0	1	1	1	0	1	0	1	1	1	0	1	0	0	0	0_2
Code	=	10	01	01	10	10	10	01	10	01	10	10	10	01	10	01	01	01	01_2

Chapter 12

12.1: *Stop and wait protocols:* These protocols are message oriented. The complete message must be transmitted to a switching computer and stored in the computer memory. Only once the complete message is stored in the memory is it then sent to the next computer. The switching computer must have a very large memory as it must serve several users simultaneously. The message transfer is slow through this system. The emphasis is on storage of the message and not on message transfer. For large messages the other users are forced to wait and hence the users are not fairly serviced.

Continuous protocols: In this system the message is broken up into small packets. The packets are continuously transferred to the switching computer. The switching computer stores the packet before transmitting it to the next computer. The emphasis here is on forwarding the data and not on storage. The memory of the switching computer is smaller and the system is a lot faster in that the users are more efficiently served.

12.2: Application layer, presentation layer, session layer, transport layer, network layer, data link layer, physical layer.

12.3: Frames are the information frame, supervisory frame, and unnumbered frame. They are discussed in Section 12.8.2.

12.4: See Section 12.6.8.

12.5: See Section 12.8.3.

12.6: See Section 12.8.4.

12.7: See Section 12.8.3.

Bibliography

The ARRL Handbook for Radio Amateurs, 73rd Edition. The American Radio Relay League, 1996.

Green, D. C., *Transmission Principles for Technicians*, 2nd Edition. Longman Scientific & Technical, 1991. ISBN 0-582-99461-6.

Helmers, S. A., *Data Communication, a Beginners' Guide*, 1st Edition. Prentice-Hall, 1989. ISBN 0-13-198870-0.

Kennedy, D., *Electronic Communication Systems*, 4th Edition. McGraw-Hill International Editions, 1992. ISBN 0-07-112672-4.

Mazda, F. (editor), *Telecommunication Transmission Principles*, 1st Edition. Butterworth-Heinemann, 1996. ISBN 0-240-51452-1.

Sklar, B., *Digital Communication Fundamentals and Applications*, 1st Edition. Prentice-Hall International Editions, 1988. ISBN 0-13-212713-X.

Reyner, J. H. and Reyner, P. J., *Radio Communication*, 2nd Edition. Pitman, 1967.

Index

Printed in the United Kingdom
by Lightning Source UK Ltd.
9803900001B/90-170